INSECT PARASITOIDS

INSECT PARASITOIDS

EDITED BY

Jeff Waage

Department of Pure and Applied Biology
Imperial College at Silwood Park

David Greathead

CAB International
Institute of Biological Control

13th Symposium of the
Royal Entomological Society of London
18–19 September 1985
at the
Department of Physics Lecture Theatre
Imperial College, London

1986

ACADEMIC PRESS
Harcourt Brace Jovanovich
London Orlando San Diego New York Austin
Boston Toronto Sydney Tokyo

ACADEMIC PRESS INC. (LONDON) LTD.
24/28 Oval Road, London NW1 7DX

United States Edition published by
ACADEMIC PRESS, INC.
Orlando, Florida 32887

British Library Cataloguing in Publication Data
Insect parasitoids.
1. Parasitic insects
I. Waage, Jeff II. Greathead, David
595.7052′49 QL463

ISBN 0–12–728900–3

Typeset by Latimer Trend & Company Ltd, Plymouth

Printed and bound in Great Britain by
T. J. Press (Padstow) Ltd., Padstow, Cornwall

Contents

10. Parasitoids in Classical Biological Control
D. J. GREATHEAD

11. Enhancing Parasitoid Activity in Crops
W. POWELL

12. Parasitoids in the Greenhouse: Successes with Seasonal Inoculative Release Systems
J. C. VAN LENTEREN

Contributors

Numbers in parentheses indicate the pages on which the authors' contributions begin.

J. J. M. VAN ALPHEN (23), Division of Ecology, Department of Population Biology, University of Leiden, Kaiserstraat 63, 2300 RA Leiden, The Netherlands

R. R. ASKEW (225), Department of Zoology, University of Manchester, Manchester M13 9PL, England

J. VAN DEN ASSEM (137), Division of Ethology, Zoological Laboratory, University of Leiden, Kaiserstraat 63, 2300 RA Leiden, The Netherlands

M. BOULETREAU (169), Laboratoire de Génétique des Populations (UA CNRS 243), Université LYON I, 69622 Villeurbanne, France

I. D. GAULD (1), Department of Entomology, British Museum (Natural History), Cromwell Road, London SW7 5DB, England

D. J. GREATHEAD (290), CAB International Institute of Biological Control, Silwood Park, Ascot, Berkshire SL5 7PY, England

M. P. HASSELL (201), Department of Pure and Applied Biology, Imperial College at Silwood Park, Ascot, Berkshire SL5 7PY, England

J. H. LAWTON (265), Department of Biology, University of York, Heslington, York Y01 5DD, England

J. C. VAN LENTEREN (342), Department of Entomology, Agricultural University, P.O. Box 8031, 6700 EH Wageningen, The Netherlands

W. POWELL (319), Department of Entomology, Rothamsted Experimental Station, Harpenden, Hertfordshire AL5 2JQ, England

M. R. SHAW (225), Department of Natural History, Royal Museum of Scotland, Chambers Street, Edinburgh EH1 1JF, Scotland

M. R. STRAND (97), Department of Entomology, College of Agricultural Sciences, Clemson University, Clemson, South Carolina 29631, USA

L. E. M. VET (23), Department of Entomology, Agricultural University, P.O. Box 8031, 6700 EH Wageningen, The Netherlands

J. K. WAAGE (63), Department of Pure and Applied Biology, Imperial College at Silwood Park, Ascot, Berkshire SL5 7PY, England. (Present address: CAB International Institute of Biological Control, Silwood Park, Ascot, Berkshire SL5 7PY, England)

Preface

The unique lifestyle of insect parasitoids continues to fascinate entomologists. The diversity of species provides a wealth of material for fundamental studies on insect behaviour, ecology and evolution, as well as an important pool of agents for the biological control of insect pests. This symposium surveys recent developments in parasitoid research along this continuum from the fundamental to the applied.

The term "parasitoid" was first used by Reuter in 1913 to describe a life history intermediate between that of predators and true parasites. Adult female parasitoids are free-living, feed on nectar, pollen or as predators and forage actively for their arthropod hosts on plants and other substrates. Usually, on locating a host, the female lays one or more eggs on or in it, and the ensuing larvae consume the host tissue, killing the host in the process. The parasitoid lifestyle is found chiefly in the orders Strepsiptera, Hymenoptera and Diptera, and is probably exhibited by over a quarter of a million species worldwide. Their sheer diversity makes identification and systematic studies difficult, which complicates many aspects of research. Therefore, the first chapter of this book begins with a consideration of parasitoid systematics, and particularly the contribution which phylogenetic systematics can make in unravelling it.

The next four chapters investigate the parasitoid lifestyle. A modern, evolutionary perspective is common to all of these chapters, and the adaptive nature of parasitoid behaviour is examined through experimental and comparative studies. In Chapter 2, foraging theory is shown to be a useful key to understanding the manner in which parasitoids find and exploit patchily-distributed hosts, while in Chapter 3, the theory of sex and progeny allocation is used to shed light on the size of parasitoid broods, which may range from one to several thousand per host, and the variable sex ratios which accompany them. Chapter 4 considers the fate of these broods as they develop in (often hostile) hosts. Concentrating on egg parasitoids, it reveals the intimate interdependence between parasitoid and host development. Finally, Chapter 5 examines the complex ethology of the courtship and mating which follows emergence, revealed in studies of the Pteromalidae and other hymenopterous parasitoids.

A further four chapters examine the close interaction between parasitoid and host at the level of populations and communities. The hypothesis that these interactions mediate a close coevolution between parasitoid and host is investigated in Chapter 6, where the role of genetic change in host-parasitoid population dynamics is demonstrated. Mathematical models for host-parasitoid dynamics are discussed in Chapter 7, with emphasis on the difference between specialist and generalist parasitoids, and the unusual dynamics of model communities which contain both. The distinction between specialist and generalist is one of several classifications used to describe parasitoid community structure. Chapter 8 introduces a novel improvement on this classification in order to explore the effect of host-related factors on the size and structure of a number of large and complex parasitoid communities. Chapter 9 turns our perspective around to investigate the possible effect of generalist parasitoids on the assembly and evolution of their host communities.

Population and community theory forms the basis of our understanding of how parasitoids are able to act as biological control agents for insect pests. The final three chapters reveal the unsatisfactory state of our understanding, and explore the potential for, and difficulties of, linking fundamental to applied research. Chapter 10 examines the pattern of success in classical biological control, in the light of new and old theories on what makes a good long-term control agent. Chapter 11 applies our understanding of adult behaviour to the problem of enhancing the impact of indigenous natural enemies on crop pests. Finally, the highly successful application of parasitoids for insect control in greenhouse systems is used in Chapter 12 as a framework for investigating questions relating to parasitoid selection, quality control and economics, which underlie any future growth in this promising method of pest management.

Throughout this book the reader will be aware that both fundamental and applied research have concentrated on the parasitic Hymenoptera, and indeed on only a relatively few species of this remarkably diverse group. We hope that this book will serve as a stimulus to expand our knowledge of parasitoids beyond this current narrow focus, to include more taxa, especially Diptera, in studies which address the interesting problems raised in these twelve chapters.

February 1986

Jeff Waage
David Greathead

Acknowledgements

It is a pleasure to thank the Speakers, the President and the Chairmen of the four sessions, Dr T. Lewis, Dr R. C. Fisher, Dr J. K. Waage and Prof. M. P. Hassell who, with the Symposium Committee, are responsible for the scientific success of the Symposium. We also thank the Registrar and his staff for its smooth organization, and staff of CIBC and Imperial College for assistance with editing the Symposium volume. In particular, we would like to thank Mrs A. H. Greathead for editorial assistance and compilation of the index.

The Society wishes to acknowledge the generous contributions of Ciba-Geigy, the British Crop Protection Council, Imperial Chemical Industries plc, Koppert B.V. and Shell Research Ltd. towards the running of the Symposium, and to thank them for their interest in research on insect parasitoids.

Opening Remarks

DR TREVOR LEWIS
President of the Royal Entomological Society of London

It is my great pleasure to welcome you to this 13th Symposium of the Royal Entomological Society of London; we are especially pleased to see representatives from 22 different countries, making it a truly international event. The Society's aim in holding these biennial symposia is to provide a forum for the exchange of the latest ideas on a chosen topic, the papers being subsequently published as a definitive account of recent developments in the subject.

The origins of this year's topic, *Insect Parasitoids*, extend far back into the history of entomology and of our Society. Probably the first recorded observation on insect parasitism was by Aldrovandi in 1602 who noted the exit of the parasitic larvae of *Apanteles glomeratus* (L.) from caterpillars of the large white butterfly, mistakenly believing the parasite's cocoons to be the eggs of the butterfly (DeBach, 1974). In 1670, M. Lister observed wasp-like larvae within scale insects on English oak branches and in 1700 Leeuwenhoek witnessed and described parasites ovipositing in aphids which died within seven or eight days. Our Dutch colleagues in the programme have clearly built on that early lead!

The first president of this Society, Rev. William Kirby, discovered the parasitoid order, Strepsiptera; hence a male of *Stylops kirbii* Leach was adopted as the Society's emblem on its foundation in 1833. In 1815 Kirby and another former president, William Spence, produced their classic monograph, *An Introduction to Entomology*, in which they not only marvelled at the ability of ichneumonids to attack hosts of all sizes and degrees of ferocity, in apparently concealed habitats, but hinted at their important role in pest control, considering them to be "sent in mercy by Heaven ... saving mankind from the horrors of famine". About 100 years later, Reuter (1913) gave these and similar insects the name "parasitoid", the term by which we know them today.

The abundance and diversity of parasitoids will no doubt be mentioned in many papers. On a world scale their numbers can only be guessed, but if the British fauna is any guide, over one-third (35·6%) of our insects are parasitic on animals, and over 90% of these are insect parasitoids; the Ichneumonidae, Braconidae and Pteromalidae are the three families with the greatest number of species in the British list. Perhaps the papers in this symposium are unduly biased towards the parasitic Hymenoptera, but this probably fairly reflects the distribution of research effort and information on the subject, and it is not too much to hope that work on other orders containing parasitoids—the Diptera, Coleoptera, Neuroptera and Lepidoptera—might be stimulated by ideas developed initially through studies on Hymenoptera.

The symposium also provides an excellent opportunity to bring together basic and applied aspects of entomology, because much of the recent drive in the subject area has come from the rising worldwide interest in biological/integrated control as a counter to environmental pollution by indiscriminate use of pesticides. Already pests on 14 million hectares of crop are supposedly controlled by the release of *Trichogramma*, whose mass rearing and release was incidentally mentioned by Mr F. Enock in 1895 at a meeting of the South London Entomological and Natural History Society—a rival but friendly organization presided over at that time by a Fellow of this Society (Turner, 1895). The work of the Commonwealth Institute of Biological Control, with whom this Society has had close and still improving links for many years, bears witness to the achievements and prospects of the applied uses in parasitoids.

In the context of our being together at this gathering it is perhaps worth noting that "parasitism" is derived from the Greek for "one who sits at the table of another" and "symposium" likewise for "fellow drinker", so this meeting seems destined to have all the elements of social conviviality as well as scientific interest.

Finally I would like to congratulate the conveners, Dr J. Waage and Dr D. Greathead and the Symposium Committee for drawing together such an excellent set of speakers, not forgetting the Registrar for much of the routine organization. With you, I look forward to the next two days and hope that you all enjoy the scientific and social programme.

REFERENCES

DeBach, P. (1974). "Biological Control by Natural Enemies." Cambridge University Press, London.

Kirby, W., and Spence, W. (1815). "An Introduction to Entomology. I." Longman, Hurst, Rees, Orme and Brown, London.

Leeuwenhoek, A. (1719). "Epistole at Societatem Regiam Anglicam et Alios Illustres Viros." John Arnold, Langerak.

Lister, M. (1671). An observation concerning certain insect husks of the Kermes kind. *Philosophical Transactions of the Royal Society* (abridged) **1,** 598.

Reuter, O. M. (1913). "Lebensgewohnheiten und Instinkte der Insekten." Friedlander, Berlin.

Turner, H. J. (1895). South London Entomological and Natural History Society—August 22nd, 1895. *Entomologist* **28,** 282–283.

1

Taxonomy, its Limitations and its Role in Understanding Parasitoid Biology

I. D. GAULD

I. INTRODUCTION

The term "parasitoid" embraces an exceedingly large number of insect species. For example, the Hymenoptera Parasitica alone probably includes more than a quarter of a million. There are more ichneumonid species than there are vertebrates, and unlike vertebrates a great majority of the ichneumonids are not even formally described, let alone readily recognizable using identification guides. Such groups are far too large for any individual to be wholly familiar with. Hence it is vital that the data available about the component taxa should be ordered in a logical and informative manner. It is usual to achieve this by adopting a hierarchical classification, but few non-taxonomists are aware of how such a system is developed, or how recent advances in systematic methodology have affected classifications now coming into general usage.

There is all too often a considerable gulf between people who classify parasitoids and those who use taxonomic services, identification keys etc. This chapter attempts to bridge this gulf by outlining modern taxonomic procedures, stating some of the problems taxonomists encounter and finally by pointing out ways in which taxonomists and other biologists can mutually benefit from increased interaction.

II. THE ROLE OF CLASSIFICATION

Classifying the bewildering diversity of organisms on this planet must be one of the most ancient of human activities, since it is a necessary prerequisite for enabling man to generalize about his surroundings (Berlin, 1973). The earliest systems of classification were likely to have been simple, self-explanatory, and essentially utilitarian. Hence plants may have been classified as poisonous or edible, as weeds or as suitable for building material. Such functional classifications were invariably anthropocentric; organisms were classified according to their value to, or impact upon man. Some of the earliest entomologists adopted similar classifications—insects were classified as "those bestowed by the Creator as a pestilence on mankind" and "those bestowed by the Creator for the benefit of mankind". In such a classification parasitoids were included in the latter group along with predatory insects, bees, lac-insects, silkmoths, dung beetles and, that notable Australian delicacy, the bogong moth (Kirby and Spence, 1815). Such classifications have one major shortcoming—they tell us nothing more about the insects than the information that was originally used to classify them. In information theory terms they are minimally informative.

Biologists have attempted to devise more informative classifications, both of a hierarchical nature (that is, usually attempting to express some sort of relationship between taxa) and of an ecological nature (that is, grouping organisms on the basis of some sort of trophic, behavioural or habitat-related similarity). Ecological classifications have proved to be very valuable and are widely used; indeed this symposium has been convened to discuss just such a group, insect parasitoids. And within parasitoids, as will be seen in the chapters to follow, ecological classification has run rampant in an attempt to organize the astounding biological diversity of this group, yielding such categories as solitary or gregarious, specialist and generalist, koinobiont and idiobiont, endoparasitoid or ectoparasitoid.

If all that were needed by biologists from a hierarchical classification of parasitoids was a system of labels by which to refer to species, then a simple scheme, such as the Dewey system used in libraries, would probably have been adopted years ago. However, the hierarchical taxonomic system in current usage is not merely an index to information, it is in itself informative. A phylogenetic classification is an attempt to maximize the information implicit in such a hierarchical system (Farris, 1979).

In addition to being informative, a phylogenetic classification provides a perspective that is complementary to an ecological classification. One may observe certain similarities between parasitoids that attack spiders' egg sacs, but there are also considerable differences, for example, between baeine scelionids and geline ichneumonids (Austin, 1985). These differences cannot simply be explained by the fact that the former are endoparasitic whilst the latter are ectoparasitic. An evolutionary perspective reveals that the relatives of baeines are exclusively egg parasitoids, whilst the relatives of gelines are parasitoids of small cocoons. This is an important observation because whilst silken egg sac construction by spiders may help to protect their eggs from some scelionid egg parasitoids, the placing of the eggs in a silken sac makes them attractive to certain ichneumonids which would otherwise be unlikely to exploit egg masses. To understand why a particular organism does what it does the way it does, it is not sufficient to comprehend only the selective forces acting: it is necessary to be aware of the constraints that a species is likely to have inherited as a result of its evolutionary history. This can only be achieved if one can recognize the taxa most closely related phylogenetically to the taxon under investigation. Such information is implicit in a cladistic classification.

Despite the fact that it is now well over a century since Darwin (1859) provided a unifying theory for taxonomy, it is only recently that the full impact of evolutionary theory has begun to be expressed in taxonomic procedure. Hennig (1966) stated a reasonably clearly defined method for inferring phylogenies. Before this, taxonomic practice was mostly unchanged from the pre-Darwinian era (Stevens, 1984). Although taxonomists talked about evolutionary relationships, prior to Hennig there was no explicit method for inferring such a relationship (Felsenstein, 1982). Classical taxonomic textbooks largely ignored the subject; Mayr *et al.* (1953) devoted less than 10 pages in their 284-page book to it, and Cain (1959) was probably correct when he stated that "young taxonomists are trained like performing monkeys, almost wholly by imitation." As a result of the lack of any formal methodological basis, most taxonomy was authoritarian and generally little justification or explanation was given for a classification. The development of phylogenetic systematics has had a markedly beneficial effect on taxonomy;

published work is increasingly becoming explicit and testable hypotheses of relationship are being advanced.

III. PHYLOGENETIC SYSTEMATICS AND PARASITOID CLASSIFICATION

The theory and practice of phylogenetic systematics have been the subject of several recent works (Nelson and Platnick, 1981; Wiley, 1981; Joysey and Friday, 1982) and Patterson (1980) gave an excellent introductory account, so a detailed discussion of methodology will not be repeated here. Basically, phylogenetic analysis involves the search for homologous features that exhibit different degrees of structural development in different assemblages of taxa. The forms of the structural development are compared and one form is postulated as being derived with respect to another. A feature may be assumed to be derived (or apomorphic) for some members of a group if the other taxa in that group and taxa in a related out-group all exhibit the unspecialized (plesiomorphic) condition (Watrous and Wheeler, 1981). For example, a comparative morphological study of two groups of Apocrita (Hymenoptera) would show that the sting of a bethylid is homologous with the ovipositor of an ichneumonid. These structures differ in that whereas the ovipositor of an ichneumonid is used for both egg-laying and envenomation, the bethylid sting is not used for egg-laying at all. The ichneumonid condition is plesiomorphic because it is essentially similar to the condition in siricoids (which can tentatively be regarded as the ancestors of the Apocrita), and thus the condition in bethylids may be regarded as apomorphic. The sharing of an apomorphic feature by two or more taxa may be construed as evidence of common ancestry. A set of taxa that are united by the possession of a shared derived feature (a synapomorphy of the group) may be said to be a holophyletic group. Thus as possession of a sting is a common feature of all aculeate superfamilies, but is found in no other Hymenoptera, possession of a sting may be considered a synapomorphy of the Aculeata which provides evidence for the group's holophyly. Phylogenies are constructed by searching for nesting series of holophyletic groups of taxa.

A cladistic classification differs from the more traditional classification in that groups of organisms which are united only by possession of primitive features, the so-called paraphyletic assemblages, are not recognized as valid taxa. Hence the Aculeata may be accepted as a valid taxon (it is demonstrably holophyletic), but because no synapomorphy has been found for the "Parasitica" (all apocritans except aculeates) the latter is paraphyletic and cannot be accorded group status (Konigsmann, 1978). The following example further illustrates the difference between a classification based on

TABLE I. Characters important in classification of anomalonine genera

	1st subdiscal cell explanate	Clypeus dentate	Male tarsi specialized	Claws pectinate	Ovipositor tip guarded	Male cuspis clubbed	Coxae 1 carinate	Propodeum reticulate
Heteropelma	X	—	X	—	X	X	—	X
Therion	X	—	X	—	X	X	—	X
Trichomma	—	X	—	X	X	X	—	X
Agrypon	—	X	—	X	—	—	X	X
Phaenolabrorychus	—	X	—	X	—	—	X	X

X = presence; — = absence of feature named.

TABLE II. Data as in Table I, recoded to show apomorphic (1) and plesiomorphic (0) states

	1	2	3	4	5	6	7	8
Heteropelma	1	1	1	1	1	1	0	1
Therion	1	1	1	1	1	1	0	1
Trichomma	0	0	0	0	1	1	0	1
Agrypon	0	0	0	0	0	0	1	1
Phaenolabrorychus	0	0	0	0	0	0	1	1

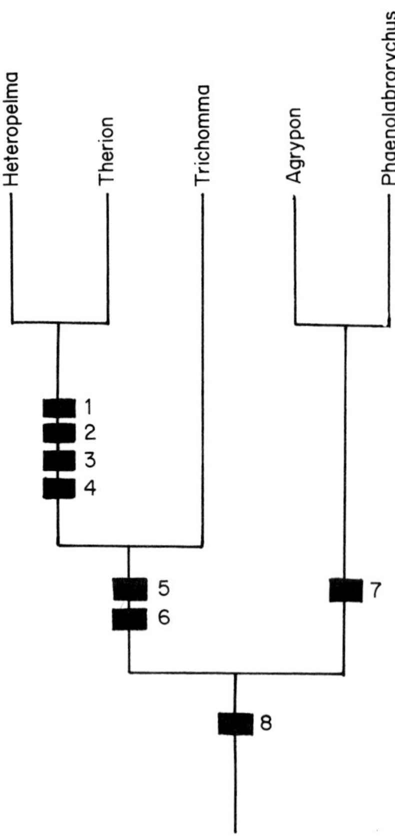

Fig. 1. Cladogram showing possible phylogenetic arrangement of anomalonine genera based on characters given in Tables I and II.

overall similarity (which recognizes paraphyletic groups), and a cladistic one which recognizes only holophyletic groups.

Consider the following five genera of anomalonine ichneumonids, *Heteropelma, Therion, Trichomma, Agrypon* and *Phaenolabrorychus*. Townes (1971), basing his classification on overall similarity, placed *Heteropelma* and *Therion* together as a group separate from the latter three genera. Gauld (1976) suggested that *Trichomma* was probably more closely related to *Therion/Heteropelma* than it was to *Agrypon* and *Phaenolabrorychus* with which it was then placed. Examination of a simple data matrix (Table I) suggests that Townes' classification is the better of the two as it is supported by four of the seven informative characters (1-4). Only three support Gauld's classification. By examining related taxa it is possible to assign evolutionary polarity to all these characters and to determine which state is apomorphic. The matrix can now be recoded, with derived character states represented by a 1 and the unspecialized plesiomorphic condition shown by a 0 (Table II). It can now be seen that there is no synapomorphy (common derived character) that supports the group *Trichomma* + *Agrypon* + *Phaenolabrorychus*, but two (5, 6) support *Therion* + *Heteropelma* + *Trichomma*, suggesting the phylogeny shown in Fig. 1. Recently discovered details of egg structure support this phylogenetic classification (Gauld, unpublished). Here then, we have essentially two classifications, one maximizing similarity and hence appearing to be more taxonomically tractable, and one more biologically meaningful which is more likely to be of use to the parasitoid researcher.

IV. PROBLEMS IN PARASITOID CLASSIFICATION

Although cladism represents a significant advance in taxonomic methodology, it has been, like many novel disciplines, embraced perhaps too enthusiastically and uncritically by some workers. Current taxonomic literature contains many spuriously authoritative evolutionary schemes appended, almost as an afterthought, to a revisionary study. As cladism offers a veil of scientific respectability for taxonomists, there is a tendency for certain of its limitations to be overlooked by some practitioners. Two of these limitations are particularly relevant to work on hymenopterous parasitoids.

A. Problems with Paraphyly

As outlined above, there are two types of monophyletic group: the holophyletic one, which contains all the descendants of an immediate common ancestor and the paraphyletic one which does not (Fig. 2). Strict cladists will

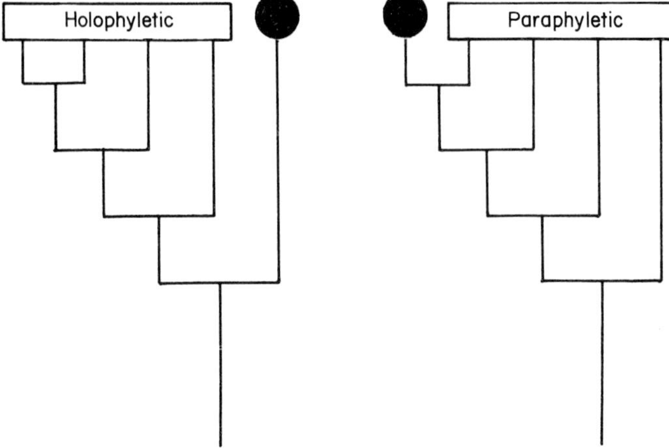

Fig. 2. Diagrams indicating the difference between holophyletic and paraphyletic groups. It can be seen that the former includes all the descendants of an immediate common ancestor, whilst the latter does not.

recognize only the former as valid taxa; indeed some workers seem to believe that, given sufficient study, all paraphyletic assemblages will be resolved as a series of holophyletic groups (Disney, 1983). In practice, in parasitoid taxonomy it is neither possible nor desirable at present to eliminate all paraphyletic groups. For example, consider the ichneumonid subfamily Pimplinae. Several of its component tribes (e.g. Pimplini, Polysphinctini, Rhyssini) are clearly definable holophyletic groups. Not only are they characterized by specialized morphological features, but each is biologically characteristic. Pimplines are endoparasitoids of lepidopterous pupae, polysphinctines are ectoparasitoids of spiders and rhyssines are parasitoids of coleopterous and hymenopterous wood-boring larvae. However, another tribe, the Ephialtini, is clearly paraphyletic. It can only be defined by its lack of the specialized features characterizing other tribes, and it is very probably the ancestral group from within which the other tribes have arisen. It is not possible either to subdivide the Ephialtini into a few holophyletic groups or to combine this tribe with one other to form a larger holophyletic group. A strict cladistic classification would have to recognize only a single group composed of the Ephialtini + Pimplini + Polysphinctini + Rhyssini. I cannot accept such an action, as a great deal of biological information would be lost. Consequently I accord the Ephialtini separate tribal status. However, it is important for the non-taxonomist to realize that a particular taxon is paraphyletic. Because of their disparate nature (i.e. in the example given the

Ephialtini comprises all the Pimplinae which cannot be placed in other tribes) such grade-groups usually contain organisms that exhibit a greater range of biological diversity than do related holophyletic groups (Table III). Therefore biological predictions about paraphyletic groups are liable to be less reliable than similar predictions about holophyletic ones.

If one accepts the saltationist view of evolution (the punctuated equilibrium of Gould and Eldredge, 1977; Stanley, 1979) then it is almost inevitable that one will have unresolvable paraphyletic groups. This seems to be a corollary of the hypothesis that most evolutionary change (i.e. development of apomorphies) is occurring rapidly in restricted populations, whilst the main lineage is barely changing at all (i.e. not acquiring any specialized features that characterize it as a lineage distinct from the rapidly evolving lineage). The existence of so many unresolvable paraphyletic groups in the Ichneumonoidea suggests that much of the evolutionary development of the superfamily may have occurred in a saltatory manner. It is interesting to note that problems with paraphyly are much less pronounced in some ancient groups whose lineages apparently diverged a very long time ago (Gauld, 1983), giving considerable time for phyletic evolution to occur, and thus for synapomorphies to accrue for each lineage.

TABLE III. Biological features of selected tribes of Pimplinae. Note the greater biological diversity of the paraphyletic Ephialtini compared with the holophyletic Pimplini, Polysphinctini and Rhyssini

Group	Biology
Pimplini	Solitary endoparasitoids of lepidopterous pupae/prepupae
Polysphinctini	Solitary ectoparasitoids of spiders
Rhyssini	Solitary ectoparasitoids of wood-boring larvae, especially Siricoidea and Cerambycidae
Ephialtini	Solitary ectoparasitoids of larvae/pupae concealed in fruit, herbaceous stems, leaves etc., including Lepidoptera, Diptera, Coleoptera, Symphyta; gregarious ectoparasitoids of cocoons of larger Lepidoptera; ectoparasitoids of wood-boring coleopterous larvae; ectoparasitoids of solitary aculeates, e.g. Sphecidae, Apidae and some social ones, e.g. Vespidae; predators in spider egg sacs

B. Problems with Homoplasy

The second major problem for phylogenetic systematics is the existence of discordance in character sets. Different combinations of apomorphic characters may be found to support different and mutually incompatible arrangements of taxa. This discordance or homoplasy results from evolutionary convergence, parallelism and character reversal (Underwood, 1982; Gauld, 1985).

Because homoplasy is reflected in many aspects of parasitoid biology and has serious implications for users of taxonomic products, it is important to be able to quantify its occurrence. For any assemblage of taxa the amount of homoplasy shown by a character set may be calculated and expressed as a coefficient (LeQuesne, 1972). Underwood and Gauld (unpublished) have expressed LeQuesne's coefficient of character state randomness as a value that varies between 0 (no homoplasy—all characters compatible and supporting a single cladogram) and 1 (character states randomly distributed across the taxa). For simplicity this coefficient is renamed the randomness ratio (RR). For a given data set, the higher the value of the RR, the more probable it is that one will find characters supporting numerous conflicting hierarchical arrangements of taxa, and that no one arrangement will be significantly better supported than several others. Character matrices for many groups of animals yield RRs of 0.50 or less, and even Kluge's (1976) classic study of homoplastic data yielded a character matrix with an RR of 0.55. In contrast, data sets for ichneumonoids usually have values between 0.70 and 0.85; only a single incidence has been found of a group with an RR of less than 0.60. As the ichneumonoid groups examined were deliberately selected because they appeared to be resolvable phylogenetically, it is probable that very high RRs are normal for ichneumonoid data sets. We may thus hypothesize that there has been considerable morphological convergence, parallelism and/or character reversal in ichneumonoids. Biological observations support this suggestion. For example, even such an apparently fundamental evolutionary step as the change from ectoparasitism to endoparasitism has occurred not only in numerous evolutionary lineages within the Ichneumonoidea, but probably has occurred more than once in related species-groups within a single genus (Shaw, 1983). There are numerous well-documented cases where members of separate lineages living in harsh habitats (e.g. deserts) have developed a series of structurally identical modifications in parallel (Gauld, 1984b; 1985).

For users of taxonomic work, high levels of homoplasy have two important implications. Firstly, phylogenetic hypotheses based on analyses of such data should be viewed with some scepticism, as there is likely to be a large number of alternative, and almost equally parsimonious, phylogenetic constructions possible for the data. Study of these numerous and almost equally plausible phylogenies will usually reveal a certain amount of common structure; hopefully, taxonomists can indicate such relatively robust parts of their preferred phylogeny, and also outline where high levels of uncertainty occur. Unfortunately, some taxonomists subjectively decide that apparent synapomorphies are homoplasy and discard them prior to undertaking phylogenetic analysis. They may then claim a spurious objectivity for their results, when in fact all that is objective is their analysis of subjectively chosen data.

C. The Construction of Identification Keys

The second and most practically important implication of a high level of homoplasy is that groups that have it are likely to be "taxonomically difficult" because subgroups are liable to be definable only in a disjunct fashion (Hull, 1965). This means that they will probably be characterized, not by any exclusive feature, but by combinations of features (Gauld and Mound, 1982). Consequently, any identification key to the taxa will usually be difficult, using long couplets with many "ifs" and "ors", or worse still, parenthetical exceptions that assume one knows the identity of the organism one is attempting to identify. This problem is partly the result of restrictions imposed by the couplet key format, and it is a problem that becomes progressively worse the larger the group. Often it is fairly easy to separate any two of the component taxa; it is just difficult to define hierarchical groups. With larger groups taxonomists generally attempt to produce a dichotomous key that segregates fairly equal sized subgroups, but paradoxically it is in large groups that it is difficult to define the subgroups.

Does this matter? Well, it is useful to attempt to produce a "symmetrical" key because this reduces the number of decisions (and consequently errors) required to make an identification. For example, consider a key to only six taxa (Fig. 3). With the asymmetrical form the average number of decisions

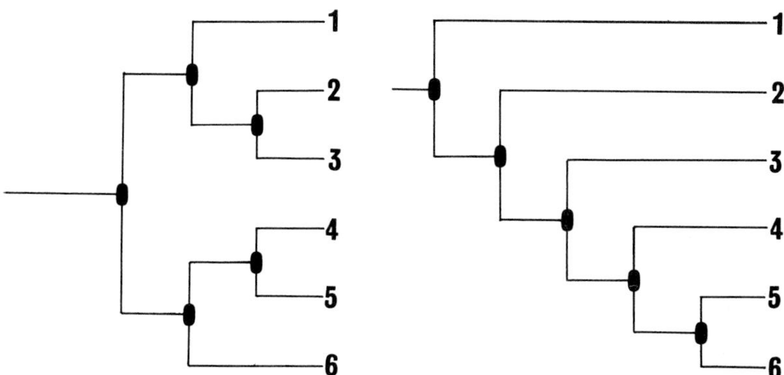

Fig. 3. Diagrams showing identification pathways for two different keys to taxa 1–6. The black oval indicates a decision point. Although the number of decision points is constant in the asymmetrical and symmetrical forms, it can be seen that the total number of decisions needed to identify all the taxa is larger in the asymmetrical form (20) than in the symmetrical form (16).

necessary to facilitate an identification is $1+2+3+4+5+5/6=3.33$, whilst in the symmetrical form $2+2+3+3+3+3/6=2.67$ average decisions suffice. The larger the key, the greater is this difference, for in a 31-couplet key the asymmetrical form necessitates 11.46 extra decisions to facilitate an average identification. In symmetrical keys to homoplastic organisms the first few couplets are likely to be very difficult and may be a major source of error. Although there are a variety of unconventional key forms that may be adopted to circumvent these problems (Pankhurst, 1978), in practice they have rarely been used for parasitoids (but see Smith and Shenefelt, 1956; Gauld, 1979).

V. CLADISTIC CLASSIFICATION AS A SCIENTIFIC HYPOTHESIS

It is important for biologists to realize that phylogenetic classifications are based on *hypotheses* of relationship and, like other hypotheses, will benefit from criticism and be improved by hypothetico-deductive reasoning (Popper, 1975). Although not directly refutable in the sense that in entomology an evolutionary sequence is likely to remain unknowable, a classification can be viewed as making testable predictions. Taxa placed together on the basis of study of a limited set of features (usually morphological) should resemble each other in features of anatomy, ecology, biochemistry etc. which were not used to formulate the classification (Crowson, 1970). The falsification of a single prediction does not necessarily refute a classification (as reversals and parallelisms do occur), but multiple failures to conform with predicted traits would strongly suggest that the classification is unsound. As a classification should involve the synthesis of all information about a group of organisms, field- or laboratory-based biologists can contribute, together with the taxonomist, to a process of reciprocal illumination, leading to the development of better classifications and to a wider understanding of the organisms concerned.

Laboratory- or field-based biologists can provide vital information that will suggest whether or not a particular classificatory hypothesis needs reexamining, or which of several competing hypotheses is preferable. For example, on the basis of adult and larval morphology, Townes (1969) considered that the ichneumonid subfamilies Ctenopelmatinae and Banchinae were closely related to the ophionoid subfamilies (Ophioninae, Campopleginae etc.) and not to the Tryphoninae and Pimplinae with which they were traditionally placed. Early but overlooked work by Pampel (1913) on the internal female anatomy, and recent studies on life histories and larval defences corroborate Townes' hypothesis and contradict other classifications.

VI. THE ROLE OF TAXONOMY IN PARASITOID RESEARCH

As taxonomists and other biologists are, to some extent, approaching the problem of information storage in different ways (group orientation versus problem orientation) each may have a reservoir of information of which the other is unaware, but which has considerable value to the other's research. For example, there have been different classifications proposed for eulophid chalcids. Work currently being undertaken on their mating behaviour (van den Assem *et al.*, 1982; in den Bosch and van den Assem, 1986; see also Chapter 5) has considerable taxonomic importance in corroborating or suggesting changes to current classifications.

A. Parasitoid Diversity

A taxonomist's group-orientated experience can be of importance in answering ecological questions. For example, Owen and Owen (1974) pointed out that certain groups of large parasitoids appeared to be no more species-rich in tropical habitats than they were in temperate ones, although their host groups were far more diverse. This paper initiated a considerable discussion in the literature concerning possible reasons for the so-called anomalous diversity (for summary see Janzen, 1981 and Chapter 8). Hypotheses were advanced and testable predictions made about the composition of tropical ichneumonid faunas, but the work was not taken further. This was because no tropical ichneumonid fauna was taxonomically well-enough studied to enable ecologists to segregate morphospecies in the plethora of very similar species encountered. Furthermore the biologies of most species were completely uninvestigated.

A recent taxonomic study of the Australian Ichneumonidae (Gauld, 1984b) has provided data which permit the testing of some of the ecological predictions about tropical ichneumonid faunas. For example, as predicted, the ratio of generalist pupal/prepupal parasitoids to more host-specific larval parasitoids is higher in tropical latitudes than in temperate ones (Fig. 4). Even though the biologies of many Australian species are unknown, taxa could be assigned to one or other biological category on the basis of evaluation of data from the phylogenetically most closely related taxa for which biological data are available (Gauld, 1986).

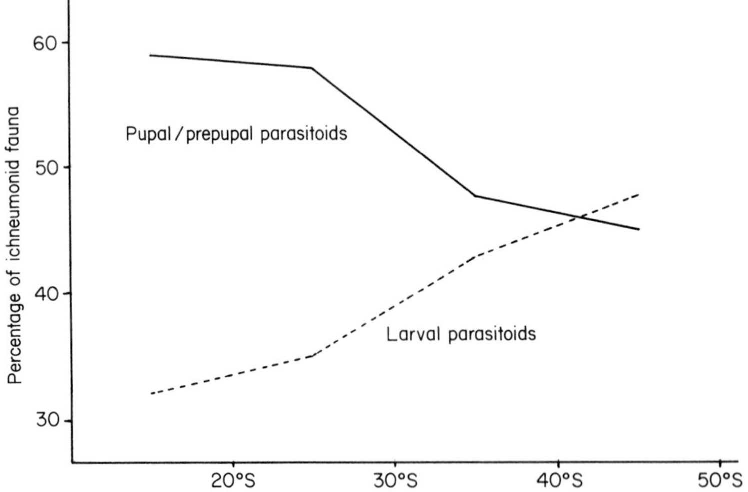

Fig. 4. Graph showing relationships between the percentage of pupal/prepupal parasitoids and larval parasitoids and latitude in the ichneumonid fauna of eastern Australia.

B. The Evolution of Life Histories

Evolutionary frameworks have great potential importance in aiding us to understand problems of co-evolution (Brooks and Mitter, 1984), but so far little progress has been made in investigating parasitoid-host systems. However, in some groups they have been of help in resolving complex and confused ranges of biology. For example, it has long been recognized that species of *Apanteles* exhibit diverse life histories; some are solitary, others gregarious; some are parasitoids of microlepidopterans, others attack butterflies etc. Mason (1981) subdivided *Apanteles* into a number of apparently holophyletic genera, each comprising a group of species that are biologically quite uniform. This phylogenetic classification may help in understanding the development of life history strategies.

As a phylogenetic classification is an explicit representation of the hypothetical evolution of a group, it is possible to develop ideas of the biological evolution and relationship of the component taxa. As many features of the biology of certain taxa are liable to be unknown, such a scenario can be viewed as a testable hypothesis because biological characteristics of these organisms are predicted (Crowson, 1982). For example, Gauld (1983), studying morphological features, recently analysed cladistically the ichneumonid subfamily Labeninae and proposed an hypothesis for the evolution of labenine life histories. Four lineages were recognized, the Labenini (*Labena*

+ *Certonotus* + *Apechoneura*), the Groteini (*Labium* + *Grotea* + *Macrogrotea*), the Poecilocryptini (*Alaothryis* + *Urancyla* + *Poecilocryptus*) and the Brachycyrtini (*Brachycyrtus* + *Habryllia* + *Adelphion* + *Monganella* + *Pedunculus*). The Labenini constituted the most primitive lineage and the Brachycyrtini the most specialized (Fig. 5). From this construct, it was possible, using the meagre details of labenine biology known and one's knowledge of evolutionary trends in a variety of related ichneumonid subfamilies, to suggest a scenario for the biological evolution of labenines (Fig. 6). Several species of the most primitive lineage are known to parasitize larvae and pupae of insects boring in wood, such as buprestids, cerambycids and siricids. Similar host ranges are known for other structurally primitive ichneumonids (e.g. some Pimplinae, Xoridinae; Townes, 1969) and have been postulated for ancestral apocritans (Konigsmann, 1978) so it is feasible that the ancestral labenine lineage had a biology similar to that of modern Labenini. Some poecilocryptines are known to oviposit on to hosts in galls and seeds (Gauld, 1984b). This behaviour is still quite similar to the Labenini as oviposition is through plant tissue to gain access to the host. Poecilocryptines differ from labenines in that they oviposit into softer, more nutritious tissue, and their larvae may well consume plant tissue to complete development (Short, 1978; Gauld, 1983). This scenario predicts that *Urancyla*, whose biology is unknown, will be found to parasitize larvae developing in nutritious plant tissue. It also suggests that the ancestors of the groteine lineage were associated with wood borers. As far as is known all groteines are cleptoparasites of bees; in Australia they attack a variety of ground nesting halictines, and thus a wood-boring ancestor seems, at first sight, to be highly speculative. However, the structurally less specialized *Grotea* and *Macrogrotea* (New World genera) are known to parasitize bees that nest in old borings in standing timber. Although oviposition is through the cap of the nest, these species are still investigating borings in wood. Our concept of a wood-boring ancestor is thus more feasible. It is further strengthened by the fact that in at least two other groups of wood-boring ichneumonids, lineages have arisen that specialize in parasitizing aculeates nesting in tunnels in timber (Townes, 1969; Jussila and Kapyla, 1975; Gupta and Gupta, 1983).

The most specialized labenine lineage, the Brachycyrtini includes only two genera with known biologies, *Brachycyrtus* (parasitoids of Hemerobiidae) and *Adelphion* (predators of spider egg sacs). These insects which are apparently biologically disparate share one important feature—both oviposit through silk. Within the Brachycyrtini these taxa are only distantly related; their most recent common ancestor is probably the ancestor of the entire tribe so it is concluded that this ancestor oviposited through silk, and it is hypothesized that all brachycyrtines will be found to be parasites in cocoons and similar small silken bags. There are no known intermediates between the

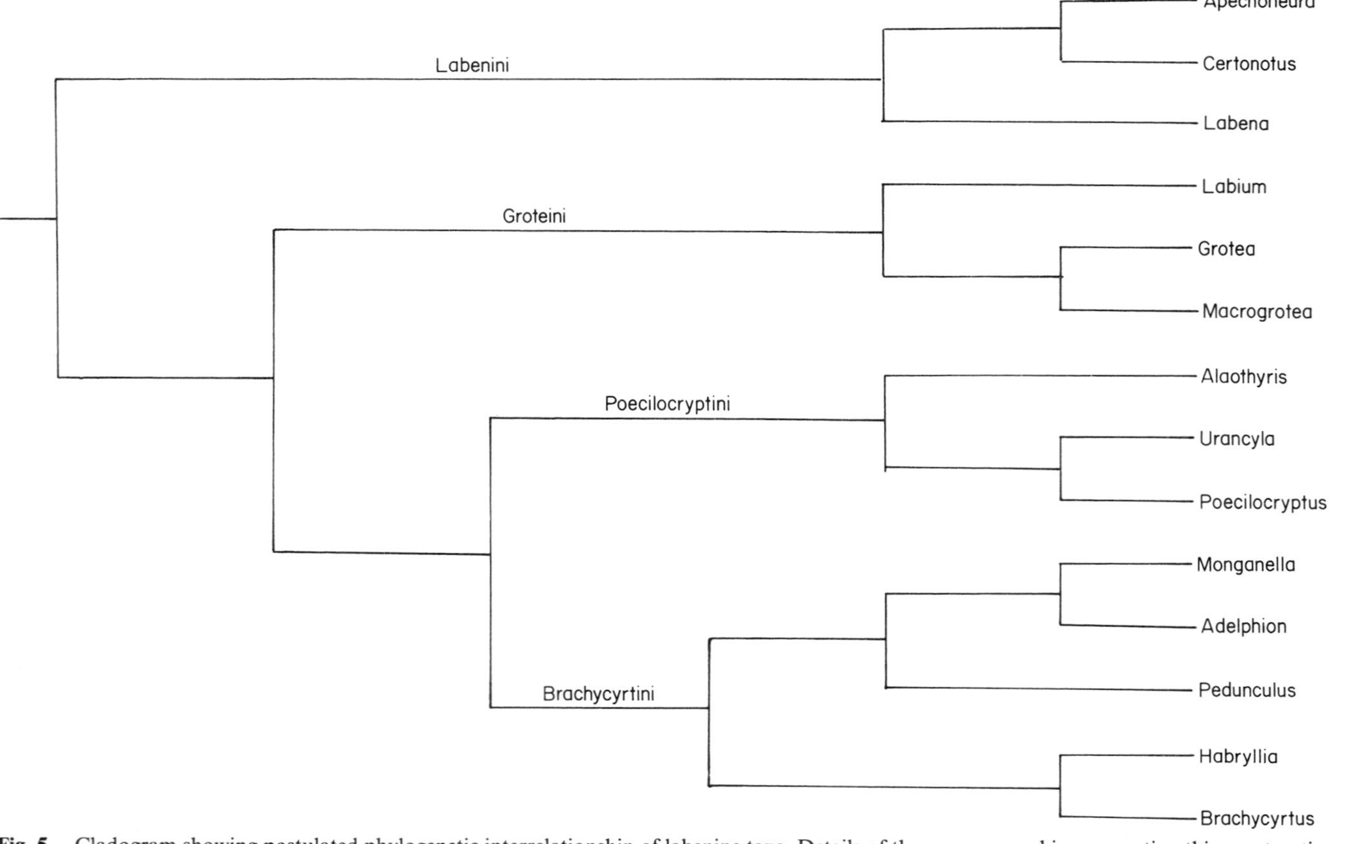

Fig. 5. Cladogram showing postulated phylogenetic interrelationship of labenine taxa. Details of the synapomorphies supporting this construction are given by Gauld (1983, 1984b).

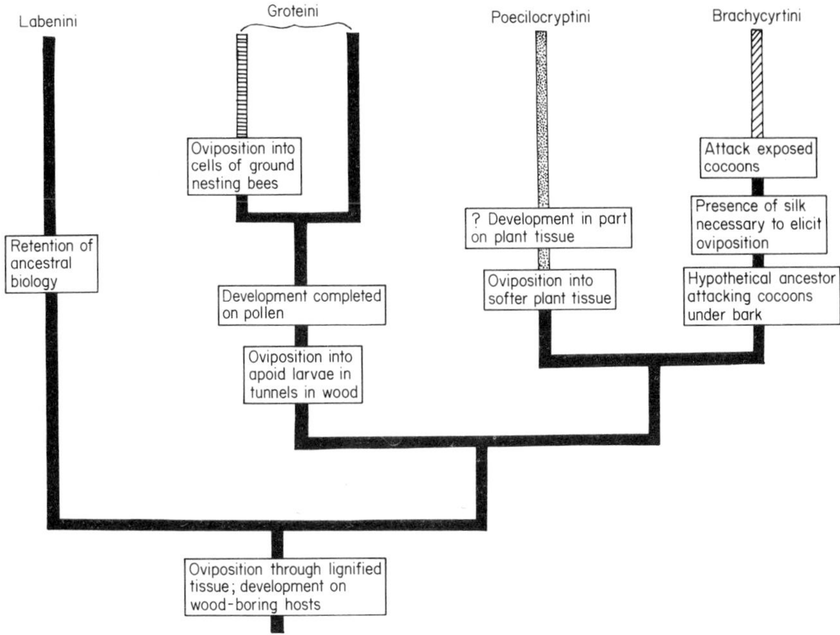

Fig. 6. Scheme showing the possible biological evolution of the Labenine superimposed over the cladogram.

hypothesized wood-boring ancestor of the Brachycyrtini + Poecilocryptini and the exposed silken bag parasitoid ancestor of the Brachycyrtini. Perhaps it is not unreasonable to postulate such an intermediate may have oviposited into silken bags in or under bark of trees. Such parasitoids are known in the Pimplinae (Gauld, 1984c) where they are structurally and biologically intermediate between the parasites of wood borers and the genera which attack exposed silken bags.

C. The Biogeography of Parasitoids

In conjunction with geological, palaeobotanical and palaeoclimatological data, phylogenies may be used to advance plausible historical zoogeographic scenarios. Attempts may be made to explain why certain species occur where they do. Although any individual pattern needs cautious consideration, certain patterns, such as the relatively recent Indo-Malayan biotic invasion of the Australian plate, are repeatedly revealed in analyses of different groups of plants and animals (Gressitt, 1961; Whitmore, 1981). Studies on two groups of ichneumonids, the Pimplinae and Ophioninae (Gupta, 1962; Gauld, 1984a) suggest that extensive radiations in New Guinea have produced relatively few species that colonize Australia. For example, the ophionine fauna of New Guinea includes about 72% endemic species, but only 16% widespread Indo-Papuan taxa. Twenty-four species of Ophioninae are known to occur in Queensland rain forests, of which eight (25%) have as their closest relatives endemic New Guinea taxa. A further 13 Queensland taxa (54%) are widespread Indo-Papuan species whose closest relatives are in south-east Asia. Further south in Australia, the majority of ophionines inhabiting drier areas belong to a single species-group whose nearest relatives are also Asian species. These data suggest that many Papuan species have a low vagility or colonizing ability compared with their Asian congeners, and that the climatic barrier between Australian everwet and xeric habitats (Taylor, 1972) has completely stopped invasion of most of Australia by Papuan species. The very few species of northern origin that have been able to cross this climatic interface have had considerable opportunity to undergo evolutionary radiation in the ichneumonid depauperate dry Australian biome (Gauld, 1984a).

VII CONCLUSIONS

The Hymenoptera Parasitica is a very large group and biologists are unlikely ever to have the time or resources to study all the species. A sound taxonomic framework will enable us to make some sense of this vast assemblage of species. It will enable us to predict biological features of poorly known species and to understand why species occur where they do; it will help us select suitable species to investigate for biological control purposes or as examples of interesting evolutionary strategies; it will enable insect ecologists to begin to study complex tropical faunas. Finally, it will provide entomologists with the means to identify species accurately, which is a vital prerequisite to the international communication of information and ideas.

Acknowledgements

I would like to thank Mike Fitton, Laurence Mound and Jeff Waage for their constructive comments on earlier drafts of the manuscript.

REFERENCES

van den Assem, J., Gijswijt, M. J., and Nubel, B. K. (1982). Characteristics of courtship and mating behaviour used as classificatory criteria in Eulophidae-Tetrastichinae, with special reference to the genus *Tetrastichus*. *Tijdschrift voor Entomologie* **125,** 205–220.

Austin, A. D. (1985). The function of spider egg sacs in relation to parasitoids and predators, with special reference to the Australian region. *Journal of Natural History* **19,** 359–379.

Berlin, B. (1973). Folk systematics in relation to biological classification and nomenclature. *Annual Review of Ecology and Systematics* **4,** 259–271.

in den Bosch, H. A. J., and van den Assem, J. (1986). The taxonomic position of *Aceratoneuromyia granularis* Domenichini (Hymenoptera: Eulophidae) as judged by characteristics of its courtship behaviour. *Systematic Entomology* **11,** 19–23.

Brooks, D. R., and Mitter, C. (1984). Analytical approaches to studying coevolution. *In* "Fungus-Insect Relationships" (Q. Wheeler and M. Blackwell, eds.), pp. 42–53. Columbia University Press, New York.

Cain, A. J. (1959). The post-Linnean development of taxonomy. *Proceedings of the Linnean Society of London* **170,** 234–244.

Crowson, R. A. (1970). "Classification and Biology." Heinemann, London.

Crowson, R. A. (1982). Computers versus imagination in the reconstruction of phylogeny. *In* "Problems of Phylogenetic Reconstruction" (K. A. Joysey and A. E. Friday, eds.), pp. 245–255. Academic Press, London.

Darwin, C. (1859). "On the Origin of Species . . ." John Murray, London.

Disney, R. H. L. (1983). A synopsis of the taxonomist's tasks, with particular attention to phylogenetic cladism. *Field Studies* **5,** 841–865.

Farris, J. S. (1979). The information content of the phylogenetic system. *Systematic Zoology* **28,** 483–519.

Felsenstein, J. (1982). Numerical methods for inferring evolutionary trees. *Quarterly Review of Biology* **57,** 379–404.

Gauld, I. D. (1976). The classification of the Anomaloninae. *Bulletin of the British Museum (Natural History)* **33,** 1–135.

Gauld, I. D. (1979). An analysis of the classification of the *Ophion* genus-group (Ichneumonidae). *Systematic Entomology* **5,** 59–82.

Gauld, I. D. (1983). The classification, evolution and distribution of the Labeninae, an ancient southern group of Ichneumonidae (Hymenoptera). *Systematic Entomology* **8,** 167–178.

Gauld, I. D. (1984a). The Australian Ophioninae (Hymenoptera: Ichneumonidae); a historical biogeographic study. *Journal of Biogeography* **11,** 269–288.

Gauld, I. D. (1984b). "An Introduction to the Ichneumonidae of Australia." British Museum (Natural History), London.

Gauld, I. D. (1984c). The Pimplinae, Xoridinae, Acaenitinae and Lycorininae (Hymenoptera: Ichneumonidae) of Australia. *Bulletin of the British Museum (Natural History)* **49,** 235–339.

Gauld, I. D. (1985). The phylogeny, classification and evolution of parasitic wasps of the subfamily Ophioninae (Ichneumonidae). *Bulletin of the British Museum (Natural History)* **51,** 61–185.

Gauld, I. D. (1986). Latitudinal gradients in ichneumonid species-richness in Australia. *Ecological Entomology* **11,** 155–161.

Gauld, I. D., and Mound, L. A. (1982). Homoplasy and the delineation of holophyletic genera in some insect groups. *Systematic Entomology* **7,** 73–86.

Gould, S. J., and Eldredge, N. (1977). Punctuated equilibria: the tempo and mode of evolution reconsidered. *Palaeobiology* **3,** 115–151.

Gressitt, J. L. (1961). Problems in the zoogeography of Pacific and Antarctic insects. *Pacific Insects Monograph* **2,** 1–127.

Gupta, S., and Gupta, V. K. (1983). The tribe Gabuniini (Hymenoptera: Ichneumonidae). *Oriental Insects Monographs* **10,** 1–313.

Gupta, V. K. (1962). Taxonomy, zoogeography and evolution of Indo-Australian *Theronia* (Ichneumonidae). *Pacific Insects Monograph* **4,** 1–142.

Hennig, W. (1966). "Phylogenetic Systematics." University of Illinois Press, Urbana.

Hull, D. L. (1965). The effect of essentialism on taxonomy: two thousand years of stasis 1. *British Journal for the Philosophy of Science* **15,** 314–326.

Janzen, D. H. (1981). The peak of North American ichneumonid species richness lies between 38° and 42°N. *Ecology* **62,** 532–537.

Joysey, K. A., and Friday, A. E. (1982). "Problems of Phylogenetic Reconstruction." Academic Press, London.

Jussila, R., and Kapyla, M. (1975). Observations on *Townesia tenuiventris* (Hlmgr) (Hym., Ichneumonidae) and its hosts *Chelostoma maxillosum* (L.) (Hym., Megachilidae) and *Trypoxylon figulus* (L.) (Hym., Sphecidae). *Annales Entomologici Fennici* **41,** 81–86.

Kirby, W., and Spence, W. (1815). "An Introduction to Entomology: or Elements of the Natural History of Insects." Longman, London.

Kluge, A. (1976). Phylogenetic relationships in the lizard family Pygopodidae: an evaluation of theory, methods and data. *Miscellaneous Publications of the Museum of Zoology, the University of Michigan* **152,** 1–72.

Konigsmann, E. (1978). Das phylogenetische System der Hymenoptera. Teil 3: Terebrants (Unterordnung Apocrita). *Deutsche Entomologische Zeitschrift* **25,** 1–55.

LeQuesne, W. (1972). Further studies on the uniquely derived character concept. *Systematic Zoology* **21,** 281–288.

Mason, W. R. M. (1981). The polyphyletic nature of *Apanteles* (Hymenoptera: Braconidae): a phylogeny and reclassification of Microgasterinae. *Memoirs of the Entomological Society of Canada* **115,** 1–147.

Mayr, E., Linsley, L. E., and Usinger, R. L. (1953). "Methods and Principles of Systematic Zoology." McGraw-Hill, New York.

Nelson, G., and Platnick, N. (1981). "Systematics and Biogeography: Cladistics and Vicariance." Columbia University Press, New York.

Owen, D. F., and Owen, J. (1974). Species diversity in temperate and tropical Ichneumonidae. *Nature* **249,** 583–584.

Pampel, W. (1913). Die weiblichen Geschlechtsorgane der Ichneumoniden. *Zeitschrift fur Wissenschaftliche Zoologie* **108,** 290–357.

Pankhurst, R. J. (1978). "Biological Identification." Edward Arnold, London.

Patterson, C. (1980). Cladistics. *Biologist* **27,** 234–240.

Popper, K. R. (1975). "The Logic of Scientific Discovery." (Revised edn). Hutchinson, London.

Shaw, M. R. (1983). On[e] evolution of endoparasitism: the biology of some genera of Rogadinae (Braconidae). *Contributions of the American Entomological Institute* **20,** 307–328.

Short, J. R. T. (1978). The final larval instars of Ichneumonidae. *Memoirs of the American Entomological Institute* **25,** 1–508.

Smith, L. K., and Shenefelt, R. D. (1956). A guide to the subfamilies and tribes of the family Ichneumonidae (Hymenoptera) known to occur in Wisconsin. *Transactions of the Wisconsin Academy of Sciences, Arts and Letters* **44,** 165–219.

Stanley, S. M. (1979). "Macroevolution: Pattern and Process." Freeman, San Francisco.

Stevens, P. F. (1984). Metaphors and typology in the development of botanical systematics 1690–1960. Or the art of putting new wine in old bottles. *Taxon* **33,** 169–211.

Taylor, R. W. (1972). Biogeography of insects of New Guinea and Cape York Peninsula. *In* "Bridge and Barrier: the Natural and Cultural History of Torres Strait" (D. Walker, ed.), pp. 213–230. Australian National University Press, Canberra.

Townes, H. (1969). Genera of Ichneumonidae 1. *Memoirs of the American Entomological Institute* **11,** 1–300.

Townes, H. (1971). Genera of Ichneumonidae 4. *Memoirs of the American Entomological Institute* **17,** 1–372.

Underwood, G. (1982). Parallel evolution in the context of character analysis. *Zoological Journal of the Linnean Society* **74,** 245–266.

Watrous, L. E., and Wheeler, Q. D. (1981). The out-group comparison method of character analysis. *Systematic Zoology* **30,** 1–11.

Whitmore, T. C. (1981). "Wallace's Line and Plate Tectonics." Clarendon Press, Oxford.

Wiley, E. O. (1981). "Phylogenetics: The Theory and Practice of Phylogenetic Systematics." John Wiley, New York.

2

An Evolutionary Approach to Host Finding and Selection

J. J. M. VAN ALPHEN AND L. E. M. VET

I. INTRODUCTION

In nature, female parasitoids seek hosts on a variety of substrates, moving on and between them in a complex non-random manner. They respond to a hierarchy of physical and/or chemical stimuli which lead them to their potential hosts. After a host is encountered, other stimuli are used to decide whether to oviposit or not.

Because of the direct link between successful searching and the production of offspring, we can expect parasitoid searching behaviour to be strongly influenced by natural selection. Therefore, parasitoid host searching and host selection behaviour is an ideal subject for testing optimization hypotheses. Although this approach is gaining popularity and has been shown to be powerful in explaining the foraging decisions of animals, thorough experimental tests of optimization hypotheses by laboratory and/or field experiments are few.

Most studies of parasitoid foraging behaviour focus on the causes of isolated steps of the host finding process (Vinson, 1981; Weseloh, 1981; Arthur, 1981). The results have contributed greatly to our understanding of *how* parasitoids find their hosts, but form only the necessary basis for answering the question *why* they behave as they do. Understanding the adaptive significance—i.e. the function—of observed behaviour is a research aim equally as necessary as trying to analyse how behaviour is caused: ideally both should go hand in hand. Furthermore, a functional approach can provide us with practical predictions about optimal parasitoid densities in mass rearing units, predictions about how parasitoids will allocate their time over areas with different host densities, and about how they will adapt host choice and sex allocation in response to host and parasitoid densities.

To explore the adaptive nature of parasitoid behaviour, we must first understand the context in which they forage. This requires a detailed knowledge of the distribution of the host population in space and time, and information on other relevant parameters such as the effect of competitors and the risk of predation. The small size of most species makes them well suited for behavioural studies in the laboratory, but hampers their observation in the field. This may be why studies on parasitoid foraging are virtually restricted to the laboratory. This lack of field data, in turn, makes it hard to design laboratory experiments with realistic patterns of host abundance and distribution. This limitation may explain why some parasitoid ecologists, after a promising start, have left the question of optimal foraging for patchily distributed hosts and taken up problems of optimal sex allocation, which can be studied more easily without detailed field knowledge.

In order to explore an optimization hypothesis, we must, of course, determine what should be optimized. In parasitoid foraging studies, it is generally assumed that offspring produced per unit of foraging time is the currency in which gains are expressed. Females should try to maximize their total genetic contribution to future generations: the maximization of oviposition rate is not the only strategy to achieve this goal. Iwasa *et al.* (1984) have considered the effect on an optimal foraging model of assuming limitation in the number of eggs, rather than limitation in searching time, but no tests of their model exist.

The costs of foraging, in terms of energy expenditure or mortality risk, have been largely neglected. Energy budgets of parasitoids have never been measured; to do so could be rewarding, because parasitoids may resorb eggs and use this energy to fly (Prokopy and Roitberg, 1984).

The most adaptive solution which the animal can realize may differ from the one predicted by an optimization model because of biological constraints. Therefore, it may be useful to define the set of possible strategies [e.g. parasitoids may use different searching modes (Vet and van Alphen, 1985)]. These have been studied in vertebrates but so far have been neglected by students of parasitoids.

Three phases can be distinguished in the host searching and host selection behaviour of parasitoids: host habitat location, host location and host selection (Doutt, 1959). We will use this division in reviewing searching behaviour.

At all these levels, the behaviour of the parasitoids may be influenced by previous experience. We will stress this throughout and suggest functions for this plasticity.

Our aim is to evaluate the state of the art with respect to evolutionary aspects of host searching and host selection by parasitoids, and to point to areas that deserve more attention. We hope that this review will encourage research effort on rigorous tests of the rapidly growing body of theory.

II. HABITAT LOCATION

In the last decade, several reviews have been written on causal aspects of host finding behaviour, including habitat location (Vinson, 1976, 1981) but there is little or no information on how parasitoids locate macrohabitats (e.g. forests or grassland communities). Most literature deals with orientation to microhabitats (i.e. potential sources of host food, such as plants). In many cases, it is difficult to distinguish between microhabitat and host location as parasitoids may use cues emitted by the host rather than the host habitat in long distance orientation. Different species may use different sensory modes in locating a microhabitat. There are examples of visual orientation: van Alphen (unpublished) showed that *Diaparsis truncatis* Gravenhorst, an ichneumonid parasitoid of the 12-spotted asparagus beetle, is attracted to wooden models of asparagus berries (Table I), while Glas and Vet (1983) showed that an opiine parasitoid of the apple maggot is attracted to

TABLE I. Visual orientation by *Diaparsis truncatis* Gravenhorst, a larval parasitoid of the 12-spotted asparagus beetle. Antennal contacts with green asparagus berries and green wooden beads. The wooden beads are larger than the berries and attract the parasitoid significantly more often (χ^2 test: $P<0.001$)

	Asparagus berries	Green wooden beads
Number on plant	170	69
Number of contacts by *Diaparsis*	275	241
Expected number of contacts	367	149

hawthorn berries. The tachinid *Eucarcelia rutilla* Villeneuve responds more strongly to twigs of *Pinus silvestris* L., the main food plant of its host, than to twigs of deciduous trees (Herrebout, 1969).

Cade (1975) gave an example of orientation to sound: the tachinid *Euphasiopterix ochracea* (Bigot) is attracted by the calls of its hosts, male crickets.

However, olfaction seems to be the most common sensory mode. The literature provides many examples of responses to microhabitat odours, particularly for hymenopterous (Vinson, 1976, 1981), but also for tachinid parasitoids (Monteith, 1955; Herrebout and van der Veer, 1969).

In order to discuss the function of host habitat location in more detail, we now focus on a group of parasitoid species that attack larval Drosophilidae.

A. Habitat Location by Parasitoids of *Drosophila* spp.

Parasitoids of *Drosophila* spp. are faced with a choice of potential microhabitats such as fermenting fruits, decaying mushrooms, decaying plant materials and sapstreams of bleeding trees (Janssen *et al.*, unpublished). Field data suggest microhabitat specificity in the majority of larval parasitoids. This ecological specialization is, to some extent, also found in the host larvae (Shorrocks, 1977). Some parasitoid species such as *Asobara tabida* Nees attack larvae only in fermenting fruits and sap streams on bleeding trees (Janssen *et al.*, unpublished); a number specialize in attacking hosts in decaying fungi, e.g. the eucoilids *Leptopilina clavipes* (Hartig) and *Kleidotoma bicolor* (Giraud), while others such as *Leptopilina fimbriata* (Kieffer), *Kleidotoma dolichocera* Thomson and *Asobara rufescens* (Förster) attack drosophilids in decaying plant materials (Vet and van Alphen, 1985). Two species, *Leptopilina heterotoma* (Thomson) and *Aphaereta scaptomyzae* Fischer parasitize hosts in all these habitats (Janssen *et al.*, unpublished).

Laboratory experiments with an airflow olfactometer support the hypothesis of microhabitat specialization. In the olfactometer, parasitoids show a specific innate preference for the odour of the microhabitat in which they occur in nature (Vet, 1983, 1985a; Vet *et al.*, 1983, 1984), and also for the stage of decay of the microhabitat. *Leptopilina clavipes*, for example, is only attracted by the odour of mushrooms which have opened their caps and started to decay, a stage most likely to contain young host larvae suitable for oviposition (Vet, 1983). *Asobara tabida* is attracted to freshly fermenting fruit media, whereas its potential competitor, *L. heterotoma*, prefers older and more decayed media (Vet *et al.*, 1984; Vet and van Opzeeland, in press). This difference is also reflected in their respective high and low alcohol tolerance (Boulétreau and David, 1981). Thus some degree of niche segregation appears to occur at an early stage of the searching process.

This clear innate preference for a specific microhabitat odour was the clue to the discovery of two sibling *Asobara* species in the same macrohabitat (i.e. in the same field), one of which attacks Drosophilidae in fruits, the other in decaying leaf material (Vet *et al.*, 1984). The adaptive value of such preference is linked to larval survival: each microhabitat harbours a different drosophilid fauna and survival in the most common species in the preferred microhabitat was significantly higher than in the most common drosophilid in the non-preferred habitat.

These findings have led to speculations about speciation in the genus *Asobara* (Vet and Janse, 1984). The evolutionary importance of habitat preference and especially of genetic changes in this preference has been emphasized before by proponents of both sympatric (Bush, 1969, 1974; Bush and Diehl, 1982) and allopatric speciation (Mayr, 1970).

The olfactometer experiments showed that species often differ only in the pattern of odour preference. Thus, some are attracted to odours of other microhabitats when the most preferred odour is absent (Vet, 1983; Vet *et al.*, 1984). The strength and persistence of these preferences are likely to be different for habitat specific and habitat generalist species.

B. Learning and Habitat Location

Several of the odour responses in *Drosophila* spp. parasitoids have proven to be variable and to some extent dependent on prior experience as an adult and, perhaps, even as a larva.

If we define learning quite broadly as any change in behaviour due to experience, there is considerable evidence that parasitoids can learn and learning has been reported for ichneumonids (Arthur, 1966, 1971; Taylor,

1974; Sandlan, 1980; Wardle and Borden, 1985), braconids (Vinson *et al.*, 1977; Dmoch *et al.*, 1985) and tachinids (Monteith, 1963).

A detailed study of conditioning to hosts has been made for *L. clavipes* (Vet, 1983). This species has a preference for the odour of decaying fungi, but shows intraspecific variation in its attraction to the odour of yeast (which represents another potential microhabitat: fermenting fruits). A fraction of the population is consistently attracted to yeast odour. By culturing *L. clavipes* on hosts feeding in a yeast medium this fraction can be increased, although mushroom odour is still significantly preferred in tests. This increased attraction to yeast odour is at least partly due to conditioning in the larval stage. Conditioned responses to yeast odour are, however, much stronger in the adult stage, when adult females are given oviposition experience with host larvae feeding on yeast. Following this experience, the attraction to the odour of yeast is so strong that females now repudiate the odour of decaying fungi and prefer the odour of yeast in choice experiments.

Leptopilina clavipes attacks mainly *Drosophila phalerata* Meigen, the most common fungivorous *Drosophila* species in western Europe. Both fly and parasitoid are found in several species of fungi. On some occasions, *D. phalerata* was found to breed in fruit baits from which *L. clavipes* was reared (Janssen *et al.*, unpublished). The parasitoids must have responded to the odours of the fruit bait as they do not make use of volatile kairomones (Vet, 1983). Similar behavioural plasticity is reported for several other parasitoids of fly larvae (Vet, 1985b; Vet and van Opzeeland, 1984, 1985a), and may be the rule rather than the exception in polyphagous or oligophagous parasitoids. The ability to learn cues in searching (as opposed to genetically fixed searching behaviour) may be more adaptive for polyphagous, habitat generalists, where successive generations may be exposed to different host species, living in different microhabitats. Learning may also be adaptive for longlived species, because it allows them to switch to new microhabitats as they become profitable.

Much present knowledge on learning in parasitoids is anecdotal and greatly in need of further investigation. As yet we have no idea of the reversibility or persistence of the learned behaviour, nor do we know exactly the mechanisms involved. Learning ability may differ strongly between species and there are indications that it can change with age (Vet, 1983; Wardle and Borden, 1985).

C. Long Distance Host Location

So far we have neglected the influence of the host on microhabitat location. Some egg parasitoids, e.g. *Trichogramma* spp., have been shown to respond to pheromones of their adult hosts (Lewis *et al.*, 1982; Noldus and van Lenteren, 1985). They may be attracted in this way to areas where adult moths are present and host eggs are likely to occur. For larval parasitoids, there are numerous examples of anemotactic responses to host-associated or host-produced volatiles (see Vinson, 1981; Weseloh, 1981 for examples).

An analysis of the olfactory responses of larval parasitoids of *Drosophila* spp. showed that none of seven eucoilid and only two of five alysiine species could distinguish between the odours of host-infested and uninfested substrates (Dicke *et al.*, 1984; Vet, 1983; Vet, 1985b). Vet and van Opzeeland (1984) found that the two alysiine species which did use host cues acquired this ability through conditioning during oviposition. Naive females showed no differential response.

We cannot give an adaptive explanation for these differences. All species respond to host habitat odours, and fermenting or decaying substrates are seldom without drosophilid larvae. Responses to host-derived cues therefore seem superfluous. More research is needed on habitat choice and the host spectrum successfully attacked by each parasitoid species.

Summarizing then, studies on host habitat location in *Drosophila* spp. parasitoids have shown that most are microhabitat specialists. They are either attracted to the odour of only one microhabitat, or show an innate preference for one habitat. Only two species were found to be generalists with respect to habitat choice, although preference could be modified by conditioning in a number of others.

III. HOST LOCATION

A. Contact Kairomones

Once parasitoids have found a microhabitat, the sequence of responses is often continued with a reaction to chemical stimuli deposited by the host or caused by the host's presence. There is a considerable variety in the sources of these chemicals (Waage, 1978). The use of such contact kairomones is a common phenomenon (Jones, 1981; Weseloh, 1981), and there is a variety of ways in which parasitoids respond to them (Waage, 1978). In several species a response is characterized by a change in walking speed (orthokinesis) and an increase in the rate of turning (random in orientation: klinokinesis; directed: klinotaxis). A combination of a reduction in walking speed when a kairomone is encountered and a klinotactic response when contact is lost is the underlying mechanism for so-called "arrestment". Like all other responses in the searching sequence, the responses to kairomones are likely to differ between different species, depending on the distribution of hosts and kairomones. For example, Vet and van der Hoeven (1984) compared the response of two *Leptopilina* species which differ in their microhabitat and host range. Upon encountering kairomone in a substrate, both reduce their walking speed and increase their probing frequency. However, *L. fimbriata* is arrested more strongly, stops more often and walks for shorter distances, whereas *L. heterotoma* walks more and stops less often, thereby covering more surface area. When offering these parasitoids a host-infested patch over which a kairomone solution is spread, *L. fimbriata* cannot find individual host larvae, whereas *L. heterotoma* finds most of them (Vet and Bakker, 1985). These differences are possibly adaptations to differences in host distribution and density in natural habitats. *L. fimbriata* finds its host, *Scaptomyza pallida* (Zetterstedt), in decaying plants. The host lays its eggs singly under wilted leaves, and each larva feeds on its own piece of leaf tissue. Kairomone concentrations are likely to be highest where a larva is actually feeding, and lower or zero in other parts of the substrate. The presence of the kairomone then indicates that a host is nearby, which makes strong arrestment functional. This contrasts with the more mobile and gregariously feeding *Drosophila* spp. larvae in the microhabitats which *L. heterotoma* visits.

Besides restricting the area of search and intensifying searching behaviour through orthokinetic and klinokinetic/tactic responses, several other functions can be associated with kairomones. Parasitoids may use them as a trail to guide them to an individual host (Klomp, 1981). Alternatively, kairomones can give information on host density (Corbet, 1973; Galis and van Alphen, 1981). Finally, parasitoids can obtain information on the identity of host species through species specific kairomones, before hosts have been encountered (Gardner *et al.*, unpublished; van Alphen *et al.*, unpublished). In some cases recognition of specific kairomones is acquired by learning, through the association of the host and its kairomone during oviposition (Strand and Vinson, 1982; Vet, 1985b; van Alphen *et al.*, unpublished). This property can be used in biological control to condition polyphagous parasitoids to target host species (see Chapter 12).

B. A Comparative Approach to Host Location

Different species have evolved in response to different environmental factors, and by comparing species we may find a correlation between a behavioural response and an environmental factor indicative of adaptation (Lack, 1971; Schoener, 1974; Ridley, 1982; Wanntorp, 1983; Thornhill, 1984). So far, this approach has not been used in parasitoid-host location research.

Vet and van Alphen (1985) and Vet and Bakker (1985) compared the way parasitoids of Diptera search for host larvae within patches. This study shows how comparative research on closely related and less related species can help to reveal whether a behavioural trait is a recent adaptation or whether it has been passed on from ancestral species (of course the trait can still be adaptive). Their largest comparison involved 32 species of Alysiinae (Braconidae) and 25 species of Eucoilidae from a variety of microhabitats and host species. Three main searching modes were distinguished: (1) vibrotaxis (movement of the host larva is the cue to detection), (2) ovipositor searching (larvae are detected by regular probing of the substrate while the parasitoid walks) and (3) antennal searching (larvae are detected through antennal contact). Some species use only one of these, while others show a more complex behaviour that consists of a combination of two or three modes.

The species can be classified according to the role these different searching modes play in host detection and this classification can then be compared with phylogenetic classification based on a cladistic analysis of morphological characters (Nordlander, 1982; van Achterberg, personal communication). The two show a fair amount of congruence. In most cases species belonging to one genus have a similar behaviour pattern. Although all searching modes are present in both families, vibrotaxis seems to be the most common in the Alysiinae, whereas ovipositor searching is more general in the Eucoilidae. Among the eucoilids, *Leptopilina* spp. spend most of their searching time probing the substrate with their ovipositors. The percentage is lower in *Kleidotoma* spp., and practically zero in *Ganaspis* spp., which spend most of their time standing still (Fig. 1). The correlation between time spent standing still and percentage larvae detected by vibrotaxis (Fig. 2a) suggests that the wasps detect host movement while standing still. From Fig. 2b we can conclude that hosts not found by vibrotaxis are detected through probing behaviour. All the eucoilids used in this study were *Drosophila* spp. parasitoids. A number of them may be found searching together on the same fruit. It is likely that different searching modes will produce different encounter rates when parasitoids search in the same patch. However, even within one fermenting fruit there may be local differences in the thickness and structure of the substrate. Also the parasitoids may search for different host stages or species. Ovipositor searching may be a better mode when searching at high host density, while vibrotaxis may be more economical at low densities. These hypotheses should be tested before we can understand the function of the different searching modes. This may be an important key to explain how closely related parasitoids using the same hosts in the same microhabitat can coexist.

Some exceptions were found to the general rule that species within a genus

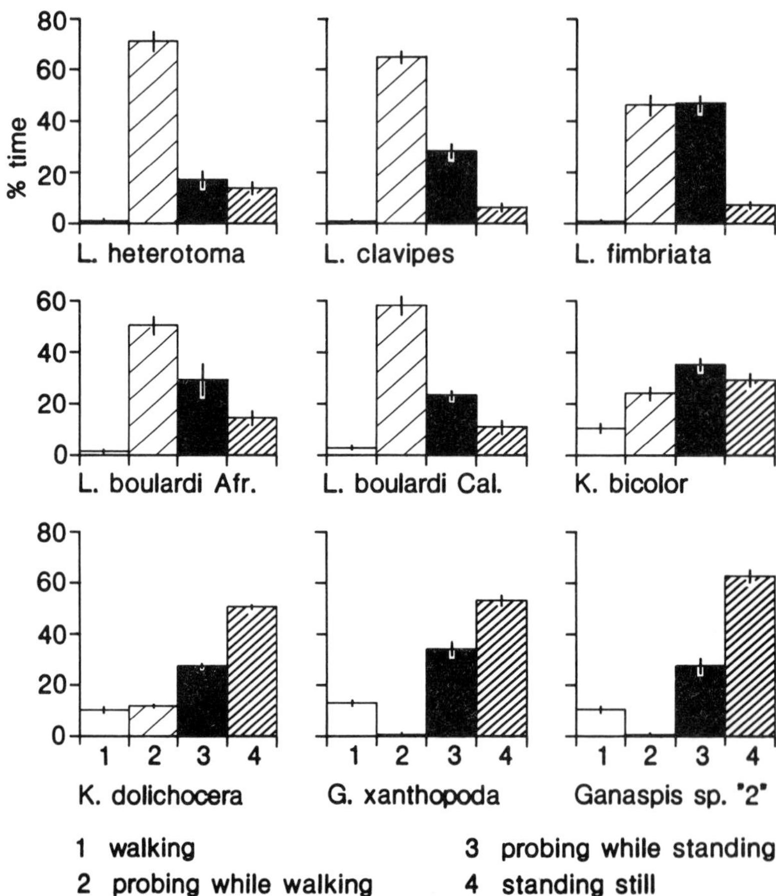

Fig. 1. The average time allocation (± s.e.) of searching parasitoids of *Drosophila* spp. to four different behaviours: walking, probing the substrate with their ovipositor while walking, probing while standing, and standing still. L = *Leptopilina*; K = *Kleidotoma*; G = *Ganaspis* (after Vet and Bakker, 1985).

have the same searching mode. These are likely to be caused by adaptive radiation, for example, two species of *Tanycarpa* (Alysiinae) that attack drosophilid larvae in different microhabitats. Based on outgroup comparison with other Alysiinae we can consider the antennal searching behaviour of *T. punctata* Achterberg, a derived trait (most alysiines use vibrotaxis to locate their hosts). This species prefers older hosts, unlike all the other larval parasitoids of drosophilid flies which we studied. The females drum the fruit substrate with the tips of their antennae to detect the hind spiracles of the

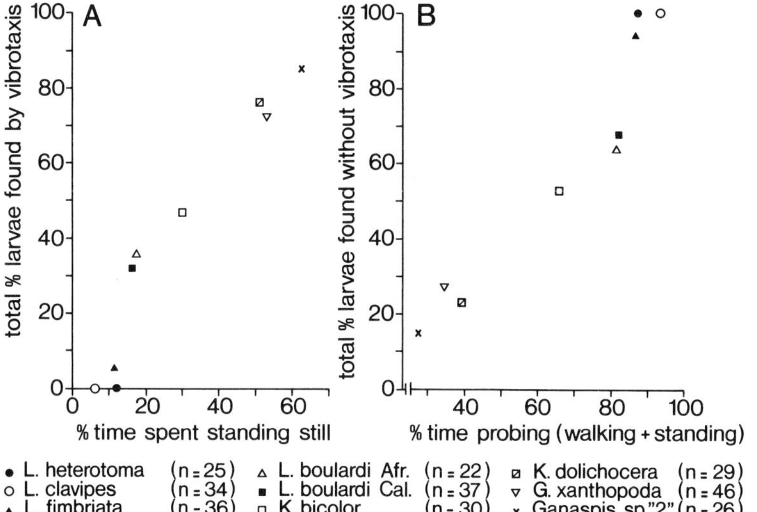

Fig. 2. (a) Relationship between the average percentage of time spent standing still by searching females and the total percentage of larvae found by vibrotaxis ($r = 0.95$), i.e. percentage of encounters with host larvae preceded within 60 s by a behavioural reaction of the female to a host moving in her vicinity. L = *Leptopilina*; K = *Kleidotoma*; G = *Ganaspis* (after Vet and Bakker, 1985). (b) Relationship between the average percentage of time spent probing (while walking and while standing still) by searching females and the total percentage of larvae found without vibrotaxis ($r = 0.98$). L = *Leptopilina*; K = *Kleidotoma*; G = *Ganaspis* (after Vet and Bakker, 1985).

older larvae which at this stage lie in the substrate, with their hind spiracles protruding from the surface. Unlike its relative *T. bicolor* and other alysiine *Drosophila* spp. parasitoids which attack younger larvae, *T. punctata* does not react to young larvae crawling over the surface.

An analogous example is reported by Glas and Vet (1983) in closely related opiine parasitoids which attack the apple maggot. The species that attacks the eggs uses the oviposition deterring pheromone of the adult fly as well as the fly's oviposition puncture to detect the host egg (Prokopy and Webster, 1978), while the larval parasitoid finds its host by vibrotaxis.

The different searching modes of larval parasitoids of *Drosophila* spp. may have different consequences for genetic variation in the behaviour of host larvae. Carton and David (1985) studied the effect of *Leptopilina boulardi* (Barbotin, Carton & Kelner-Pillault) on two types of *D. melanogaster* Meigen larvae, of which one penetrated deep in the host medium and the other remained on the surface. The larvae which remained on the surface suffered more parasitism than those living deep in the substrate. Sokolowski

(1980, 1982) and Bauer and Sokolowski (1984) identified another genetically determined behavioural polymorphism. One genotype called "rovers" is highly mobile; the other, called "sitters" spends less time crawling. It is likely that *Asobara tabida* and other parasitoids searching by vibrotaxis will find "rovers" more easily than "sitters".

IV. PATCH TIME ALLOCATION BY PARASITOIDS

Because of the importance of parasitoids for the regulation of host populations and their use in the biological control of insect pests, there is a strong and persistent interest in the population dynamics of parasitoid-host systems. A successful approach to understanding these systems has been to explain population dynamics from the foraging behaviour of individual parasitoids (Hassell, 1978). A central question is how individuals allocate their searching time over different parts of the habitat. We discuss the factors which influence such decisions and review them in the light of current foraging theory. Population dynamical consequences are discussed in Chapter 7.

We use the term "patch" for a limited area in which a parasitoid searches for hosts. Its boundaries should be defined by the searching behaviour of the parasitoid: a patch is what a parasitoid perceives as a patch. Sometimes it is easy to define a patch this way, e.g. when chemical cues tell the parasitoid to restrict its search within a certain area (Waage, 1978), but at other times it is impossible to define a patch unambiguously, because no clear boundaries exist. Also, there may be different hierarchical levels of patchiness (Hassell and Southwood, 1978). It will be clear that for this reason it may also be difficult to define local host density. Even if we know the patch size exactly, and are able to calculate density, we should realize that a parasitoid experiences encounter rates, and in many situations it will be unaware of absolute density. We use "patch" and "density" where our examples allow us, or when we discuss general models of parasitoid behaviour.

A. Patch Time Allocation by Individual Parasitoids

Following the seminal papers by MacArthur and Pianka (1966) and Emlen (1966), many ecologists have explored the foraging behaviour of animals assuming that natural selection will favour those individuals which exploit their resources most efficiently. If this assumption is true, animals can be expected to allocate their foraging time in an optimal way and hence to maximize their encounter rate with prey or hosts. For a known distribution

of hosts or prey within a habitat, the optimal allocation of foraging time can be calculated from a simple model (Charnov, 1976). This model, known as the marginal value theorem, predicts that each patch within a habitat should be exploited until the encounter rate within the patch has decreased to a marginal value which is the same for all patches. Because animals cannot be expected to know the exact abundance and distribution of prey or hosts in the habitat, it is generally assumed that they use simple rules to decide how long to stay in a patch.

Three rules for patch time determination have been suggested in the literature relating to the impact of predators and parasitoids on population dynamics: (1) number expectation (Krebs, 1973), (2) time expectation (Gibb, 1962) and (3) giving-up time (Hassell and May, 1974; Murdoch and Oaten, 1975). We describe these rules and explore the circumstances in which they may result in an adaptive, if not optimal, allocation of foraging time.

The number expectation rule assumes that a predator or parasitoid leaves a patch after it has encountered a fixed number of hosts. Such a rule, on its own, would doom a parasitoid to stay forever in a patch without hosts if brought there by chance, but in patches where hosts were found, it could provide a useful mechanism. Strand and Vinson (1982) found that *Cardiochiles nigriceps* Viereck leaves a tobacco leaf immediately after it has completed oviposition in a host larva. This could be explained as the most simple case of a number expectation rule. Here the rule may be adaptive because of the low variation in host density between patches.

The time expectation rule assumes that a parasitoid leaves a patch after it has spent a fixed amount of searching time. Weis (1983) suggested a fixed time rule for *Torymus capite* (Huber), because it leaves a gall before all hosts are parasitized and time spent per gall is not correlated with the number of hosts present. However, the time spent per gall was not constant, which should be the case if the parasitoids really used a fixed time rule. Weis also suggested that the fixed time rule can explain the inverse density dependent parasitism found by Morrison *et al.* (1980) and Morrison and Strong (1981).

Iwasa *et al.* (1981) have shown that both the fixed time rule and the number expectation rule are adaptive when the variance in patch densities is small and may result in an optimal allocation of time.

The fixed giving-up time rule assumes that a parasitoid leaves an area if it has not encountered unparasitized hosts after a fixed time. This rule has been used to test Charnov's (1976) marginal value theorem, assuming that giving-up times are inversely related to the threshold rate of encounter (Cook and Hubbard, 1977; Krebs *et al.*, 1974). There exists no strong evidence that parasitoids do use fixed giving-up times to determine patch time. Cook and Hubbard's (1977) optimal foraging model, based on fixed giving-up times, predicted patch times which were at least in qualitative agreement with those found in their experiments, but this does not prove that parasitoids do indeed use a fixed giving-up time.

Density dependent variation in optimal giving-up times have been proposed by McNair (1982), who showed that to increase giving-up time with host density could be an optimal strategy. Van Alphen and Galis (1983) found that giving-up times of *Asobara tabida*, while highly variable at any density, did increase with host density. However, patch time in *A. tabida* is determined in a more complex way, as discussed below.

Waage (1978, 1979), in a set of elegant experiments, showed that patch time in *Venturia canescens* (Gravenhorst) is determined by several interacting factors: the wasps react to a contact kairomone associated with the silk webbing produced by the host larvae. The parasitoid turns sharply when the edge of the patch is reached, and so remains on the patch until this response to the patch edge has gradually waned by habituation or sensory adaptation. Whenever a host is found, oviposition increases the responsiveness to the patch edge. A simple behavioural model in which the time increment added by an oviposition depends on the interval since the last oviposition, predicted patch times in a way similar to Cook and Hubbard's (1977) optimal foraging model. There is an interesting similarity between Waage's (1979) model and that of Iwasa *et al.* (1981), who developed a function for the estimator of the number of unparasitized hosts remaining in a patch: when hosts follow a negative binomial distribution the shape of this function is very similar to Waage's behavioural model. This supports the idea that the way in which patch time is determined in *V. canescens* is adaptive, assuming that hosts in nature have a clumped distribution.

Waage's experiments gave no support for a fixed giving-up time as the determinant of patch time since times were highly variable and not related to host density, as his model predicted. As Oaten (1977) and Green (1979, 1984) have argued, Charnov's (1976) and Cook and Hubbard's (1977) models are deterministic and assume omniscience in the foraging parasitoid. It is more realistic to consider the foraging process as stochastic and to assume that the parasitoid must use past encounters with hosts or other clues to obtain an expectation of the future encounter rate, and Waage's behavioural model does this.

Thus, there are a number of factors which may affect patch time in parasitoids and simple rules may not be adequate to predict their behaviour.

1. Responses to Patch Structure

Some factors relate to the form of the patch itself. Thus, *Leptopilina heterotoma* increases searching time on empty yeast patches with increasing patch size (van Lenteren and Bakker, 1978), and searches longer on patches with trails of crawling larvae, than on patches of similar size and kairomone concentration with a smooth surface (van Alphen and Bakker, in press).

2. *Responses to Chemical Stimuli*

Other factors are related to the presence of kairomones in the patch. Waage (1978, 1979), Galis and van Alphen (1981) and Dicke *et al.* (1985) found increasing patch times with increasing kairomone concentrations, at least over part of the range of concentrations. If kairomone concentration is proportional to host density (Corbet, 1973) then the parasitoid can obtain an estimate of host density before it encounters a host. A response to chemical stimuli may indicate the profitability of a patch in other ways. *A. tabida* females spend more time in patches with a mixture of host species when these contain a higher fraction of their preferred host, *D. subobscura* Collin (van Alphen and van Harsel, 1982). Galis and van Alphen (1981) found that yeast suspensions in which high densities of *D. melanogaster* larvae had fed for 18 h lose their attractivity for *A. tabida*, even though these substrates contain high concentrations of kairomone. In choice experiments the hosts left these patches and migrated to fresh yeast suspensions. Patch quality therefore indicates the probability of finding hosts in a patch.

3. *Response to Unparasitized Hosts*

Encounters with unparasitized hosts will affect patch time allocation. As noted, oviposition increases patch time in *V. canescens* (Waage, 1979) but induces patch leaving in *Cardiochilis nigriceps* (Strand and Vinson, 1982). These differences can be explained as adaptations to host distribution since increases in patch time after an oviposition are only functional when more hosts can be expected in the same patch. *A. tabida* (van Alphen and Galis, 1983) and *L. heterotoma* (van Lenteren and Bakker, 1978) both increase searching time in a patch after an oviposition and host distributions in these species are generally highly clumped.

4. *Response to Parasitized Hosts*

Waage (1979) and van Lenteren (1981) suggested that encounters with parasitized hosts also affect patch time. This may be true for two reasons. While exploiting a patch, the ability to discriminate between parasitized and unparasitized hosts allows the parasitoid to perceive the decreasing rate of encounter with unparasitized hosts and thus provides it with information on the level of exploitation. Also, encounters with parasitized hosts may decrease the motivation to continue search in an analogous way to that by which ovipositions cause an increase. This was shown by van Alphen and Bakker (in press) who found that *Leptopilina heterotoma* searches longer when exploiting patches from which the parasitized hosts were removed immediately after oviposition than when the parasitoid exploited patches with the same density from which parasitized hosts were not removed. In both sets of experiments the wasps experienced a decreasing rate of encounter with unparasitized hosts. They spent more time on patches from which the parasitized hosts had been removed (Table II). Hence, encounters with parasitized hosts in the other set of experiments caused a decrease in the motivation to continue search. Other experiments corroborated this finding (van Alphen and Bakker, in press) but experiments by van Alphen and Galis (1983) failed to show such an effect in *Asobara tabida*.

TABLE II. The effect of encounters with parasitized hosts on the motivation to continue search. Data from experiments with *Leptopilina heterotoma* searching on a patch with 32 *Drosophila melanogaster* larvae (from van Alphen and Bakker, in press)

	Parasitized hosts			
	Left in patch		Removed	
Mean searching time (s)	2995	s.d. 402	4091	s.d. 1398*
Mean GUT (s)	275	s.d. 222	446	s.d. 502
Number of ovipositions	25.9	s.d. 4.7	26.7	s.d. 6.1
Number of rejections	10.7	s.d. 4.8	1.8	s.d. 2.2

*$P \leqslant 0.05$; Mann-Whitney U-test.

5. *Encounters with Parasitoids or Their Traces*

Encounters with other parasitoids, or with their traces, are known to affect patch time. Observations by Hassell (1971) and Beddington (1975) suggest that searching parasitoids tend to leave the patch after encountering one another. Waage (1979) and van Lenteren and Bakker (1978) found that successive visits of *V. canescens* and *L. heterotoma* to the same patch were shorter than the first visit. They suggested patch marking as a possible explanation, but shorter return visits may also be caused by a reduced motivation to search, by habituation to kairomone concentration or because the parasitoid recognises the patch from visual landmarks. Dicke *et al.* (1985) later showed that *L. heterotoma* does not mark previously visited areas, but Harrison (1985) found that *V. canescens* indeed leaves a deterrent pheromone trail. Price (1970), Greany and Oatman (1972) and Galis and van

Alphen (1981) demonstrated that marking of searched areas does occur in *Pleolophus basizonus* (Gravenhorst), *Orgilus lepidus* Muesebeck and *A. tabida*, respectively. All these studies show that parasitoids spend less time in areas previously visited by other females.

According to Price (1970), "this is adaptive since searching becomes more efficient when pre-searched areas can be avoided". The question why females would provide marks which can be used by other females to increase searching efficiency remains to be answered. Is the benefit to the marking female itself enough to allow such a trait to evolve or is kin selection the driving force in the evolution of these marks? The benefit to other females may be reduced because the marks may decay in a few hours: Price (1970) showed that the effect of the odour mark of *P. basizonus* decreased with time. Another question that remains is why *A. tabida* leaves a patch mark and *L. heterotoma* does not, although both parasitoids exploit similar patches.

6. Previous Experience on Other Patches

Patch time may be influenced by previous experience. Waage (1979) studied the effect of a previous patch visit by *V. canescens* on the duration of a visit to another patch. No such effect was found when the host density in the first patch differed from that in the second, but when the density in both patches was the same, the visit to the second patch was shorter. This surprising result may be caused by habituation to a specific concentration of kairomone. Van Alphen and van Harsel (1982) showed that when *A. tabida* was conditioned the previous day on mixtures of *Drosophila melanogaster* and *D. subobscura*, it searched for less than 10 min on patches with 32 *D. melanogaster* larvae as hosts, but when conditioned on *D. melanogaster* alone, it stayed for more than 40 min in such patches. These results clearly show that *A. tabida* can retain information for at least 24 h, and that previous experience sets the level of the motivation to search the patch.

7. Travel Time between Patches

Finally, patch time may be influenced by the travel time between patches. It follows from the marginal value theorem that longer patch times are expected when more time is spent travelling from one patch to the next. Surprisingly, this important factor has not been studied in insect parasitoids. The small size of most parasitoids makes it hard to follow them on interpatch flights, but mark-recapture experiments in the field could provide an estimate of the mean proportion of time parasitoids spend outside patches. One might use this information to design realistic laboratory experiments to investigate the effect of travel time on patch times.

The relative importance of all these factors and the way they interact should be studied before we can decide whether they should be included in particular patch time models.

B. From Individuals to Populations: Mutual Interference

So far, we have discussed patch time allocation by individual parasitoids. We will now consider the way in which a parasitoid population may distribute its searching time over host patches. As we have seen, a parasitoid may respond to the presence of other parasitoids or their trail marks and to parasitized hosts. These responses play an important role in the theory of mutual interference.

Hassell and Varley (1969) introduced this concept into parasitoid ecology by showing that in a number of studies the searching efficiency of parasitoids decreased with increasing parasitoid density. Starting from these empirical data they proceeded to show that such an effect can stabilize the otherwise unstable Nicholson-Bailey model. Because of this property mutual interference has elicited much theoretical interest (Royama, 1971; Rogers and Hassell, 1974; Beddington, 1975; Free *et al.*, 1977; Sutherland, 1983), but behavioural tests have been scarce (Ridout, 1981). Sutherland and Parker (1985) assume that mutual interference is unimportant in determining dispersion in parasitoids, while others (Hassell and May, 1974) assume that it is an important effect of parasitoid aggregation. There is no general agreement on the importance of interference in parasitoids, partly because behavioural aspects of mutual interference have been confused with the effects of exploitation (therefore called pseudo-interference), and because research on interference behaviour of parasitoids has been insufficient. We now review the existing literature, discuss recent work in which true mutual interference has been found and suggest a functional explanation for it.

Hassell (1978) proposed that the behavioural mechanisms which cause searching efficiency to decrease with increasing parasitoid densities are firstly, an increased tendency to leave a patch after encounters with other individuals and secondly an increased tendency to leave a patch after encounters with hosts which have already been parasitized. Rogers and Hassell (1974) also mention that some parasitoids avoid odour trails left by other females.

Because behaviour that decreases searching efficiency relative to that of conspecifics will be culled by natural selection, it is unlikely that searching parasitoids, after encountering each other, temporarily abandon their search for some time, as Rogers and Hassell (1974) and Beddington (1975) assume in their models. Abandoning search for a while—when hosts are still available—would benefit other parasitoids which continue to search and would therefore be detrimental for the wasp that retreats. Therefore, the evolutionary basis of mutual interference models is weak. The only plausible evolutionary explanation of the three behavioural mechanisms cited above is that all are responses to local overcrowding which forces the parasitoid to leave and search for more profitable patches. As such, they do not necessarily lead to a decrease in searching efficiency. At the population level, they may cause a rapid redistribution of the parasitoids over the available patches. Hence, the behavioural mechanisms invoked to explain mutual interference could also be instrumental in maintaining or increasing searching efficiency.

Often, in laboratory studies concerning mutual interference, different parasitoid densities are exposed to a particular number of hosts for a constant period (Burnett, 1958; Hassell, 1971; Ridout, 1981). Parasitoid searching efficiency is calculated from the Nicholson-Bailey model, using the number of ovipositions observed, the number of eggs later dissected from the hosts, or the number of parasitoids which emerge from the host and the time the experiments lasted. The decrease in the instantaneous attack rate with increasing parasitoid densities found in such experiments is at least partly caused by exploitation of the available hosts. At higher parasitoid densities, the parasitoids will have found and parasitized most hosts long before the experimental time has passed, and they will spend the rest of the time in attempting dispersal or resting on the wall of the cage. Alternatively the parasitoids may revisit the patch because they cannot leave and go to other unexploited patches. This effect has been called pseudo-interference by Free *et al.* (1977), who argued that it plays an important role at high levels of parasitoid aggregation and at high parasitoid densities. There are, as far as we know, no studies on interference which discriminate between true interference and pseudo-interference. The only study in which a detailed analysis was made of the changes in behaviour due to encounters with other parasitoids is that completed by Ridout (1981). She found that *Venturia canescens* females change their behaviour after encountering a conspecific, but these changes did not result in a decrease in the time spent probing with increasing parasitoid density. Ridout's observations lasted only 30 min, while the parasitoids were exposed to the hosts for 24 h. The decrease in searching efficiency over the 24-h period could therefore be caused by pseudo-interference, or, as she suggests, by avoidance reactions of the hosts. An hypothesis that the decrease in parasitoid efficiency is caused by time wasted in the rejection of parasitized hosts was not corroborated by her data. Besides, such an effect would be caused by exploitation of the patches and therefore be part of pseudo-interference.

To show whether behavioural interference does occur or whether all observed interference effects are due to pseudo-interference, encounter rates with hosts of parasitoids searching at different parasitoid densities should be measured in experiments in which the parasitoids can search until they give up and leave.

C. The Ideal Free Distribution

Optimal foraging models in which the distribution of parasitoid populations over patchily distributed hosts is considered (Royama, 1971; Comins and Hassell, 1979) predict that parasitoids will distribute themselves in an ideal free manner (Fretwell and Lucas, 1970). The ideal free distribution is that which would minimize interference. In parasitoids with low transit time between patches, it would be relatively easy to achieve an optimal distribution which minimizes interference, and ideal free distribution theory would be a good predictor of foraging time distribution. As transit times increase, an optimal foraging argument analogous to the marginal value theorem would predict that parasitoids should be more persistent in remaining in patches and so experience interference, simply because the cost of leaving would increase relative to the cost of staying. When high density host patches are more easily found than low density patches, and parasitoids stay longer in such patches, aggregation of parasitoids in high density patches will occur.

D. Superparasitism and Interference

An important attribute of parasitoids is that they leave parasitized hosts vulnerable to further attack. When the probability of raising offspring by superparasitism is greater than zero, it may pay to oviposit in hosts already parasitized by other females (van Alphen and Nell, 1982; Charnov and Skinner, 1984, 1985; Parker and Courtney, 1984; Iwasa *et al.*, 1984; Bakker *et al.*, 1985; see also Chapter 3). Therefore, parasitoids may choose to stay in a patch and superparasitize. Although many parasitoids are able to discriminate between unparasitized and parasitized hosts (van Lenteren, 1981), superparasitism is often found in nature. If parasitoids do not interfere, do not superparasitize, and encounter their hosts randomly, then the product of parasitoid density (P) and searching time (Ts) should be constant for different values of P, or the sum of the Ts values should be constant. Van Alphen and Nell (unpublished) tested whether superparasitism in *Leptopilina heterotoma* increased with parasitoid density, when the parasitoids were free to leave the patch. *L. heterotoma* strongly avoids superparasitism when searching alone (van Lenteren, 1981). It encounters its hosts at random (van Batenburg *et al.*, 1983 and Fig. 3) but in the experiments the sum of the

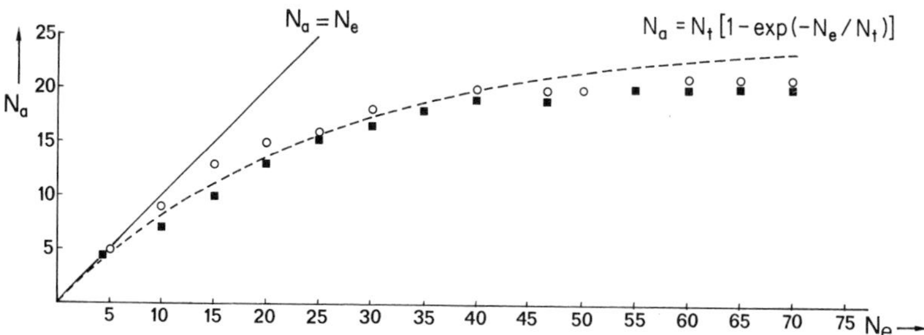

Fig. 3. The relation between the total number of encounters and the number of hosts parasitized: data of two experiments with *Leptopilina heterotoma* (Thomson) at density 25, compared with a random search model (– –), and with a model for systematic search (——).

searching times increased with parasitoid density, and encounter rates (with unparasitized as well as parasitized hosts) decreased. Superparasitism occurred more often at higher parasitoid densities (Table III). Thus these experiments demonstrate mutual interference (Figs 4 and 5). They also show that superparasitism increases in the presence of other females, which suggests that interference occurs as a consequence of the parasitoid's decision to stay longer and superparasitize in the presence of competing conspecifics. When transit times are long, or dispersal to other patches is associated with high mortality risks such a decision could be adaptive. We should therefore measure the costs of interpatch movement in order to understand parasitoid aggregation and to detect true interference.

Mutual interference is generally expressed as a decrease in the number of

TABLE III. Total time spent searching by one, two or four females of *Leptopilina heterotoma* on patches with 32 *D. melanogaster* larvae and the resulting degree of superparasitism (from van Alphen and Nell, unpublished)

	Parasitoid density		
	1	2	4
Mean no. eggs/host	1.0	1.2	1.8
Mean time spent searching (s)	2523.5	4214.0	6417.2

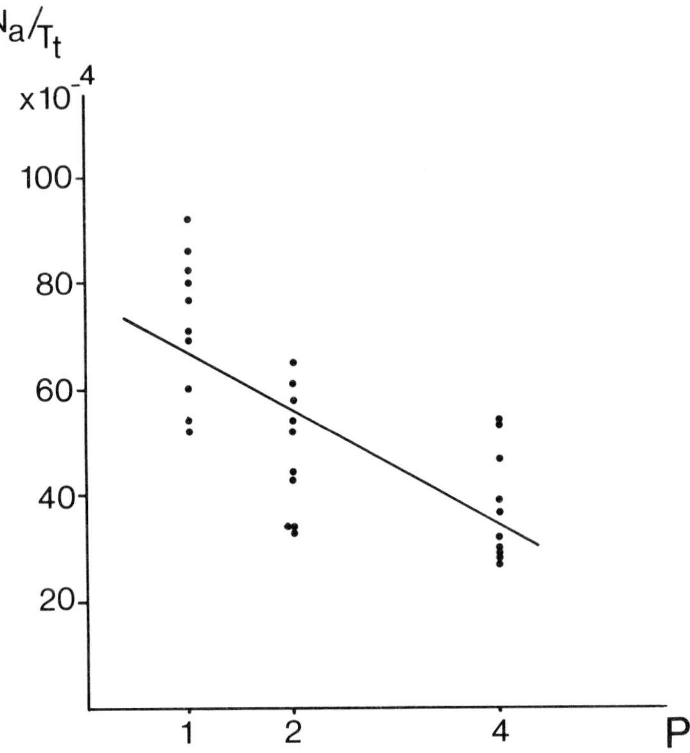

Fig. 4. Attack rate (expressed as number of ovipositions/total patch time × 10^{-4}) as a function of parasitoid density in experiments with one, two or four females of *Leptopilina heterotoma* (Thomson) searching for 32 hosts. N_a = number of ovipositions; P = parasitoid density; T_t = total patch time.

eggs laid per unit of time. In local mate competition (Werren, 1980; see also Chapter 3), the genetic contribution to future generations could be large when only a small number of male producing eggs are added to a host already parasitized or to a patch already exploited by another female. A reduction in search efficiency under such circumstances, expressed as the number of eggs laid per unit of time, does not necessarily mean that the fitness gain per unit of time also decreases.

E. Patch Time Allocation and the Distribution of Parasitism

A number of parasitoid species vary in percentage of parasitism with host density in a density dependent way (van Lenteren and Bakker, 1978; van

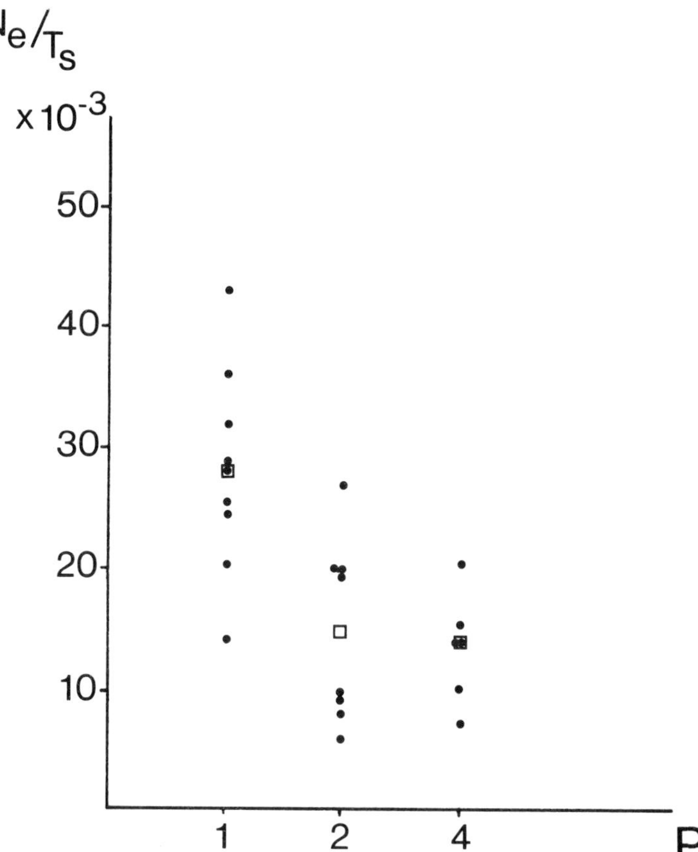

Fig. 5. Encounter rate (expressed as number of encounters/time spent searching $\times 10^{-3}$) as a function of parasitoid density in experiments with one, two and four females of *Leptopilina heterotoma* (Thomson), searching for 32 hosts. N_e = number of encounters; T_s = time spent searching; P = parasitoid density.

Alphen and Galis, 1983), but density independent and even inverse density dependent relations have been found (Lessells, 1985). Lessells investigated whether inverse density dependence could be caused by optimally foraging parasitoids either when time is lost during recognition of parasitized hosts, or when mutual interference makes high density patches unattractive, and concluded that these factors are unable to cause inverse density dependency.

Weis (1983), discussing his own work on *Torymus capite*, as well as that of Morrison *et al.* (1980) and Morrison and Strong (1981) on egg parasitoids, suggests that the observed negative density dependent relationship between parasitism and host density could be generated by the parasitoids using a fixed time rule to determine patch time. It is not certain that the parasitoids did use this rule, because their behaviour was not observed, but it is clear that such a rule applied to patches containing different host densities would cause a decreasing percentage of parasitism with increasing host density. The same reasoning can be applied to the fixed number rule. An alternative explanation of these results is that with increasing host density, a larger fraction of hosts escapes the parasitoid. Van Alphen and van Batenburg (unpublished) showed that a fixed giving-up time rule produces density dependent parasitism (Fig. 6) over a range of host densities. Other rules, by which parasitoids are able to respond to a threshold rate (e.g. the marginal value theorem) will do likewise since time spent on patches increases with host density. This agrees with Sih's (1984) general conclusion that, if maximization of the

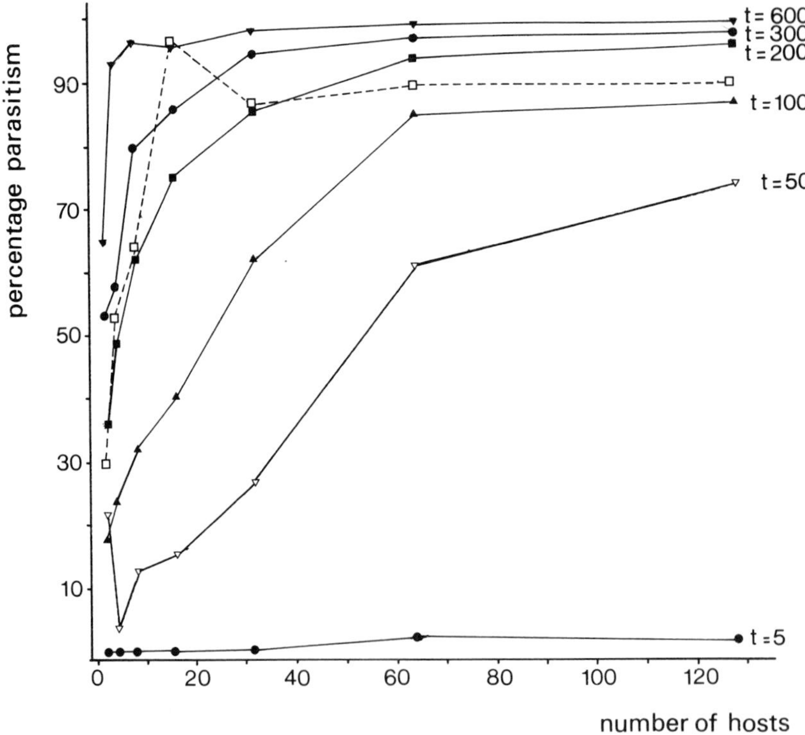

Fig. 6. The relation between percentage parasitism and host density for different fixed giving-up times. Results from simulations of the searching behaviour of *Leptopilina heterotoma* (Thomson) using the model by van Batenburg *et al.* (1983). Broken line shows data from experiments by van Lenteren and Bakker (1978).

encounter rate is the foraging goal, density dependent parasitism or predation can be expected.

Other aspects of searching behaviour which may cause density dependent parasitism are an increase in instantaneous attack rate, or a decrease in handling time. Galis and van Alphen (1981) found that *A. tabida* increases its search intensity with increasing kairomone concentrations. Collins *et al.* (1981) found that handling time decreases with host density in *Aphelinus thomsoni* Graham. Both *Asobara tabida* (van Alphen and Galis, 1983) and *Aphelinus thomsoni* show density dependent parasitism.

V. HOST SPECIFICITY AND HOST SPECIES SELECTION

Host selection often equates with habitat selection (for both generalists and specialists) and only in a few circumstances does selection need to operate at the lowest level of foraging, when parasitoids are actually moving between adjacent hosts. Hence, host habitat choice and the olfactory cues used may rule out potential host species. Situations in which a parasitoid may encounter different host species within a patch are not common (larval and pupal parasitoids of flies which live in dung, decaying leaves, fermenting fruits or fungi and some aphid parasitoids are examples). Thus, for many polyphagous parasitoids host species selection is a matter of habitat and/or patch choice. Also, no real choice may exist; the host species occur at different times during the season, viz. the ichneumonid *Poecilostictus cothurnatus* (Gravenhorst) (van Veen, 1981), or the biology of the parasitoids precludes selection, as in the tachinid *Cyzenis albicans* Fallén which releases its microtype eggs in response to general cues; they may subsequently be swallowed by defoliators which are unsuitable as hosts (Embree and Sisojevic, 1965).

A. Host Specificity and Habitat Choice

A particular habitat may be selected because it harbours a preferred host species. On the other hand, hosts may be attacked, not because they are preferred, but because they are accessible in a particular habitat (Townes, 1962; Zwolfer, 1962; Haeselbarth, 1967; Askew and Shaw, 1978; van Achterberg, 1984; Vinson, 1984). There are species which use a wide range of habitats and hosts, such as the braconid ectoparasitoid *Colastes braconius* Haliday, which attacks a broad range of leafmining insect larvae, belonging to Diptera, Coleoptera and Lepidoptera, in both herb and tree zones (Shaw, 1983). Species in the braconid genus *Coeloides* select a particular habitat (tree species), but attack therein larvae of many different hosts which are not closely related (Haeselbarth, 1967). A similar case is reported by van Achterberg (1984) for the braconid *Gnaptodon breviradialis* Fischer which parasitizes any nepticulid in a selected vegetation zone, even species mining in mistletoe. It is more likely that ectoparasitoids attack a wider range of phylogenetically unrelated hosts than endoparasitoids, which are more likely to be limited in their host species range by physiological constraints (see Chapter 8). By attacking a wider range of hosts in one habitat, polyphagous parasitoids can economize on transit times to other habitats.

For some parasitoids, habitat choice is unimportant, like the pteromalid *Pachycrepoideus vindemiae* (Rondani) which attacks *Drosophila* spp. pupae in urban environments, woods and open landscape, or the aphelinids *Encarsia tricolor* Förster, *E. partenopea* Masi and *E. formosa* (Gahan) which attack whiteflies on many different species of host plants.

We concentrate now on functional aspects and restrict ourselves to the problem of host selection within patches.

B. Host Species Selection

Two theoretical approaches are found in the literature which are applicable to host or prey selection by parasitoids or predators:

1. Switching

The tendency of predators or parasitoids to concentrate on the most abundant host species has been called switching by Murdoch (1969). Whether switching will occur depends on the preference of a parasitoid for one of the host species. Preference is usually measured as a deviation in the proportion of the prey attacked from the proportion of the prey available in the environment. Such a definition does not discriminate between the different behavioural components that cause the disproportionate attack. Preference defined in such a manner may result from differential searching rates for the different host species, from active rejection of a host species following encounter, or from differing abilities of the hosts to escape after encounter. The adaptive value of switching could be an increase in searching efficiency because the parasitoid forms a search image of the dominant host species (Cornell, 1976).

2. Optimal host selection

Emlen (1966), McArthur and Pianka (1966), Charnov (1976), Krebs *et al.* (1977), Hughes (1979), Houston *et al.* (1980) and Iwasa *et al.* (1984) have provided the theoretical basis for optimal host selection in parasitoids. Their models predict that when a parasitoid has a choice between two host species, it should always accept the best one in terms of fitness gain and that the second should only be accepted if it would increase number of offspring produced per time unit. Depending on the encounter rate with the best host species (or, when recognition time plays a role, depending on the encounter rates with both host species and the amount of time spent to recognize them), the less profitable species should either be accepted or rejected. Hence, compared to Murdoch's switching, optimal host selection theory predicts a variable response to the less profitable host species alone.

The difference between these two approaches is not fundamental: switching theory stresses the importance of adaptive changes in encounter rates while assuming that profitability of the two host species does not differ much; optimal host selection models stress the difference in profitability while assuming constant searching rates for the two host species. These two aspects could easily be included in one model. Considering the clear-cut predictions of the models, and the attention which prey choice by predators has received in optimal foraging literature, the number of studies on host selection in parasitoids based on these theories is small. This may be because of the relative scarcity of parasitoids which encounter more than one host species in the same patch.

Cornell and Pimentel (1978) studied switching in *Nasonia vitripennis* Walker, using three host species: *Musca domestica* L., *Phormia regina* (Meigen) and *Lucilia sericata* (Meigen). The parasitoid has an innate preference for *P. regina* rather than the other two species, but concentrates its attack on the most abundant host species. Preference for a host species increased when the parasitoids were allowed to feed on pupae of this species before testing. Hence, learning influences host selection, and contributes to switching. Cornell (1976) assumes that the adaptive significance of this behaviour is that the parasitoid increases its encounter rate by forming some kind of a search image. However, the authors did not measure the consequences of the parasitoid's behaviour in terms of fitness.

Rouault (1979) found that *Leptopilina boulardi* preferred *D. melanogaster* to *D. simulans* Sturtevant as host. Correspondingly, survival in *D. melanogaster* was higher than in *D. simulans*.

Van Alphen and Janssen (1982) showed that *Asobara tabida* is able to rank five *Drosophila* species according to the probability of survival of their offspring in these hosts: a positive correlation ($r=0.95$; $P<0.05$) between survival and percentage parasitism was found. In experiments with the most profitable host species (*D. subobscura*) and *D. melanogaster*, *A. tabida* always accepted *D. subobscura* larvae. At first, it also accepted *D. melanogaster*, but

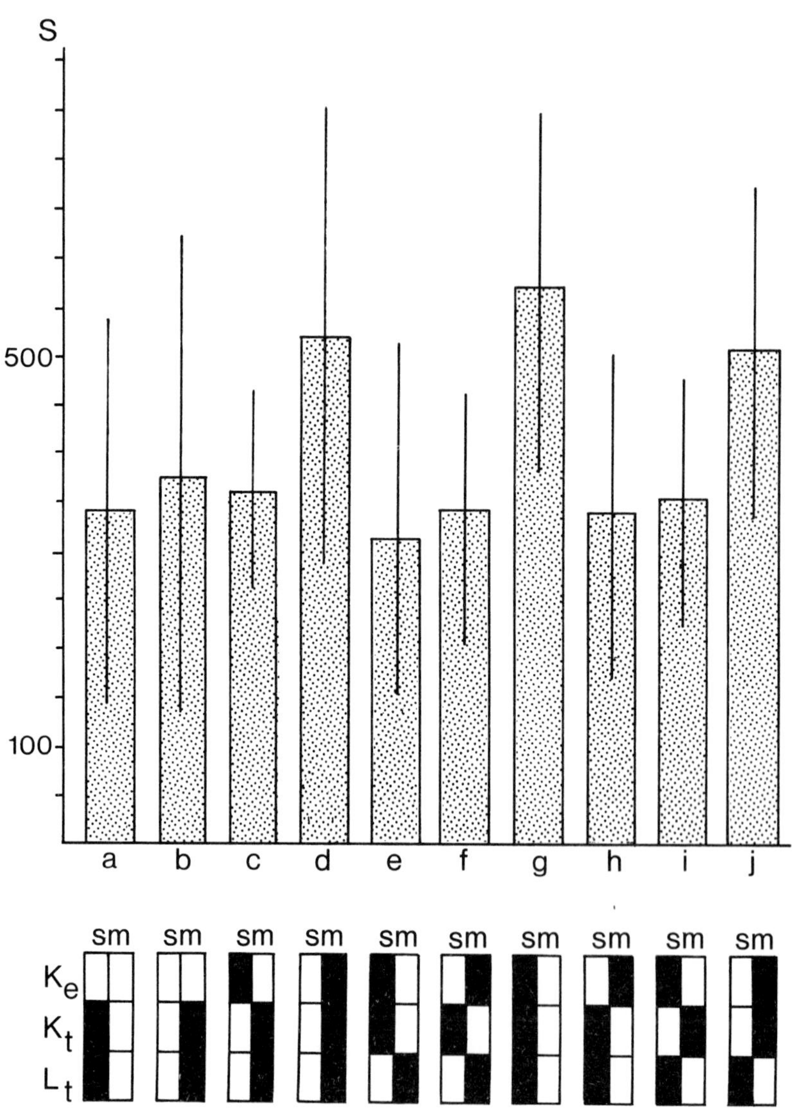

Fig. 7. Mean patch times (in s) of 10 *Asobara tabida* Nees females (bars); the lines indicate lowest and highest values. In the lower part of the figure the black area indicates whether larvae and/or kairomone was used from *Drosophila melanogaster* Meigen (m; right half of each bar) or *D. subobscura* Collin (s; left half of each bar). The lowest row (L_t) indicates the species of larvae used during training, the middle row (K_t) indicates the kairomone used during training and the top row (K_e) indicates which kairomone was used in the experiment, for example, A = (control) parasitoids trained on patches with *Drosophila subobscura* larvae and *D. subobscura* kairomone. Patch time measured on patches containing clean yeast; D = parasitoids trained on patches with

rejected this species after it had searched for a while. This suggested that the decision to reject the less profitable species follows a period of sampling. However, in some experiments, rejection occurred before a *D. subobscura* larva was encountered. Recent experiments (van Alphen *et al.*, unpublished) revealed that *A. tabida* is able to distinguish kairomones produced by *D. melanogaster* from those produced by *D. subobscura*, when it has had the opportunity to associate the kairomone with the host species during previous experience (Fig. 7). Recognition of kairomones could therefore shorten the sampling period necessary to decide whether to reject or accept a host species. Vet (1985b) showed that the more generalist species *L. heterotoma* is able, after experience with one host species and its kairomone, to recognize kairomones of other Drosophilidae, in contrast to *A. tabida* which needs experience with specific kairomones before it is able to recognize them. This may indicate that *L. heterotoma* does not discriminate between kairomones of different host species.

Van Alphen and van Harsel (unpublished) studied host species selection in *A. tabida* using different ratios of *D. melanogaster* and *D. subobscura*. Three different laboratory strains of the parasitoid were studied: one reared on *D. melanogaster* for about 50 generations, another reared on *D. melanogaster* for eight generations and a third reared for six generations on *D. subobscura*. The strain reared on *D. subobscura* consistently rejected *D. melanogaster* larvae. The other two also preferred *D. subobscura* over *D. melanogaster*, but accepted an increasing proportion of *D. melanogaster* larvae with increasing relative frequency of this species. The probability of survival of the parasitoid eggs in *D. subobscura* was much higher than in *D. melanogaster* for all three strains. In agreement with optimal host selection theory, *D. subobscura* larvae were always accepted when encountered.

As in *N. vitripennis*, previous experience influenced host selection behaviour in *A. tabida*: wasps conditioned on *D. melanogaster* the day before the experiment accepted more *D. melanogaster* larvae than wasps conditioned on mixtures of both host species. This shows that foraging decisions are based on more information than can be obtained from the patch which is presently exploited. For the two strains which had been collected recently from the field, patch times increased in proportion to the most profitable host species.

D. melanogaster larvae and *D. melanogaster* kairomone. Patch times measured on patches containing *D. melanogaster* kairomone; J = parasitoids trained on *D. subobscura* larvae and *D. melanogaster* kairomone. Patch times measured on patches containing *D. melanogaster* kairomone.

The data show firstly, that experience with kairomone of one host species results in longer patch times on patches containing this kairomone (D), (G) and (J); for a possible explanation of exception (E) see van Alphen *et al.* (unpublished); and secondly, that the kairomones of *D. melanogaster* and *D. subobscura* are species-specific (after van Alphen *et al.*, unpublished).

In the field, the ratio of *D. melanogaster* and *D. subobscura* larvae varies in space and time. It thus appears adaptive that the willingness of *A. tabida* to accept *D. melanogaster* larvae is influenced by the experience of the wasps. Both switching and optimal host selection may result in density dependent parasitism. In the case of switching this may be found for both host species; in the case of optimal host selection, the most profitable species is always taken in proportion to its availability, and density dependent parasitism may only be found for the less preferred species.

C. Host Stage Selection

Parasitoids may be classified into guilds according to the stage which they attack (egg, larval, pupal or adult parasitoids). Within these guilds, parasitoids may show preference for younger or older developmental stages. The function of this host stage preference is sometimes obvious: for example, the other stages may be invulnerable to parasitoid attack, like sexually mature red scale to *Aphytis melinus* DeBach (Luck *et al.*, 1982). Alternatively host defence systems decrease the survival of parasitoid eggs in older hosts (see also Chapter 4), like *D. melanogaster* larvae, which are able to encapsulate almost all the eggs of *A. tabida* when attack occurs in late second instar or older larvae. In choice experiments, *A. tabida* parasitizes early second instars more than older ones (van Alphen and Drijver, 1982). Liu *et al.* (1984) and Liu (1985) found that the aphid parasitoid *Aphidius sonchi* Marshall prefers larger apteriform nymphs which produce larger and more fecund offspring than less preferred instars.

Parasitoids may kill some hosts by feeding on them, and oviposit in others. For instance, the eulophid *Tetrastichus asparagi* Crawford uses young eggs for host feeding, but oviposits in eggs with a developed embryo (van Alphen, 1980), and the aphelinid *Encarsia formosa* oviposits in whitefly larvae of the third and fourth instar and prepupae, but feeds mainly on second instars and pupae (van Alphen *et al.*, 1976).

In other situations the function of preference is less clear; handling time of older hosts may be much longer because of defence by the host, as in *Tetrastichus asparagi*, which oviposits normally in eggs, but will attack larvae when eggs are scarce (van Alphen, 1980). When attacked, the host larva lifts its head and exudes a brown liquid from the mouth parts with the result that the parasitoid coming into contact with this fluid stops attacking and spends some hours preening itself. On the other hand, the probability of encounter with older and therefore larger hosts may be higher than with young hosts (van Alphen and Drijver, 1982) and may counteract the prolonged handling time.

Intraspecific competition may be the selective force favouring the attack of early stages, when the first egg laid in a host has the best chance to survive competition after superparasitism.

It has been suggested that by attacking different stages parasitoids partition the resource, and that interspecific competition is one of the selective forces which has shaped parasitoid complexes (Force, 1975; Askew, 1975; Price, 1975; see also Chapter 8). Some parasitoids attack early host stages, taking advantage of the fact that they are more numerous than older hosts; others, usually good competitors, attack later and are able to eliminate the parasitoids which attack the early stages. Hence, interspecific competition may be one of the selective forces operating on host stage selection.

Encounter probability, handling time and profitability of the hosts in terms of offspring survival, as well as intra- and interspecific competition may all determine host stage selection. Optimal host selection models can be used to test the importance of the first three factors, but this has not been done so far.

VI. CONCLUSIONS

There are probably over a quarter of a million species of parasitoids, all developing as larvae feeding on a host insect, but differing enormously in behaviour and ecology. We have tried to show how these differences and similarities can be understood when the function of the observed behaviour is explained.

The methods used to answer questions about function are comparison of observed behaviour with the predictions of optimal foraging models, and comparison of different species in the hope of finding trends which correlate with trends in their ecology. This last method has hardly been applied to the searching behaviour of parasitoids (Waage, 1982), but has been successfully used in other areas of ecology (Ridley, 1982).

Searching behaviour is divided into three phases: host habitat location, host location and host selection. These phases are usually studied in the laboratory as isolated steps. For some parasitoids this practice reflects the way in which they themselves experience the searching process, but for others such a rigid hierarchy does not exist. For the latter, cues produced by the host (sounds or kairomones) attract the parasitoid and host species selection may occur before entering a patch.

Patch time allocation may be determined by simple rules, such as number or time expectation in some species, or in a more complex manner. These different mechanisms can all be adaptive, depending on the host distribution pattern.

There exists a wealth of data on how parasitoids find their host's habitat, but no quantitative data exist on the costs of interpatch travel, in terms of time, energy or mortality risk. Because travel costs influence other foraging decisions, this is an important lacuna. We suggest that for parasitoids which aggregate, transit costs in nature are high, and this could explain why parasitoids are reluctant to leave a patch once they have found it. Because parasitoids leave parasitized hosts vulnerable to further attack, to stay and superparasitize can be adaptive when the probability of producing additional offspring is higher than leaving and searching for other patches. This would cause parasitoid aggregation and mutual interference. If travel costs were negligible, ideal free distributions rather than aggregation would be found.

Our understanding of parasitoid foraging behaviour is largely based on observations of individual wasps. In population models it is often assumed that all individuals are equal and constant throughout their lives. However, phenotypical differences such as size differences, changes in behaviour caused by learning or genetical differences in host or host habitat preferences may produce immense variation within populations. These differences partly cause the substantial variance in experimental data used to test deterministic foraging models. Optimal foraging theory is based on the notion that individuals try to maximize their genetic contribution to future generations, but we normally use average performance to test this hypothesis. It could therefore be rewarding to look for conditional or mixed strategies. Game-theory approaches so far have only been applied to the problem of sex allocation, but not to other foraging decisions.

The fact that detailed quantitative data are needed to answer functional questions about parasitoid foraging means that most data must be obtained from laboratory experiments. We should however take care that our experimental techniques reflect situations similar to those a parasitoid may encounter in nature. Therefore, field work is essential to learn which selective forces are the main determinants of observed behaviour.

Acknowledgements

We thank the following people for reading and commenting on the manuscript: Kees Bakker, Arne Janssen, Gerard Driessen and Hannah Nadel. Many valuable suggestions for improvement of the first draft were given by Jeff Waage. We thank Jan Herzberg and Martin Brittijn for preparing the figures.

REFERENCES

van Achterberg, C. (1984). The preference of zoophagous Hymenoptera for certain types of plants as shown by the subfamily Gnaptodontinae (Braconidae). *Proceedings of the 10th International Symposium of Entomofauna of Central Europe.*

van Alphen, J. J. M. (1980). Aspects of the foraging behaviour of *Tetrastichus asparagi* Crawford and *Tetrastichus* spec. (Eulophidae), gregarious egg parasitoids of the asparagus beetles *Crioceris asparagi* L. and *C. duodecimpunctata* L. Chrysomelidae. *Netherlands Journal of Zoology* **30,** 307–325.

van Alphen, J. J. M. and K. Bakker (in press). Reduction of patch time by encounters with parasitized hosts in *Leptopilina heterotoma* (Thomson), a larval parasitoid of *Drosophila. Netherlands Journal of Zoology* **36.**

van Alphen, J. J. M. and van Batenburg, F. H. D. (unpublished). Sigmoid functional responses in insect parasitoids and predators.

van Alphen, J. J. M., and Drijver, R. A. B. (1982). Host selection by *Asobara tabida* Nees (Braconidae; Alysiinae) a larval parasitoid of fruit inhabiting *Drosophila* species. I. Host stage selection with *Drosophila melanogaster* as host. *Netherlands Journal of Zoology* **32,** 215–231.

van Alphen, J. J. M., and Galis, F. (1983). Patch time allocation and parasitization efficiency of *Asobara tabida* Nees, a larval parasitoid of *Drosophila. Journal of Animal Ecology* **52,** 937–952.

van Alphen, J. J. M., and van Harsel, H. H. (1982). Host selection by *Asobara tabida* Nees (Braconidae; Alysiinae), a larval parasitoid of fruit inhabiting *Drosophila* species. III. Host species selection and functional response. *In* "Foraging Behaviour of *Asobara tabida*, a Larval Parasitoid of Drosophilidae" (J. J. M. van Alphen, ed.), pp. 61–93. PhD Thesis, University of Leiden.

van Alphen, J. J. M. and van Harsel, H. H. (unpublished). Host selection, switching and functional response in *Asobara tabida* Nees, a larval parasitoid of *Drosophila.*

van Alphen, J. J. M., and Janssen, A. R. M. (1982). Host selection by *Asobara tabida* Nees (Braconidae, Alysiinae), a larval parasitoid of fruit inhabiting *Drosophila* species. II. Host species selection. *Netherlands Journal of Zoology* **32,** 215–231.

van Alphen, J. J. M., and Nell, H. W. (1982). Superparasitism and host discrimination by *Asobara tabida* Nees (Braconidae, Alysiinae), a larval parasitoid of Drosophilidae. *Netherlands Journal of Zoology* **32,** 232–260.

van Alphen, J. J. M., and Nell, H. W. (unpublished). Mutal interference and superparasitism by *Leptopilina heterotoma* (Thomson), a larval parasitoid of *Drosophila.*

van Alphen, J. J. M., Nell, H. W., and Sevenster-van der Lelie, L. A. (1976). The parasite-host relationship between *Encarsia formosa* (Hymenoptera: Aphelinidae) and *Trialeurodes vaporariorum* (Homoptera: Aleyrodidae). VII. The importance of host feeding as a mortality factor in greenhouse whitefly nymphs. *Bulletin O.I.L.B./S.R.O.P.* **4,** 165–169.

van Alphen, J. J. M., van Megen, D., and Simbolotti, G. (unpublished). Recognition of species specific kairomones by *Asobara tabida* Nees, a larval parasitoid of *Drosophila.*

Arthur, A. P. (1966). Associative learning in *Itoplectis conquisitor* (Say) (Hymenoptera: Ichneumonidae). *Canadian Entomologist* **98,** 213–223.

Arthur, A. P. (1971). Associative learning by *Nemeritis canescens* (Hymenoptera: Ichneumonidae). *Canadian Entomologist* **103,** 1137–1141.

Arthur, A. P. (1981). Host acceptance by parasitoids. *In* "Semiochemicals, their Role in Pest Control" (D. A. Nordlund, R. L. Jones and W. J. Lewis, eds.), pp. 97–120. J. Wiley, New York.

Askew, R. R. (1975). The organisation of chalcidoid-dominated parasitoid communities centered upon endophytic hosts. *In* "Evolutionary Strategies of Parasitic Insects and Mites" (P. W. Price, ed.), pp. 130–153. Plenum, New York.

Askew, R. R., and Shaw, M. R. (1978). Account of Chalcoidea parasitizing leaf-mining insects of deciduous trees in Britain. *Biological Journal of the Linnean Society* **6,** 289–335.

Bakker, K., van Alphen, J. J. M.. van Batenburg, F. H. D., van der Hoeven, N., Nell, H. W., van Strien-van Liempt, W. F. T. H., and Turlings, T. C. J. (1985). The function of host discrimination and superparasitization. *Oecologia* **67,** 572–576.

van Batenburg, F. H. D., van Lenteren, J. C., van Alphen, J. J. M., and Bakker, K. (1983). Searching for and parasitation of *Drosophila melanogaster* (Dipt.: Drosophilidae) by *Leptopilina heterotoma* (Hym.: Eucoilidae): a Monte Carlo simulation model and the real situation. *Netherlands Journal of Zoology* **33,** 306–336.

Bauer, J. S., and Sokolowski, M. B. (1984). Larval foraging behaviour in isofemale lines of *Drosophila melanogaster* and *D. pseudobscura. Journal of Heredity* **75,** 131–134.

Beddington, J. R. (1975). Mutual interference between parasites or predators and its effect on searching efficiency. *Journal of Animal Ecology* **44,** 331–340.

Boulétreau M., and David, J. R. (1981). Sexually dimorphic habitat response to host habitat toxicity in *Drosophila* parasitic wasps. *Evolution* **35,** 395–399.

Burnett, T. (1958). Dispersal of an insect parasite over a small plot. *Canadian Entomologist* **90,** 279–283.

Bush, G. L. (1969). Sympatric host race formation and speciation in frugivorous flies of the genus *Rhagoletis* (Diptera, Tephritidae). *Evolution* **23,** 237–351.

Bush, G. L. (1974). The mechanism of sympatric host race formation in the true fruit flies (Tephritidae). *In* "Genetic Analysis of Speciation Mechanisms" (M. J. D. White, ed.), pp. 3–23. Australia and New Zealand Book Co.

Bush, G. L., and Diehl, S. R. (1982). Host shifts, genetic models of sympatric speciation and the origin of parasitic insect species. *Proceedings of the 5th International Symposium on Insect–Plant Relationships*, 297–305.

Cade, W. (1975). Acoustically orienting parasitoids: fly phonotaxis to cricket song. *Science* **190,** 1312–1313.

Carton, Y., and David, J. R. (1985). Relation between the genetic variability of digging behaviour of *Drosophila* larvae and their susceptibility to a parasitic wasp. *Behavior Genetics* **15,** 143–154.

Charnov, E. L. (1973). "Optimal Foraging: Some Theoretic Explorations." PhD Thesis, University of Washington.

Charnov, E. L. (1976). Optimal foraging, the marginal value theorem. *Theoretical Population Biology* **9,** 129–136.

Charnov, E. L., and Skinner, S. W. (1984). Evolution of host selection and clutch size in parasitoid wasps. *Florida Entomologist* **67,** 5–21.

Charnov, E. L., and Skinner, S. W. (1985). Complementary approaches to the understanding of parasitoid oviposition decisions. *Environmental Entomology* **14,** 383–391.

Collins, M. D., Ward, S. A., and Dixon, A. F. G. (1981). Handling time and the functional response of *Aphelinus thomsoni*, a predator and parasite of the aphid *Drepanosyphon platanoides. Journal of Animal Ecology* **50,** 479–488.

Comins, H. N., and Hassell, M. P. (1979). The dynamics of optimally foraging predators and parasitoids. *Journal of Animal Ecology* **48,** 335–351.

Cook, R. M., and Hubbard, S. F. (1977). Adaptive searching strategies in insect parasites. *Journal of Animal Ecology* **46,** 115–125.

Corbet, S. A. (1973). Concentration effects and the response of *Nemeritis canescens* to a secretion of its host. *Journal of Insect Physiology* **19,** 2119–2128.

Cornell, H. (1976). Search strategies and the adaptive significance of switching in some general predators. *American Naturalist* **110,** 317–320.

Cornell, H., and Pimentel, D. (1978). Switching in the parasitoid *Nasonia vitripennis* and its effect on host competition. *Ecology* **59,** 297–308.

Dicke, M., van Lenteren, J. C., Boskamp, G. J. F., and van Dongen-van Leeuwen, E. (1984). Chemical stimuli in host-habitat location by *Leptopilina heterotoma* (Thomson) (Hymenoptera: Eucoilidae), a parasite of *Drosophila. Journal of Chemical Ecology* **10,** 695–712.

Dicke, M., van Lenteren, J. C., Boskamp, G. J. F., and van Voorst, R. (1985). Intensification and elongation of host searching in *Leptopilina heterotoma* (Thomson) (Hymenoptera:

Eucoilidae) through a kairomone produced by *Drosophila melanogaster*. *Journal of Chemical Ecology* **11,** 125–136.

Dmoch, J., Lewis, W. J., Martin, B. P., and Nordlund, D. A. (1985). Role of host produced stimuli and learning in host selection behavior of *Cotesia* (= *Apanteles*) *margiventris* (Cresson) (Hymenoptera, Braconidae). *Journal of Chemical Ecology* **11,** 453–463.

Doutt, R. L. (1959). The biology of the parasitic Hymenoptera. *Annual Review of Entomology* **4,** 161–182.

Embree, D. G., and Sisojevic, P. (1965). The bionomics and population density of *Cyzenis albicans* (Fall.) (Tachinidae: Diptera) in Nova Scotia. *Canadian Entomologist* **97,** 631–639.

Emlen, J. M. (1966). The role of time and energy in food preference. *American Naturalist* **100,** 611–617.

Force, D. C. (1975). Succession of r and K strategists in parasitoids. *In* "Evolutionary Strategies of Parasitic Insects and Mites". (P. W. Price, ed.), pp. 112–129. Plenum, New York.

Free, C. A., Beddington, J. R., and Lawton, J. H. (1977). On the inadequacy of simple models of mutual interference for parasitism and predation. *Journal of Animal Ecology* **46,** 543–554.

Fretwell, S. D., and Lucas, H. L. Jr. (1970). On territorial behavior and other factors influencing habitat distribution in birds. I. Theoretical Development. *Acta Biotheoretica* **19,** 16–36.

Galis, F., and van Alphen, J. J. M. (1981). Patch time allocation and search intensity of *Asobara tabida* Nees (Hym.: Braconidae). *Netherlands Journal of Zoology* **31,** 701–712.

Gardner, S. M., Dissevelt, M., and van Lenteren, J. C. (unpublished). Behavioural adaptations in host finding by *Trichogramma evanescens* Westwood: the influence of oviposition experience on response to host contact kairomones.

Gibb, J. A. (1962). Tinbergen's hypothesis of the role of specific searching images. *Ibis* **104,** 106–111.

Glas, P. C. G., and Vet, L. E. M. (1983). Host-habitat location and host location by *Diachasma alloeum* Muesebeck (Hym.: Braconidae) a parasitoid of *Rhagoletis pomonella* Walsh (Dipt. Tephritidae). *Netherlands Journal of Zoology* **33,** 41–54.

Greany, P. D., and Oatman, E. R. (1972). Analysis of host discrimination in the parasite *Orgilus lepidus* (Hymenoptera: Braconidae). *Annals of the Entomological Society of America* **65,** 377–383.

Green, R. F. (1979). Baysian birds: a simple example of Oaten's stochastic model of optimal foraging. *Theoretical Population Biology* **15,** 244–256.

Green, R. F. (1984). Stopping rules for optimal foragers. *American Naturalist* **123,** 30–43.

Haeselbarth, E. (1967). Zur Kenntnis der palearktischen Arten der Gattung *Coeloides* Wesmael (Hym., Braconidae). *Mitteilungen Münchener entomologischen Gesellschaft* **57,** 20–53.

Harrison, E. (1985). "Oviposition Behaviour of *Venturia canescens*: a Study of the Effect of a Pheromone". PhD thesis, University of London.

Hassell, M. P. (1971). Mutual interference between searching insect parasites. *Journal of Animal Ecology* **40,** 473–486.

Hassell, M. P. (1978). "The Dynamics of Arthropod Predator-Prey Systems." Princeton University Press, Princeton.

Hassell, M. P., and Varley, G. C. (1969). New inductive population model for insect parasites and its bearing on biological control. *Nature* **223,** 1133–1136.

Hassell, M. P., and May, R. M. (1974). Aggregation in predators and insect parasites and its effect on stability. *Journal of Animal Ecology* **43,** 567–594.

Hassell, M. P., and Southwood, T. R. E. (1978). Foraging strategies of insects. *Annual Review of Ecology and Systematics* **9,** 75–98.

Herrebout, W. M. (1969). Habitat selection in Eucarcelia rutilla Vil. (Diptera: Tachinidae) II. Experiments with females of known age. *Zeitschrift für angewandte Entomologie* **63,** 336–349.

Herrebout, W. M., and van der Veer, J. (1969). Habitat selection in *Eucarcelia rutilla* Vill.

(Diptera: Tachinidae). III. Preliminary results of olfactometer experiments with females of known age. *Zeitschrift für angewandte Entomologie* **64,** 55–61.

Houston, A. I., Krebs, J. R., and Erichson, J. T. (1980). Optimal prey choice and discrimination time in the great tit (*Parus major* L.). *Behavioural Ecology and Sociobiology* **6,** 169–175.

Hughes, R. N. (1979). Optimal diets under the energy maximization premise: the effect of recognition time and learning. *American Naturalist* **113,** 209–221.

Iwasa, Y., Higashi, M., and Yamamura, N. (1981). Prey distribution as a factor determining the choice of optimal foraging strategy. *American Naturalist* **117,** 710–723.

Iwasa, Y., Suzuki, Y., and Matsuda, H. (1984). Theory of oviposition strategy of parasitoids. I. Effect of mortality and limited egg number. *Theoretical Population Biology* **26,** 205–227.

Janssen, A. R. M., Driessen, G. J. J., de Haan, M., and Roodbol, N. (unpublished). Microhabitats of larval parasitoids of temperate woodland *Drosophila.*

Jones, R. L. (1981). Chemistry of semiochemicals involved in parasitoid-host and predator-prey relationships. *In* "Semiochemicals, Their Role in Pest Control" (D. A. Nordlund, R. L. Jones and W. J. Lewis, eds), pp. 239–250. J. Wiley, New York.

Klomp, H. (1981). Parasitic wasps as sleuthhounds: response of an ichneumon wasp to the trail of its host. *Netherlands Journal of Zoology* **31,** 762–772.

Krebs, J. R. (1973). Behavioral aspects of predation. *In* "Perspectives in Ethology" (P. P. G. Bateson and P. H. Klopfer, eds), pp. 73–111. Plenum Press, New York.

Krebs, J. R., Ryan, J. C., and Charnov, E. L. (1974). Hunting by expectation or optimal foraging? A study in patch use by chicadees. *Animal Behaviour* **22,** 953–964.

Krebs, J. R., Erichsen, J. T., Webber, M. I., and Charnov, E. R. (1977). Optimal prey selection in the great tit, *Parus major. Animal Behaviour* **25,** 30–38.

Lack, D. (1971). "Ecological Isolation in Birds." Blackwell Scientific Publications, Oxford.

van Lenteren, J. C. (1981). Host discrimination by parasitoids. *In* "Semiochemicals: Their Role in Pest Control" (D. A. Nordlund, R. L. Jones and W. J. Lewis, eds), pp. 153–179. John Wiley, New York.

van Lenteren, J. C., and Bakker, K. (1978). Behavioural aspects of the functional response of a parasite (*Pseudocoila bochei* Weld) to its host (*Drosophila melanogaster*). *Netherlands Journal of Zoology* **28,** 213–233.

Lessells, C. M. (1985). Parasitoid foraging: should parasitism be density dependent? *Journal of Animal Ecology* **54,** 27–41.

Lewis, W. J., Nordlund, D. A., Gueldner, R. C., Teal, P. E. A., and Tumlinson, J. H. (1982). Kairomones and their use for management of entomophagous insects. XIII. Kairomonal activity for *Trichogramma* spp. of abdominal tips, excretion, and a synthetic sex pheromone blend of *Heliothis zea* (Boddie) moths. *Journal of Chemical Ecology* **8,** 1323–1331.

Liu Shu-sheng (1985). Development, adult size and fecundity of *Aphidius sonchi* reared in two instars of its aphid host, *Hyperomyzus lactucae. Entomologia Experimentalis et Applicata* **37,** 41–48.

Liu Shu-sheng, Morton, R., and Hughes, R. D. (1984). Oviposition preference of a hymenopterous parasite for certain instars of its aphid host. *Entomologia Experimentalis et Applicata* **35,** 249–254.

Luck, R. F., Podoler, H., and Kfir, R. (1982). Host selection and egg allocation behaviour by *Aphytis melinus* and *Aphytis lignanensis*: comparison of two facultatively gregarious parasitoids. *Ecological Entomology* **7,** 397–408.

MacArthur, R. H., and Pianka, E. R. (1966). On optimal use of a patchy environment. *American Naturalist* **100,** 603–609.

Mayr, E. (1970). "Populations, Species and Evolution." Harvard University Press, Cambridge, Massachusetts.

McNair, J. N. (1982). Optimal giving up times and the marginal value theorem. *American Naturalist* **119,** 511–529.

Monteith, L. G. (1955). Host preference in *Drino bohemica* Mesn. (Diptera: Tachinidae), with particular reference to olfactory responses. *Canadian Entomologist* **87,** 509–530.

Monteith, L. G. (1963). Habituation and associative learning in *Drino bohemica* Men. (Diptera: Tachinidae). *Canadian Entomologist* **95,** 418–426.

Morrison, G., and Strong, D. R., Jr. (1981). Spatial variation in egg density and the intensity of parasitism in a neotropical chrysomelid (*Cephaloleia consanguinea*). *Ecological Entomology* **6,** 55–61.

Morrison, G., Lewis, W. J., and Nordlund, D. A. (1980). Spatial differences in *Heliothis zea* egg density and the intensity of parasitism by *Trichogramma* spp.: an experimental analysis. *Environmental Entomology* **9,** 79–85.

Murdoch, W. W. (1969). Switching in general predators: experiments on predator specificity and stability of prey populations. *Ecological Monographs* **39,** 335–354.

Murdoch, W. W., and Oaten, A. (1975). Predation and population stability. *Advances in Ecological Research* **9,** 2–132.

Noldus, L. P. J. J. and van Lenteren, J. C. (1985). Kairomones for the egg parasite *Trichogramma evanescens* Westwood. I. Effect of volatile substances released by two of its hosts, *Pieris brassicae* L. and *Mamestra brassicae* L. *Journal of Chemical Ecology* **11,** 781–791.

Nordlander, G. (1982). Systematics and phylogeny of an interrelated group of genera within the family Eucoilidae (Insecta: Hymenoptera, Cynipoidea). Doctoral dissertation, University of Stockholm.

Oaten, A. (1977). Optimal foraging in patches: a case for stochasticity. *Theoretical Population Biology* **12,** 263–285.

Parker, G. A., and Courtney, S. P. (1984). Models of clutch size in insect oviposition. *Theoretical Population Biology* **26,** 21–48.

Price, P. W. (1970). Trail odours: recognition by insects parasitic in cocoons. *Science* **170,** 546–547.

Price, P. W. (1975). Reproductive strategies of parasitoids. *In* "Evolutionary Strategies of Parasitic Insects and Mites" (P. W. Price, ed.), pp. 87–111. Plenum, New York.

Prokopy, R. J., and Roitberg, B. D. (1984). Foraging behaviour of true fruit flies. *American Scientist* **72,** 41–49.

Prokopy, R. J., and Webster, R. P. (1978). Oviposition deterring pheromone of *Rhagoletis pomonella*, a kairomone for its parasitoid *Opius lectus*. *Journal of Chemical Ecology* **4,** 481–494.

Ridley, M. (1982). How to explain organic diversity. *In* "Darwin up to Date" (J. Cherfas, ed.), pp. 42–44. IPC, London.

Ridout, L. M. (1981). Mutual interference: behavioural consequences of encounters between adults of the parasitoid wasp *Venturia canescens* (Hymenoptera: Ichneumonidae). *Animal Behaviour* **29,** 897–903.

Rogers, D. J., and Hassell, M. P. (1974). General models for insect parasite and predator searching behaviour: interference. *Journal of Animal Ecology* **43,** 239–253.

Rouault, J. (1979). Rôle des parasites entomophages dans la competition entre espèces jumelles de Drosophiles: approche experimentale. *Compte Rendu Hebdomadaire Séances Académie Sciences* **289,** 643–646.

Royama, T. (1971). Evolutionary significance of a predator's response to local differences in prey density: a theoretical study. *In* "Dynamics of Populations" (P. J. den Boer and G. R. Gradwell, eds.), pp. 344–357. Centre for Agricultural Publishing and Documentation, Wageningen.

Sandlan, K. (1980). Host location by *Coccygomimus turionellae* (Hymenoptera: Ichneumonidae). *Entomologia Experimentalis et Applicata* **27**, 233–245.

Schoener, T. W. (1974). Resource partitioning in ecological communities. *Science* **185**, 27–39.

Shaw, M. R. (1983). On evolution of endoparasitism: the biology of some genera of Rogadinae (Braconidae). *Contributions to the American Entomological Institute* **20**, 307–328.

Shorrocks, B. (1977). An ecological classification of European *Drosophila* species. *Oecologia* **26**, 335–345.

Sih, A. (1984). Optimal behaviour and density dependent predation. *American Naturalist* **123**, 314–326.

Sokolowski, M. B. (1980). Foraging strategies of *Drosophila melanogaster*: a chromosomal analysis. *Behavior Genetics* **10**, 291–302.

Sokolowski, M. B. (1982). Rover and sitter larval foraging patterns in a natural population of *Drosophila melanogaster*. *Drosophila Information Service* **58**, 138–139.

Strand, M. R., and Vinson, S. B. (1982). Behavioral response of the parasitoid *Cardiochiles nigriceps* to a kairomone. *Entomologia Experimentalis et Applicata* **31**, 308–315.

Sutherland, W. J. (1983). Aggregation and the "ideal free" distribution. *Journal of Animal Ecology* **52**, 821–828.

Sutherland, W. J., and Parker, G. A. (1985). Distribution of unequal competitors. *In* "Behavioural Ecology. Ecological Consequences of Adaptive Behaviour" (R. M. Sibly and R. H. Smith, eds.), pp. 255–273, Blackwell Scientific Publications, Oxford.

Taylor, R. J. (1974). Role of learning in insect parasitism. *Ecological Monographs* **44**, 89–104.

Thornhill, R. (1984). Scientific Methodology in Entomology. *Florida Entomologist* **67**, 74–96.

Townes, H. K. (1962). Host selection patterns in some Nearctic Ichneumonids. *Proceedings of the 2nd International Congress of Entomology* **2**, 738–744.

van Veen, J. C. (1981). The biology of *Poecilostictus cothurnatus* (Hymenoptera, Ichneumonidae) and endoparasite of *Bupalus piniarius* (Lepidoptera, Geometridae). *Annals of Entomology* **47**, 77–93.

Vet, L. E. M. (1983). Host-habitat location through olfactory cues by *Leptopilina clavipes* (Hartig) (Hym: Eucoilidae), a parasitoid of fungivorous *Drosophila*: the influence of conditioning. *Netherlands Journal of Zoology* **33**, 225–248.

Vet, L. E. M. (1985a). Olfactory microhabitat location in some eucoilid and alysiine species (Hymenoptera), larval parasitoids of Diptera. *Netherlands Journal of Zoology* **35(3)**, 486–496.

Vet, L. E. M. (1985b). Response to kairomones by some alysiine and eucoilid parasitoid species (Hymenoptera). *Netherlands Journal of Zoology* **35(4)**.

Vet, L. E. M., and van Alphen, J. J. M. (1985). A comparative functional approach to the host detection behaviour of parasitic wasps. I. A qualitative study on Eucoilidae and Alysiinae. *Oikos* **44**, 478–486.

Vet, L. E. M., and Bakker, K. (1985). A comparative functional approach to the host detection behaviour of parasitic wasps. II. A quantitative study on eight eucoilid species. *Oikos* **44**, 487–984.

Vet, L. E. M., and van der Hoeven, R. (1984). Comparison of the behavioural response of two *Leptopilina* species (Hymenoptera: Eucoilidae), living in different microhabitats, to kairomone of their host (Drosophilidae). *Netherlands Journal of Zoology* **34**, 220–227.

Vet, L. E. M., and Janse, C. J. (1984). Fitness of two sibling species of *Asobara* (Braconidae: Alysiinae), larval parasitoids of Drosophilidae in different microhabitats. *Ecological Entomology* **9**, 345–354.

Vet, L. E. M., and van Opzeeland, K. (1984). The influence of conditioning on olfactory microhabitat and host location in *Asobara tabida* (Nees) and *A. rufescens* (Foerster) (Braconidae: Alysiinae), larval parasitoids of *Drosophila*. *Oecologia* **63**, 171–177.

Vet, L. E. M., and van Opzeeland, K. (1985). Olfactory microhabitat selection in *Leptopilina heterotoma* (Thomson) (Hym.: Eucoilidae), a parasitoid of Drosophilidae. *Netherlands Journal of Zoology* **35,** 497–504.

Vet, L. E. M., van Lenteren, J. C., Heymans, M., and Meelis, E. (1983). An airflow olfactometer for measuring olfactory responses of hymenopterous parasitoids and other small insects. *Physiological Entomology* **8,** 97–106.

Vet, L. E. M., Janse, C. J., van Achterberg, C., and van Alphen, J. J. M. (1984). Microhabitat location and niche segregation in two sibling species of drosophilid parasitoids: *Asobara tabida* (Nees) and *A. rufescens* (Foerster) (Braconidae: Alysiinae). *Oecologia* **61,** 182–188.

Vinson, S. B. (1976). Host selection by insect parasitoids. *Annual Review of Entomology* **21,** 109–133.

Vinson, S. B. (1981). Habitat location. *In* "Semiochemicals, their Role in Pest Control (D. A. Nordlund, R. L. Jones and W. J. Lewis, eds.), pp. 51–77. John Wiley, New York.

Vinson, S. B. (1984). Parasitoid-host relationship. *In* "Chemical Ecology of Insects" (W. J. Bell and R. T. Carde, eds.). Chapman and Hall, London.

Vinson, S.B., Barfield, C. S., and Henson, R. D. (1977). Oviposition behaviour of *Bracon mellitor*, a parasitoid of the boll weevil (*Anthonomis grandis*). II. Associative learning. *Physiological Entomology* **2,** 157–164.

Waage, J. K. (1978). Arrestment responses of the parasitoid, *Nemeritis canescens*, to a contact chemical produced by its host, *Plodia interpunctella. Physiological Entomology* **3,** 135–146.

Waage, J. K. (1979). Foraging for patchily-distributed hosts by the parasitoid, *Nemeritis canescens. Journal of Animal Ecology* **48,** 353–371.

Waage, J. K. (1982). Sib mating and sex ratio strategies in scelionid wasps. *Ecological Entomology* **7,** 103–112.

Wanntorp, H. E. (1983). Historical constraints in the adaptation theory: traits and non-traits. *Oikos* **41,** 157–160.

Wardle, A. R., and Borden, J. H. (1985). Age-dependent associative learning by *Exeristes roborator* (F.) (Hymenoptera: Ichneumonidae). *Canadian Entomologist* **117,** 605–616.

Weis, A. E. (1983). Patterns of parasitism by *Torymus capite* on hosts distributed on small patches. *Journal of Animal Ecology* **52,** 867–878.

Werren, J. H. (1980). Sex ratio adaptations to local mate competition in a parasitic wasp. *Science* **208,** 1157–1159.

Weseloh, R. M. (1981). Host location by parasitoids. *In* "Semiochemicals, their Role in Pest Control" (D. A. Nordlund, R. L. Jones and W. J. Lewis, eds.), pp. 79–95.

Zwölfer, H. (1962). Die Orientierung entomophager Parasiten als Problem der angewandten Entomologie. *Zeitschrift für angewandte Entomologie* **50,** 93–98.

3

Family Planning in Parasitoids: Adaptive Patterns of Progeny and Sex Allocation

J. K. WAAGE

I. INTRODUCTION

This chapter is concerned with how parasitoids allocate offspring to the hosts which they encounter while foraging. This problem involves two main decisions: how many eggs to lay per host (and this may be none if the host is not accepted), and what sex ratio to produce amongst the eggs laid (Waage and Ng, 1984).

Variation between parasitoid species in patterns of progeny and sex allocation is considerable. At present, such variability is of interest to most parasitoid researchers only as a means of classifying their particular species, for instance as solitary or gregarious. Some workers, however, have recognized that the "pure" problem of progeny and sex allocation is relevant to several "applied" problems in parasitoid rearing, such as superparasitism and the change in fitness or sex ratio with crowding. Most work on these phenomena has been strongly empirical in that it has sought solutions by manipulating rearing conditions, without considering the behaviour of the parasitoid involved.

In this chapter, I will show how an evolutionary perspective towards progeny and sex allocation can lead to a better understanding of this variable character of parasitoids. This approach utilizes an impressive armoury of recent evolutionary models to generate predictions about oviposition behaviour under particular conditions. These models have so far been tested on only a small number of parasitoid species, but results suggest that parasitoids do approach the predicted patterns, and are therefore making adaptive decisions in progeny and sex allocation. To bridge the gap between this evolutionary understanding and solutions to the practical problems already mentioned, it is important to know *how* parasitoids achieve these adaptive responses, and this mechanistic problem will also be considered, with emphasis on the insects studied most by our research group, the hymenopterous egg parasitoids of the families Trichogrammatidae and Scelionidae. This chapter represents just one of several recent attempts to review and interpret theories on progeny and sex allocation for parasitoid researchers; somewhat different approaches may be found in papers by Charnov and Skinner (1984, 1985) and Waage and Godfray (1985).

II. OPTIMAL PROGENY ALLOCATION

Evolutionary models usually aim at a level of generality, and the prediction of simultaneous optimal progeny and sex allocation in a single general model is difficult and has rarely been attempted (but see Waage and Godfray, 1985; Godfray, 1986a). Instead, the problem has usually been approached along two paths: models for progeny allocation and models for sex allocation. This approach may prove more comprehensible to the reader as well as to the modeller, but the simultaneous nature of these decisions should not be forgotten in what is to follow, and I will return to this in Section VI.

A. Maximizing Fitness per Host Attacked

How many eggs should a female parasitoid lay in a host which represents a limited amount of resource for her developing offspring? Given that the amount of resource a larva gets determines its survival and fitness as an adult, a fitness function, $f(c)$ can be defined to describe the fitness of an offspring in a clutch of size c. The fitness realized *per host attacked* is then simply the product of the clutch size, c, and this function, $f(c)$. The value of c where this function is maximized is the optimal clutch size (Lack, 1947; Charnov and Skinner, 1984; Waage and Godfray, 1985).

Data sets on fitness per individual offspring at different clutch sizes have been obtained for only a few parasitoid species, mostly gregarious, and these vary considerably in the measure chosen to estimate fitness. Waage and Godfray (1985), for instance, have extracted fitness functions from the literature for six species of gregarious parasitoid, using the probability of survival from egg to adult as a crude measure of fitness. Three of these functions are illustrated in Fig. 1 for the gregarious egg endoparasitoids *Trichogramma evanescens* Westwood (Pallewatta, 1986) and *Telenomus fariai* Lima (Escalante and Rabinovich, 1979), and for the gregarious pupal ectoparasitoid, *Dahlbominus fuliginosus* (Nees) (Wilkes, 1963). These illustrate two methods by which fitness functions can be obtained. For the egg parasitoids, data were obtained by dissection of part of a cohort, to determine initial egg density, and rearing of the other part, to determine survival (hence, sampling error can lead to overestimates of survival as in Fig. 1b). For the ectoparasitoid, a more precise method was possible, whereby eggs were transferred from one host to another, to create the desired

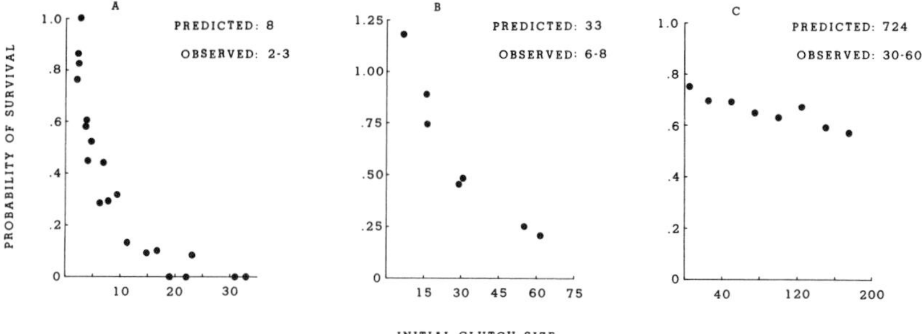

Fig.1. Fitness ($f(c)$) functions for three gregarious parasitoids, estimated by the probability of survival in initial broods of different size: (a) *Trichogramma evanescens* Westwood on eggs of *Mamestra brassicae* L. (Pallewatta, 1986); (b) *Telenomus fariai* Lima on eggs of *Triatoma phyllosoma pallidipennis* Stål (Escalante and Rabinovich, 1979) and (c) *Dahlbominus fuliginosus* (Nees) on pupae of *Neodiprion lecontei* (Fitch) (Wilkes, 1963).

initial broods. If larval mortality does not occur until late in development, and dead larvae are not consumed, fitness functions may be obtained by a third method, which is the simple comparison of emerged and unemerged offspring (Takagi, 1985; Ikawa and Okabe, 1985).

More accurate fitness functions can be obtained by measuring not only offspring survival, but the fitness of surviving parasitoids. This may involve measurement of size/fecundity (Charnov and Skinner, 1984; Waage and Ng, 1984; Takagi, 1985) or weight (Shiga and Nakanishi, 1968a; Takagi, 1985; Ikawa and Okabe, 1985).

Most available data sets, like those in Fig. 1, indicate a continuous decline in individual fitness with clutch size, presumably as a result of larval competition. Only in a few studies (Takagi, 1985; Ikawa and Okabe, 1985) is there evidence for an Allee effect, where fitness initially rises and then falls. With more study, such dome-shaped fitness functions may prove common in gregarious endoparasitoids. For these parasitoids, very small broods often fail entirely, because of their inability to overcome host defences (DeLoach and Rabb, 1972; Kitano and Nakatsuji, 1978; Ikawa and Okabe, 1985) or to consume all of the host tissues, a prerequisite in some cases for successful pupation and emergence (Schieferdecker, 1969; Strand and Vinson, 1985). Allee effects are less likely to occur for ectoparasitoids (a large proportion of existing data sets), where host defence and excess food do not limit parasitoid survival so strongly, and of course for those endoparasitoids whose normal clutch size has been taken as the minimum, and larger clutches are obtained by encouraging superparasitism, as in Fig. 1a and b.

We can test the optimal clutch size model by comparing observed clutch sizes with those predicted by the calculation of $cf(c)$ from available data sets. This yields support for a general prediction of the model: species with "shallower" fitness functions allocate larger clutches to hosts (Waage and Godfray, 1985; see Fig. 1). Similarly, in intraspecific studies of egg allocation by *Trichogramma* spp., host species on which fitness functions are shallower (larger hosts) are allocated relatively larger clutches (Charnov and Skinner, 1984).

However, when predicted and observed clutch sizes are compared for a particular parasitoid and host, predicted clutch size is often substantially larger (Fig. 1). This discrepancy may occur for several reasons (Charnov and Skinner, 1984). It may simply reflect a poor measure of individual fitness. For instance, the predicted clutch size for *Trichogramma evanescens* attacking *Mamestra brassicae* (L.), is about eight when a fitness function based on larval survival is used (Fig. 1a). When fitness is measured instead by the fecundity of surviving females, the predicted clutch size is two to three (for all-female broods), which is within the range of clutch size normally produced on that host (Waage and Ng, 1984; Waage and Godfray, 1985).

An increase in clutch size is often associated with a decrease in the size of surviving parasitoids. This in turn may reduce their fitness because smaller parasitoids often have reduced fecundity, longevity, and searching rates (see Chacko, 1969; Pak and Oatman, 1982; Waage and Ng, 1984 as examples for *Trichogramma* spp.). Accurate fitness functions should take all these components of fitness into account. Measures based on a single component, or on just a few, are therefore likely to underestimate the rate of decline of fitness at larger clutch sizes, and therefore overestimate optimal clutch size.

Another reason for the discrepancy between observed clutch sizes and those predicted by the optimal clutch size model may lie in the assumption that lifetime reproductive success is maximized by maximizing fitness per host. Is this correct?

B. Other Criteria for Optimizing Clutch Size

A number of recent models show that maximizing fitness per host may not always be the best strategy for an ovipositing parasitoid. If, for instance, there is a cost in time to laying an egg, and if the fitness realized per egg decreases with the number of eggs already present in a host (i.e. as we approach the peak of the domed $cf(c)$ function where $f(c)$ is decreasing), it may benefit a parasitoid to cease adding more eggs to a host and allocate the time saved to finding another host (Parker and Courtney, 1984; Skinner, 1985). The advantage of leaving hosts and seeking new ones will increase as the rate of finding new hosts increases. Hence, in this model which maximizes fitness per unit time, wasps will maximize fitness per host attacked only when hosts are very scarce. As hosts become more abundant, they should leave each host sooner and hence lay smaller clutches per host.

Similar predictions are obtained from a quite different model which assumes that a parasitoid has a limited number of eggs to lay (Iwasa *et al.*, 1984; Parker and Courtney, 1984; Godfray, 1986b). If this number is very large relative to the number of hosts available, then she could do no better than lay the clutch size which maximized fitness per host. But if her eggs are limited relative to hosts, either as a species character or because she has laid most of them, then she would do better to spread out her eggs between hosts, in order to maximize fitness per egg. Assuming a monotonically decreasing fitness function, as in Fig. 1, this would lead ultimately to a strategy of placing one egg per host. Parasitoid survival and longevity will also play a role here. Whatever her fecundity, as the probability of a female surviving to lay all her eggs decreases, her optimal clutch size will increase to a maximum value, which is that which maximizes fitness per host attacked (Weiss *et al.*,

1983: Parker and Courtney, 1984; Iwasa *et al.*, 1984; Waage and Godfray, 1985; Ikawa and Okabe, 1985).

The risk of mortality has somewhat different theoretical consequences if it is associated with the act of oviposition: any cost to oviposition which would reduce subsequent reproductive success should decrease optimal clutch size (Charnov and Skinner, 1984; Iwasa *et al.*, 1984).

These various models make different specific predictions for different initial conditions, but when considered together they point to some general, expected patterns of egg allocation. A parasitoid should never exceed the clutch size which maximizes fitness per host attacked, and this clutch size will be determined by the fitness function, $f(c)$, which in turn is determined by the amount of resource in the host and the number of parasitoid larvae competing for it (the "vital space" calculation of Rabinovich, 1971). This maximal clutch size will be favoured when parasitoids are relatively host limited, either because hosts are scarce or because parasitoid survival is low. As parasitoids become less limited by host abundance (and perhaps more limited by the eggs which they have to lay), optimal clutch sizes should decrease, ultimately to the extreme where only one egg is laid per host.

Thus, these models may help to explain the finding of the last section, that observed clutch sizes are often less than those predicted by a model which maximizes only fitness per host.

C. Mechanisms for Adaptive Progeny Allocation

Models for optimal clutch size suggest that parasitoids should base clutch size decisions on a variety of endogenous and exogenous stimuli. These include the value of the "host at hand", previous foraging experience and the parasitoid's physiological state (e.g. egg load and age). What evidence is there that parasitoids do respond to such stimuli by adjusting clutch size?

Host size may be one indication of the value of an encountered host, particularly for idiobiotic parasitoids (Haeselbarth, 1979; see Chapter 8), whose larvae develop on the precise host resource encountered (and usually killed) by the mother. *Trichogramma* spp. are typical idiobionts, and studies have shown that wasps make a physical inspection of the host egg to determine the clutch size laid (Klomp and Teerink, 1962, 1967). This inspection may involve an estimation of egg volume from a measure of both curvature and surface area of the egg (Schmidt and Smith, 1985a). For some other gregarious idiobionts which lay fewer eggs in smaller hosts, surface area alone has been shown to be a poor predictor of clutch size (Wylie, 1967; Takagi, personal communication).

Host size, however, may not be as good an estimate of host value for a koinobiotic parasitoid, whose host continues to grow after parasitism (Waage, 1982b). Nonetheless, a positive correlation of clutch size with host instar has been found for some parasitic Hymenoptera (Neser, 1973; Sato and Tanaka, 1984) and Diptera (DeLoach and Rabb, 1971; Danks, 1975b) attacking growing lepidopterous larvae. In these cases, older hosts may provide, ultimately, more larval resource, and there is evidence that host instar itself, and not host size (weight), is the stimulus for changing clutch size (Sato and Tanaka, 1984).

Another aspect of host value for many species is host age: survival on older hosts is often less, and some parasitoids respond by laying smaller clutches on these hosts (Pak and Oatman, 1982; Juliano, 1982). Similar patterns are found when host value is reduced by previous parasitism (see Section III). In Chapter 4, Strand discusses in detail the role of physiological and structural changes in the host, and of host markers in the discrimination of age and previous parasitism by egg parasitoids.

Models which consider fitness gain per unit time (Parker and Courtney, 1984; Skinner, 1985) predict that parasitoids should vary clutch size on the same host in response to the rate of host finding (specifically the travel time between hosts). While these models have not had specific tests, their general predictions conform to observations for gregarious parasitoids that larger clutches are laid at lower rates of oviposition (Jackson, 1966; Glas *et al.*, 1981; Ikawa and Suzuki, 1982; Waage and Ng, 1984). *Trichogramma evanescens* lays larger clutches when encountering solitary host eggs at intervals of 30 min than when placed on a patch of eggs, where ovipositions follow at roughly 5-min intervals (Waage and Ng, 1984). This difference in clutch size between solitary and clumped hosts has been observed for other species (Hirose *et al.*, 1976; Vu, 1985). Waage and Godfray (1985) interpret this phenomenon as evidence that wasps base clutch size on a measurement of time between ovipositions. Schmidt and Smith (1985b) propose quite a different mechanism for this shift in clutch size on solitary versus clumped eggs: because host surface area is used to determine clutch size (Schmidt and Smith, 1985a), hosts packed in groups will on average have less exposed surface and hence will be allocated fewer eggs. They suggest, in effect, that the distribution of hosts, rather than the rate of host finding, might provide information on host abundance.

By their very nature, endogenous stimuli which may affect progeny allocation are difficult to measure, and have received little study. Models predict that egg load may be an important factor determining clutch size. The number of eggs in the oviducts may be used as an indication of the rate of oviposition, independent of or together with a time-based measure, but this "egg pressure" effect remains to be investigated in the context of progeny allocation.

III. SUPERPARASITISM

From the point of view of an ovipositing parasitoid, superparasitism is the allocation of an egg or eggs to a host which has eggs from a previous oviposition by that or another conspecific parasitoid (van Dijken and Waage, unpublished). Given fitness functions such as those in Fig. 1, laying an egg in a parasitized host is likely to confer less fitness than laying an egg in an unparasitized host. In solitary parasitoids, comparable fitness functions would usually be even more steep and hence the value of superparasitism would be even less obvious. This kind of reasoning is the basis for the widespread belief that superparasitism is disadvantageous and must therefore represent a "failure to discriminate" (Fiske, 1910; van Lenteren, 1981).

In fact, superparasitism need not always be disadvantageous. As long as an egg has a finite probability of survival in competition with a previous brood, conditions may arise (e.g. the complete absence of unparasitized hosts) where this probability would make it advantageous to superparasitize. Furthermore, even when the probability of survival is zero, superparasitism may be advantageous for a parasitoid which is not egg-limited, if it would take her more time and energy to discriminate and reject the host than to lay an egg in it (Hughes, 1979). To consider more specific advantages of superparasitism, it is first important to recognize two distinct kinds of superparasitism: one in which a female lays eggs into hosts which she has attacked previously (self superparasitism), and one in which a female lays eggs into hosts attacked by other conspecifics (conspecific superparasitism; van Dijken and Waage, in press; Bakker *et al.*, 1985).

A. Self Superparasitism

When an individual parasitoid is confined in the laboratory with a few hosts, it will often superparasitize. This can be seen in Fig. 2a for *T. evanescens* exposed to hosts for varying periods. In such experiments, as exposure to hosts continues, parasitoids encounter fewer and fewer unparasitized hosts. Theory suggests that when the rate or chance of finding unparasitized hosts falls, optimal clutch size should increase, up to that which maximizes fitness per host (Parker and Courtney, 1984; Skinner, 1985; Waage and Godfray, 1985). In this context, returning to lay more eggs in a host can be viewed as an adaptive, retrospective adjustment of clutch size, in the light of foraging experience. Indeed, after some superparasitism in *T. evanescens*, clutch size levels off at around three eggs per host, close to the value predicted earlier as maximizing fitness per host attacked.

Adjustment of clutch size occurs quite frequently in *T. evanescens*, even

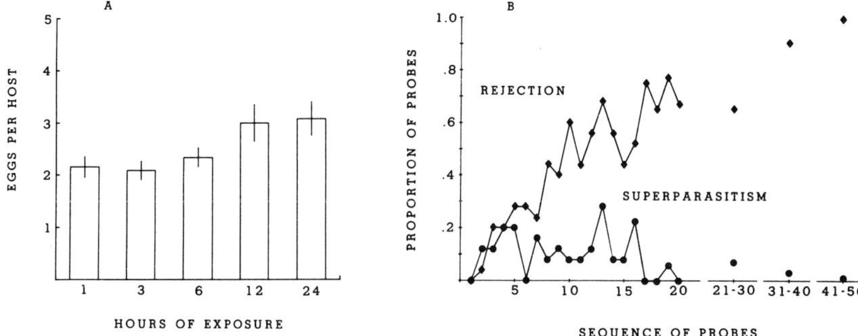

Fig. 2. Self-superparasitism in *Trichogramma evanescens* Westwood: (a) mean eggs laid per host by an experienced female exposed to 10 hosts for different periods, with 95% confidence intervals; (b) frequency of successive probes which lead to rejection or superparasitism by an experienced female exposed to a mass of 10 unparasitized eggs (probes into unparasitized hosts are not shown).

early in an oviposition bout when only a few hosts have been attacked (van Dijken and Waage, in press). Fig. 2b shows the frequency with which parasitized hosts are attacked or rejected as a wasp begins to exploit a patch of 10 host eggs. Initially this frequency is the same, but soon the rate of superparasitism falls to a very low level. After an exposure of about 12 h (Fig. 2a) it rises slightly. In *T. evanescens*, smaller broods are allocated in self superparasitism than in attacks on unparasitized hosts (see below).

We might expect solitary parasitoids to be much more reluctant to self superparasitize. Superparasitism might be advantageous when two or more eggs increase the probability of any offspring surviving in the egg. This hypothesis is supported by limited evidence that superparasitism in solitary endoparasitoids reduces the effectiveness of the host immune response (Askew, 1968; Puttler, 1974; Bakker *et al.*, 1985), permitting greater parasitoid survival per host.

B. Conspecific Superparasitism

When a parasitoid encounters a host parasitized by another individual, quite a different evolutionary problem is posed. These parasitized hosts are essentially hosts of a lower quality, in terms of fitness gained per egg laid. This has been shown elegantly by Bakker *et al.* (1985), using wild type and mutant strains of the solitary endoparasitoid, *Leptopilina heterotoma* (Thomson). The eggs of a mutant had a 94% chance of survival on an unparasitized hosts, an 8% chance on a host parasitized 1 h previously by a wild type female, and a 3% chance on a host parasitized 3 h previously.

For conspecific superparasitism, the decision to oviposit can be approached by optimal diet or optimal patch use models (Iwasa *et al.*, 1984; Parker and Courtney, 1984; Skinner, 1985). These have been formulated with respect to the fitness gained by a parasitoid per unit time, and predict generally that superparasitism should increase at lower rates of encounter with unparasitized hosts, and decrease as parasitized hosts become relatively less suitable hosts. This latter change may occur as the time from first oviposition increases (as in the example above), or as hosts receive more eggs. Bakker *et al.* (1972) have shown that supernumerary eggs of *L. heterotoma* are distributed more regularly than predicted by random allocation, suggesting that females can estimate the number of eggs in a host, and superparasitize hosts with less eggs.

Smaller clutches are also predicted in conspecific superparasitism, as long as the first and second wasps to find a host have a similar oviposition experience (Skinner, 1985). This prediction is supported by a variety of studies on gregarious parasitoids (Wylie, 1965; Holmes, 1972; Ikawa and Suzuki, 1982; van Dijken and Waage, in press).

Clearly, models for self and conspecific superparasitism make very similar predictions. Both predict that superparasitism will increase when unparasitized hosts become scarce, and both predict that clutches in superparasitism will be reduced. However, all things being equal, a parasitoid given a choice between a host which it has parasitized and one parasitized by a conspecific should usually choose the latter. Evidence that parasitoids can make this distinction is limited. When *T. evanescens* was allowed to self superparasitize, van Dijken and Waage (in press) found that it allocated fewer eggs on average to parasitized than to unparasitized hosts. The same was true for wasps allowed to parasitize hosts attacked by other females under identical conditions, but no significant difference was found in clutch size between self and conspecific superparasitism. Using a very similar experimental method, Hubbard *et al.* (in press) found clear evidence that the solitary endoparasitoid, *Venturia canescens* (Gravenhorst), laid eggs in parasitized hosts less frequently during self than during conspecific superparasitism. Finally, no difference between self and conspecific superparasitism was found for the solitary endoparasitoid, *Asobara tabida* Nees (van Alphen and Nell, 1982).

These studies have looked for individual-specific markers, but parasitoids may use other, less precise, kinds of information to estimate the probability that a parasitized host contains their eggs. These may include the presence of other ovipositing parasitoids or the time elapsed or distance moved between last oviposition and the encounter of a parasitized host (van Alphen and Nell, 1982; van Dijken and Waage, in press) or a change in markers (Chapter 4). Thus, Bakker *et al.* (1985) report that *Leptopilina heterotoma* superparasitizes more in the presence of other wasps than when alone (see also Chapter 2). In an interesting development of this argument, van Alphen and Nell (1982) suggest that the tendency of inexperienced wasps to superparasitize may reflect, firstly, that their experience of finding better, unparasitized hosts is nil and, secondly, that they can be certain that parasitized hosts do not contain their eggs. This hypothesis challenges the traditional explanation for this phenomenon, that wasps must learn to avoid superparasitism (van Lenteren and Bakker, 1975; van Lenteren, 1976; Klomp *et al.*, 1980).

IV. THE EVOLUTION OF GREGARIOUSNESS

At this point, the reader may feel uncomfortable about the discrepancy between the *theory* of parasitoid progeny allocation, wherein optimal clutch size models predict a continuous range of clutch sizes with varying conditions, and the *reality* of parasitoid progeny allocation, namely that there are gregarious species and there are solitary species. Figure 3 illustrates this discontinuity in the oviposition behaviour of real parasitoids, for the genus

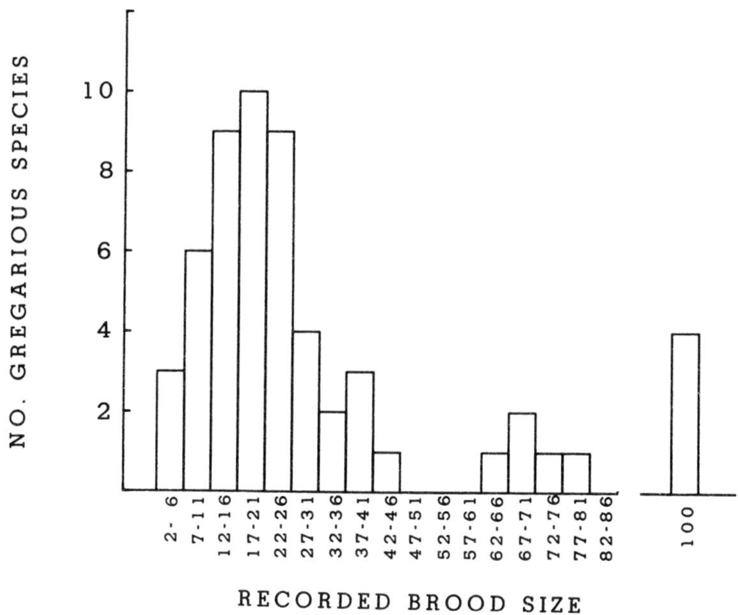

Fig. 3. The distribution of brood size for 56 gregarious species of the braconid genus *Apanteles* (*sensu lato*), collected from the literature (Le Masurier, unpublished).

Apanteles (*sensu lato*) (Le Masurier, unpublished). This analysis of brood size for 276 species reveals that *Apanteles* either develop alone in hosts, or in fairly large groups of about 20, 70 or several hundred.

This clear discontinuity between solitary and gregarious lifestyles is found in many parasitoid taxa. In order to understand it, we must first appreciate that clutch size is the evolutionary consequence of two processes: selection acting on adults to optimize progeny allocation and selection acting on larvae to maximize their own adult lifetime fitness. As a result, the evolution of clutch size may involve a degree of parent-offspring conflict. For instance, a glance at Fig. 1 suggests that, for these three species, the fitness of individual offspring is maximized (from the fitness function) at clutch sizes below those which maximize fitness per host for the mother.

Under conditions where it greatly benefits a larva to develop alone in a host, we would expect it to evolve the capacity to eliminate competitors, even siblings placed with it by the mother. The larvae of solitary parasitoids eliminate competitors by physical attack, usually in an early instar, and by physiological suppression (Salt, 1961; Vinson and Iwantsch, 1980). Gregarious and facultatively gregarious species, in contrast, rarely show either the weaponry or the inclination to be aggressive.

If we assume for the sake of argument that gregariousness is a secondary, derived character in parasitoids [and this is not unreasonable, given that it is widely distributed but generally uncommon in parasitoid taxa (Clausen, 1940)], then how could it evolve in the face of parent-offspring conflict? Godfray (in press) has developed a simple genetical model for the evolution of larval tolerance as a first step towards gregariousness. The model predicts that a mutant gene for tolerance will spread in a population of aggressive, solitary parasitoids only when fitness per larva in a clutch of one is less than that in larger clutch sizes. Thus, only with an Allee effect does fitness increase with clutch size, favouring gregariousness.

A gene for tolerance suffers the disadvantage, when rare, of frequently sharing a host with aggressive siblings, and this acts as a brake to its initial spread. Factors such as the cost of fighting apparatus (e.g. mandibles) and the chance of two aggressive larvae killing each other can make the spread of tolerance more likely, while the possibility of superparasitism by non-siblings is a potent selective force for the retention of aggressive behaviour and the solitary habit.

The requirement of an Allee effect is a fairly stringent limitation to the evolution of gregariousness, but if satisfied and larvae evolve a tolerant, gregarious habit, clutch size might be expected to evolve further to the parental optimum. Parent-offspring conflict does not necessarily disappear, however, and a killer gene which ensures solitary development may spread. Of particular interest in Godfray's model is the finding that a gregarious species is most likely to revert to a solitary species when clutch sizes are small, which might help explain the sharp discontinuity in clutch size between some solitary and gregarious species (Fig. 3).

The observation that gregarious species in largely solitary taxa are often associated with particularly large host species conforms to the prediction of this model, if large hosts represent a resource which must be shared to be utilized successfully or efficiently (e.g. if all of the host must be consumed for successful pupation, see Section II. A). Le Masurier (unpublished) has found for the genus *Apanteles* (*sensu lato*) that very small hosts support only solitary species, very large hosts only gregarious species, but hosts of intermediate size support both. Thus, host size can explain some but not all of the distribution of these lifestyles in a particular taxon.

A number of other factors may increase the likelihood that larval tolerance will evolve, perhaps the most interesting being increased relatedness of larvae in the host. An extreme case is that of polyembryonic parasitoids: where larvae are genetically identical, a larva's own fitness is maximized in a clutch size which maximizes fitness per host. It is noteworthy that polyembryonic species show some of the largest brood sizes of any parasitoid: broods of 1000–3000 wasps per host are not uncommon in some polyembryonic Encyrtidae (Clausen, 1940). Polyembryonic species also reveal the potential for evolutionary switching between solitary and gregarious forms. In *Pentalitomastix* sp., two successive larval forms occur in a single host. The first to develop is mandibulate and aggressive, and serves to protect the host from multiparasitism, while the second form has reduced mandibles and is tolerant of its many siblings (Cruz, 1981). These superficially "solitary" and "gregarious" larvae are produced by the same female, indeed perhaps by the same genome!

A more subtle case of kin selection may be implicated in the evolution of gregariousness, involving female-biased broods in the parasitic Hymenoptera. Because, in these haplodiploid insects, sisters are more related to each other than they are to their brothers, mean relatedness increases in broods with relatively more daughters. It is generally believed that gregariousness favours local mating between siblings (a form of local mate competition, see Section V. C), which favours female-biased broods. Perhaps the reverse is also true: female-biased broods favour larval tolerance and hence larger broods, generating a system of positive feedback which could accelerate the evolution of gregariousness in a particular species.

There are undoubtedly other, more ecological factors which have influenced the evolution of gregariousness and benefited parents and larvae alike. These may include the advantage in some environments of producing many small offspring instead of a few large offspring or the ability of gregarious larvae to develop more rapidly than their solitary counterparts, and the advantage this may give in multiparasitism (Conde and Rabinovich, 1979). At present, we do not have the information to determine how important these advantages may be, or even how general they are among gregarious species.

V. OPTIMAL SEX ALLOCATION

The great inter- and intraspecific variation found in parasitoid sex ratios indicates that understanding sex allocation is important to understand reproductive strategies of parasitoids. In the parasitic Hymenoptera, sex ratios commonly range from a slight male bias to entirely female broods. Less is known about other parasitoid taxa; in the Tachinidae a sex ratio of about 0.5 (proportion males) seems most common, although female-biased ratios are reported (Clausen, 1940).

A major problem in the study of sex allocation by parasitoids is the measurement of primary sex ratios. Sex ratios are usually obtained through rearing or, less commonly, by sampling of adults in the field. These secondary sex ratios may differ from those allocated to hosts because of differential larval mortality or differences in adult behaviour, respectively. In only a few cases has it been possible to measure the sex of eggs or young larvae. Sometimes male and female eggs may be laid in different positions on the host (Flanders, 1950; Luck *et al.*, 1982), or larvae may be sexually dimorphic (Tooke, 1955). Sexes can be distinguished at oviposition by differences in the oviposition movements of females for *Trichogramma* spp. (Suzuki *et al.*, 1984; see below), and for the ichneumonid, *Itoplectis maculator* (Fabricius) (Cole, 1981). This method may work for other species. In the parasitic Hymenoptera, the potential also exists to distinguish the sex of eggs or larvae by chromosome counts, but thorough studies of this method have yet to be made.

In the absence of methods to determine primary sex ratios, larval competition may be measured, to allow correction for differential mortality in our estimate of sex ratio. This has permitted the analysis of secondary sex ratios as primary sex ratios in some studies on gregarious parasitoids (Werren, 1983). Wellings *et al.* (in press) have recently developed a useful statistical model for predicting primary sex ratios from secondary sex ratios in solitary parasitoids which corrects for sex-specific mortality.

In the discussion to follow, it should be borne in mind that some results are based on measurement of secondary sex ratios.

A. Fisherian Sex Ratios

The starting point for a consideration of the evolution of sex allocation is Fisher's (1930) model which predicts that, in panmictic (randomly mating) populations, investment in the production of males and females should be equal. One way in which this may be achieved is for parents to invest equally in the production of sons and daughters. If sons and daughters are equally expensive to produce, a sex ratio of 0.5 is then the predicted allocation. If we consider parental investment for parasitoids to involve the energy or time expended in producing eggs, finding the hosts for them and laying them, most parasitoids probably do invest equally in male and female offspring. A possible exception is found in heteronomous hyperparasitoids of the family Aphelinidae which lay their sexes in different kinds of hosts or even different host species (Walter, 1983). Some species of *Encarsia*, for instance, lay female eggs in unparasitized hosts and male eggs in (probably rarer) parasitized hosts. Furthermore, male eggs can require a much greater oviposition time than female eggs (Alomar, personal communication). Hence, the investment in time per female offspring is relatively less than per male, and we might expect a sex ratio biased towards the cheaper sex (Colgan and Taylor, 1981)—which is what occurs.

Given equal investment in both sexes, selection for sex ratios other than 0.5 may arise if the assumptions underlying Fisher's models do not apply. Before discussing alternate models which cause these assumptions to vary, a comment should be made on the ability of natural selection to change sex ratio. Virtually all studies of parasitoid sex allocation have been made on arrhenotokous Hymenoptera, where males generally develop from unfertilized eggs and females from fertilized eggs. Control of insemination may therefore control sex ratio, and this might involve regulating the release of sperm from the spermatheca (Flanders, 1939). In principle, selection might act quite easily on such a behavioural response, although success in selecting for changes in wasp sex ratios has in fact been limited (Simmonds, 1947; Wilkes, 1947; Ram and Sharma, 1977). In diploid species, in contrast, sex appears usually to be determined by Mendelian assortment of sex chromosomes. Williams (1979) has suggested that this essential process of assortment restricts the potential for sex ratios to evolve away from 0.5. Thus we might expect, for the same selection pressures, the evolution of more sex ratio variation in haplodiploid parasitoids than in diploid parasitoids, such as tachinids.

B. The Effect of Differences Between Male and Female Larvae

Differential mortality between sexes does not influence the Fisherian prediction of 0.5, unless it results from the effect of the presence of one sex in a brood on another (Waage and Godfray, 1985; Godfray, 1986a). For instance, if the presence of males in a brood decreases the fitness of a female more or less than the presence of an equal number of females, then optimal sex ratio will be biased towards females, the sex which "loses" in a relative sense. Evidence for differential larval mortality in gregarious parasitoids is considerable. Most cases indicate that males, perhaps because of their more rapid development or lower nutritional requirements, fare relatively better as survival decreases for parasitic Hymenoptera (Narayanan and Subba Rao, 1955; Wilkes, 1963; Wylie, 1966; Suzuki *et al.*, 1984) and perhaps also for Diptera (Wylie, 1977; Ziser *et al.*, 1977). However, in only a few of these instances is a decline in fitness of a sex linked to the presence of the other sex, and not simply the presence of too many competitors. Chacko (1969), for instance, found that female *Trichogramma minutum* Riley emerging from *Corcyra cephalonica* Stainton had a lower fecundity if they had shared that egg with a male rather than with a female. In this species, theory predicts a female-biased sex ratio, which is seen.

Differential mortality and sexual asymmetries encompass a range of fascinating interactions between male and female parasitoid larvae. Pickering (1980), for instance, has noted that patterns of relatedness between siblings may influence sexual asymmetries. Thus, in parasitic wasps, female larvae may be more tolerant of sisters than of brothers, because they share on average more genes with the former. This could generate parent-offspring conflict which encourages mothers to separate sexes in different broods (Pickering, 1980; Waage and Godfray, 1985). Single-sex broods are indeed found in some gregarious parasitoids (Bryan, 1983). The response of male larvae to female siblings has been discussed by Kurosu (1985), who suggests that male *Trichogramma* sp. sacrifice their own wing production to free more nutrients to sisters, *unless* male larvae from another mother are present, whereupon they become less altruistic and develop wings.

The fact that resource limitation can affect sexes differently has another important consequence for optimal sex allocation. Charnov (1979) has pointed out that if hosts vary in the amount of resource they contain, for instance because of size, and if the incremental gain in fitness per unit of resource is greater for one sex than for the other, then females should lay that sex into the larger hosts. Charnov's model explains a very widespread observation amongst solitary parasitoids that females emerge from larger hosts and males from smaller hosts (see Charnov, 1982 and Charnov *et al.*,

1981 for review). This would be predicted if female fitness increased more rapidly per unit of host resource than male fitness. Although the clear effects of larger size on many aspects of female fitness (see Section II) make this assumption appealing, male fitness may also increase with size (Grant *et al.*, 1980), and the underlying assumption remains to be proven quantitatively for any parasitoid.

Charnov's model explains the variation in the distribution of sexes across hosts, which affects indirectly the population sex ratio. The model predicts an overall population sex ratio which is slightly male-biased (Charnov, 1982). Because the model predicts that a parasitoid should lay females in *relatively* larger hosts, lifetime sex ratios are not determined by the number of each host size attacked: a small host may become relatively large if no larger hosts are found. Studies on various parasitoids show that sex allocation often switches at a specific host size or growth stage (Jones, 1982; Luck and Podoler, 1985), but parasitoids offered only one size of host do eventually begin to produce both sexes and a more equitable lifetime sex ratio, as predicted (Sandlan, 1979; Avilla and Albajes, 1984). Van den Assem *et al.* (1984) developed a behavioural model for this pattern of sex allocation, which involves the adjustment of sex ratio on the basis of both short- and long-term experience.

Charnov's model may also explain patterns of sex allocation to hosts varying in qualities other than just size. Some parasitoids are known to allocate males preferentially to hosts at a less suitable stage for development (van Alphen and Thunnissen, 1983), which are poorer for development and may be poorer in larval resource, to parasitized hosts (see below) and even to different host species amongst a number of species which are attacked (Hails, personal communication).

C. Differences in the Mating Structure of Adult Populations

Hamilton (1967) was the first to show that optimal sex ratios could vary from 0.5 if the assumption of panmixis was relaxed. He constructed the local mate competition (LMC) model for a situation where females placed offspring in discrete patches of resource, and those offspring mated randomly within their patch before female offspring dispersed to colonize new patches. The optimal sex ratio (proportion males) predicted by this model is $(n-1)/2n$, where n represents the number of females colonizing a single patch. Thus, when n is very large, mating is between the offspring of a large part of the population (i.e. near panmixis), and a sex ratio of 0.5 is predicted. As n decreases, so does sex ratio until the point when only one female colonizes a patch, she should produce only as many sons as necessary to mate all her daughters (in the model this limit is, rather unrealistically, no males). A number of modifications have since been proposed for this model (e.g. for haplodiploids; Hamilton, 1979) which do not change its general predictions, and much effort has been spent interpreting precisely why the sex ratios change as they do (Taylor, 1981; Colwell, 1981), which cannot be reviewed here.

In his original paper, Hamilton (1967) identified gregarious parasitic Hymenoptera as a group particularly suited for testing the model, because n per host is likely to be small, and mating of progeny will often be restricted to the vicinity of the host. In many gregarious species, males emerge first and mate females as they emerge from the host or from pupal cases nearby. However, the degree to which all mating occurs before dispersal from the host is variable in parasitic wasps. At one extreme, broods of some egg parasitoids may mate before leaving the host egg (Costa Lima, 1928; Suzuki and Hiehata, 1985). In other cases, aggression between males may encourage their dispersal from the host (Waage, 1982a), and females may disperse as well, such that mating may occur elsewhere and more panmictically (Tagawa and Kitano, 1981). A preference for non-sibling males has been demonstrated in one gregarious species (Grant *et al.*, 1980), which would also encourage panmixis.

In general, gregarious Hymenoptera do have female-biased sex ratios, which supports the LMC model. The model is further supported by comparison with solitary species. When they do not attack grouped hosts, solitary parasitic wasps, because of the dispersion of hosts, are likely to be more panmictic in mating than gregarious wasps. This argument is supported by evidence that some solitary species mate in swarms (Vater, 1971; van Achterberg, 1977; Jervis, 1979) and utilize long distance sex pheromones (see Chapter 5). Sex ratios in these species tend to be near 0.5 or slightly female biased. A specific comparison has been made for species of the braconid genus *Apanteles* (*sensu lato*) by Shaw and Smith (unpublished) and Le Masurier (unpublished). In the latter study, the average sex ratio reported in solitary species was 0.50 ($n=28$), while that for gregarious species was 0.35 ($n=36$).

An interesting contrast to these patterns is found in the parasitic Diptera. Gregarious tachinids, while developing together in the host, often pupate away from host remains, show marked protandry and mate in sunny habitats far from the emergence site. Such species exhibit sex ratios of close to 0.5 (DeLoach and Rabb, 1971; Danks, 1975a), as we might predict from LMC theory, given that diploids are free to evolve adaptive sex ratios, as discussed earlier.

A more rigorous test of LMC is to compare species which vary in the average number of females colonizing a host patch. This has been done for different species of solitary scelionid wasps which attack, develop and mate on the egg masses of different host species (Waage, 1982a). In this test, it was assumed that *n* for species attacking hosts with small egg masses would be near 1, as one female could parasitize a whole mass, and females tended to avoid superparasitism. Species attacking egg masses too large for one female to exploit, on the other hand, would experience higher *n*. Data from 31 species of scelionids showed an increase in sex ratio with host egg mass size, as predicted.

For a particular species, LMC theory predicts that individual females should increase sex ratio per host patch in response to an increase in perceived *n*. Density dependent shifts in sex ratio are widely reported for hymenopterous parasitoids under laboratory conditions (Walker, 1967; Waage and Lane, 1984), but because of differential mortality and possible effects of sexual asymmetries, these cannot be taken as clear support of LMC theory. A positive association between sex ratio and *n* has been found in more controlled experiments on the parasitoids *Gregopimpla himalayensis* Cameron (Shiga and Nakanishi, 1968b), *Nasonia vitripennis* Walker (Wylie, 1966; Werren, 1983), *Muscidifurax zaraptor* Kogan & Legner (Wylie, 1979), *Trichogramma evanescens* (Waage and Lane, 1984) and *Telenomus remus* Nixon (van Welzen and Waage, unpublished). In the first three of these, independent studies indicated that differential larval mortality was unlikely, while in the last two, wasps were observed continuously and superparasitized hosts were excluded from the analysis.

Another experimental approach has been to consider the sex ratio laid by the first and second females to visit a patch, representing an *n* of 1 and 2, respectively. Werren (1980) and Suzuki and Iwasa (1980) modified Hamilton's model to predict the optimal sex ratio for the second female, and found that this depended primarily on the relative clutch sizes of both females, and to a lesser extent on the sex ratio of the first female. The general prediction is that the second female should lay a more male-biased sex ratio, and this has been shown experimentally for *Nasonia vitripennis* (Wylie, 1966; 1976a; Holmes, 1972; Werren, 1980), but was not found for *Trichogramma evanescens* (Waage and Lane, 1984, van Dijken and Waage, in press).

The production of more males when superparasitizing is not an exclusive prediction of LMC theory; it may also arise from Charnov's (1979) model if parasitized hosts are taken as providing less resource for larval development than unparasitized hosts (Waage and Lane, 1984). Taylor (unpublished) has modified the Charnov model to apply to cases where LMC theory also applies, and from this has generated means of distinguishing the predictions of the two models. Whereas the LMC model predicts a smooth rise in sex ratio with *n*, Taylor's modification of Charnov's model predicts a more discontinuous function, which resembles some found in the literature.

D. Mechanisms of Adaptive Sex Allocation

If natural selection is acting on sex ratio evolution, as suggested above, it should be reflected not only in the sex ratio produced, but in the mechanism of sex allocation itself. For parasitic Hymenoptera, a simple hypothesis for the adaptive allocation of sex would be that a fixed rate of sperm is released from the spermatheca, ensuring a fixed proportion of fertilization, and generating an optimal sex ratio which would appear as male and female eggs in random sequence. For the solitary braconid, *Asobara persimilis* Papp, male and female offspring are produced in a random sequence, supporting such a mechanism (Owen, 1983). According to King (1961), even the relatively precise, female-biased sex ratio of the gregarious *Nasonia vitripennis* might be generated at random. He suggests that an egg passing through the common oviduct will be oriented so as to obscure the micropyle (through which sperm passes to effect fertilization) about 25% of the time, generating the observed sex ratio.

When sex ratios get very female-biased (as they might under extreme LMC), and broods get small, problems may arise with random sex allocation. In order to ensure the presence of a male in every brood, a female may have to lay a higher sex ratio than predicted by LMC alone (Hartl, 1971). A more precise, non-random pattern of sex allocation would therefore be more advantageous, and this is indeed found in many species which lay relatively small, female-biased clutches (Green *et al.*, 1982; Waage, 1982a, 1982b).

Non-random patterns of sex allocation are often achieved by a particular sequence of sex allocation. In solitary scelionid egg parasitoids which attack egg masses, a male is usually laid in the first or second host egg, followed by a series of females, and then another male, more females, and so on (Waage, 1982a). This not only ensures the presence of a male on each egg mass, but it suggests a simple mechanism for adaptive sex allocation to host patches of different size (Waage, 1982a; Waage and Ng, 1984). Rather than counting the number of hosts in the patch and calculating what fraction should be sons and daughters (to ensure that all daughters are mated), the female can lay a son, then the number of daughters he can mate, then another son, and so on. Wherever this sequence stops, for lack of more hosts in that patch, an optimal, minimal number of sons will have been allocated.

Similar patterns of sex allocation are found in gregarious species. Figure 4 shows probable sequences of sex allocation for a series of 20 ovipositions by different *Trichogramma* species to different natural and artificial hosts. Ovipositing females in this genus exhibit different patterns of abdominal movements when laying male and female eggs: a pause in abdominal vibration occurs for female eggs and is probably associated with the act of fertilization. By continuous observation, the sequence of sex allocation between and within hosts can be measured. In our laboratory, this method predicts actual sex ratio sequences of *T. evanescens* with over 95% accuracy (van Dijken and Waage, in press). Errors are probably attributable to both the observer and the wasp.

Two conclusions arise from this comparison. Firstly, the pattern of sex allocation is remarkably conservative over a wide range of species and clutch sizes: whether eggs are laid one per host on adjacent hosts (Fig. 4a), two to three per host on adjacent hosts (Fig. 4b) or all together in one host (Fig. 4c and d), the second or third egg laid is most likely to be a male and males follow at intervals thereafter.

Secondly, overall sex ratio is seen to decrease with clutch size per host. This results primarily from an increase in the average number of females laid between successive male eggs (obscured here because of the superposition of male-female-male sequences of different length in the same graph). Given the diverse sources of these data, it is difficult to discuss any adaptive significance for this shift. Hosts which support only one parasitoid may receive male eggs more frequently because these hosts are of lower quality, or because emerging wasps may need to disperse more to find mates, thereby experiencing less LMC. In *Trichogramma*, intervals between oviposition bouts tend to "reset" this sequence of sex allocation. Thus, Waage and Ng (1984) found that individual *T. evanescens* laid the same pattern of males and females (similar to Fig. 4b) on three egg masses of *Mamestra brassicae* presented on three consecutive days. Furthermore, when wasps were presented with single host eggs at 30-min intervals, they always laid a male egg in each, as expected if the sex allocation sequence were reset between presentations.

Intraspecific sex ratio shifts, in response to LMC, must involve a modification of these patterns of sex allocation. Or must they? Waage and Lane (1984) found no evidence for an effect of encountering other wasps or parasitized eggs on the sex ratio laid in a particular egg. The sequence effect (Fig. 4b) was alone sufficient to explain density dependent sex ratio shifts, given that females avoided superparasitism and thus hosts per wasp decreased with wasp density. However, van Welzen and Waage (unpublished) in a similar study of *Telenomus remus* Nixon, found that wasps did respond to the presence of other females, and changed their sequence of sex allocation. The stimulus for this change may have been contact with traces of other wasps—chemicals deposited by foraging females are known to cause shifts in sex allocation in other scelionids (Viktorov, 1968; Viktorov and Kochetova, 1971). Other authors have suggested that changes in sex allocation patterns may be elicited by physical jostling of parasitoids by other individuals (Wylie, 1976a, 1979) or the encounter of parasitized hosts (Wylie, 1973, 1976b).

Not all parasitoids lay male and female eggs in the patterns discussed

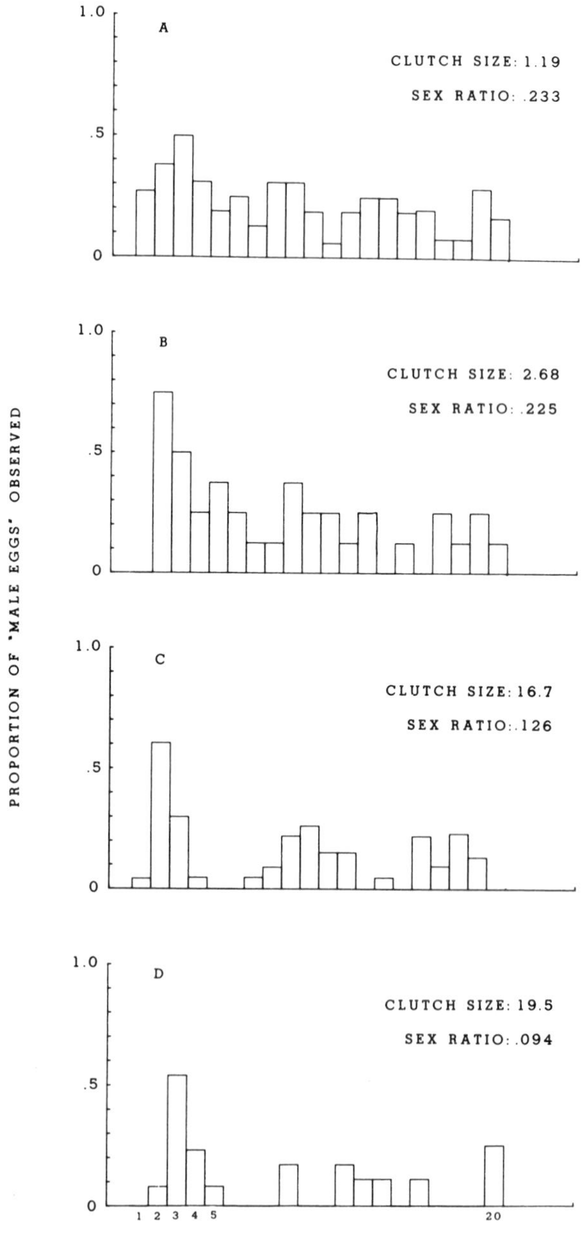

Fig. 4. Sequence of sex allocation (average frequence of male eggs per oviposition) for different *Trichogramma* spp. on different hosts: (a) *Trichogramma confusum* Viggiani on *Corcyra cephalonica* (Stainton) (Waage and Guo Ming-fang, unpublished); (b) *Trichogramma evanescens* Westwood on *Mamestra brassicae* L. (van Dijken, unpublished); (c) *Trichogramma chilonus* Ishii

above. Quite different patterns of sex allocation have been suggested for other solitary (Waage, 1982b) and gregarious (Putters and van den Assem, 1985; Feijen and Schulten, 1981) parasitoids. Whether these have an adaptive significance, or merely reflect one of many equally acceptable ways of producing an adaptive lifetime sex ratio, is a problem for future study.

VI. THE APPLICATION OF PROGENY AND SEX ALLOCATION THEORY

Do evolutionary models for progeny and sex allocation increase our understanding of parasitoids? The answer is certainly yes, if only because they help to organize the bewildering literature on parasitoid reproduction under a number of clear hypotheses, which themselves have great predictive power for parasitoids yet unstudied. Thus, when we discover a new gregarious parasitoid, we can predict it will have a female-biased sex ratio, and if it does not, then at least we know where in its complex biology to look for an explanation. Prior to sex ratio models, this new information would be just one more statistic to be added, for purely bibliographic reasons, to our empirical knowledge of parasitoids.

Despite this value, the study of parasitoid reproductive strategies is beset with a number of problems.

A. Problems with Optimization Models

We are presently at a point where there are clearly more models than studies which test them. Worse, these models often make predictions which are too broad for confident testing: e.g. parasitoids should lay smaller clutches when attacking hosts more rapidly, parasitoids should lay more males when on patches with other females. The more general a prediction, the more likely it will be shared by other hypotheses, adaptive and non-adaptive.

Rarely, this can be informative. The similarity between predictions of models for self and conspecific superparasitism, for example, tells us that a single behavioural mechanism might allow a parasitoid to respond adaptively to both. More often, however, this redundancy weakens the power of this approach. In the previous section, for instance, I have successively

---

on *Papilio xuthus* L. (data from Suzuki *et al.*, 1984) and (d) *Trichogramma confusum* Viggiani on a large, artificial, wax coated egg (Waage and Guo Ming-fang, unpublished). *T. confusum* and *T. chilonus* are considered synonymous by some taxonomists: the data above come from geographically distinct populations originating from different natural hosts. Average clutch size and sex ratio are for laboratory conditions similar to those in which sex allocation was measured.

claimed that a female-biased sex ratio under certain conditions constitutes support for Fisher's model of equal investment, Godfray's model of sexual asymmetries, Charnov's model of host quality, and the LMC model. Species can be found (species of *Trichogramma*, for instance) whose female-biased sex ratios may be explained by invoking several different models and their general predictions.

This problem does not bother many modellers. For them, generality, and the mathematical elegance which it courts, is a primary goal. If parasitoid researchers want to use these models to understand progeny and sex allocation by particular parasitoids, they must take it upon themselves to develop from them unambiguous, testable predictions by which the different models may be distinguished. Some progress has already been made in distinguishing between the precise predictions of different models (Taylor, unpublished) and in combining models to generate more precise predictions: e.g. host quality and LMC models (Werren, 1984), sexual asymmetry and LMC models (Godfray, 1986a), and optimal foraging and host quality models (Green, 1982).

Most existing theory deals with progeny allocation *or* sex allocation; few general models have been able to incorporate both (e.g. Godfray, 1986a). But wasps must decide on both every time they lay an egg. Can separate analysis of these two components tell us what a parasitoid should do? The answer is probably yes for a solitary species, where sibling larvae do not interact, and for "very" gregarious species, where the addition of a male or female egg will not have a large effect on fitness per clutch (Werren, 1984; Godfray, 1986a).

However, independent analysis of progeny and sex allocation will be inappropriate for small-brooded, "facultatively-" or "semi-gregarious" species. This has been shown by Waage and Ng (1984) for *Trichogramma evanescens*. They used a simulation model to explore optimal clutch size and sex ratio for this species. Assuming (1) an optimal clutch size model which maximized fitness per host and (2) complete LMC, they showed that optimal clutch size is critically influenced by sex allocation: the need to include one or more male eggs in a clutch to ensure that all daughters are mated affects the optimal number of female eggs to allocate in order to maximize the collective fitness of those daughters when mated. The simulation model was found to predict what wasps do more accurately than separate progeny and sex allocation models.

Another circumstance in which it may be misleading to separate the analysis of sex and progeny allocation is in superparasitism where even for large-brooded parasitoids the allocation of a second clutch may greatly reduce larval survival, with consequent effects of sex ratio via differential mortality and sexual asymmetries.

B. Parasitoids in Laboratory Rearing—an Opportunity for Practical Application

Given that the problems discussed above can be solved, the most valuable future application of models for progeny and sex allocation will be in understanding why parasitoid populations in laboratory culture are different from those which we collect in the field. This underlies a real problem in biological control, often referred to as "quality control". Laboratory populations often yield parasitoids with reduced fitness and undesirable population sex ratios. Even in their present simple state, models for progeny and sex allocation can tell us that some of these changes may not represent behavioural aberrations under abnormal conditions, but rather a very natural, adaptive response of parasitoids to particular host and parasitoid densities.

For instance, we may predict from theory that increasing ratios of parasitoid to host, by decreasing the rate of finding unparasitized hosts, should cause wasps to make adaptive increases in clutch size. This they would do through larger first clutches and superparasitism. The increase in clutch size through superparasitism would be greater as the absolute number of wasps increased, as this increases the frequency of conspecific relative to self superparasitism. The sex ratio allocated to hosts should also increase adaptively with the ratio of parasitoid to host, because of LMC effects on first clutches and in superparasitism. With low levels of superparasitism, this should not exceed 0.5, but with high levels, allocation might be considerably above this.

These predicted changes in allocation correspond to what we find in many crowded laboratory cultures of parasitic wasps: low productivity, small, unfit wasps and highly male-biased sex ratios. Adaptive changes in progeny and sex allocation will not be the only factors affecting the quality of cultures, and may not be the most important. Male-biased sex ratios in wasp cultures, for instance, may result not only from shifts in sex allocation, but from differential mortality in superparasitized hosts, insufficient mating, excessive mating, inbreeding, high temperatures, and even the rapid spread of sex ratio diseases (Waage *et al.*, 1984).

However, to the extent that undesirable shifts in progeny and sex allocation do occur in laboratory rearing, evolutionary theory does provide a basis for understanding why they occur, and more importantly, what stimuli parasitoids may be using to modify clutch size and sex ratio. These stimuli should, for instance, be associated with the rate of encounter of unparasitized and parasitized hosts, and with other wasps or their traces. This knowledge then serves as a guide for modifying rearing systems to improve quality.

Without this insight, improving culture quality is a laborious matter of trial and error, but of course it can be achieved. For example, detailed studies led researchers in China to discover that the male bias in sex ratios of the mass-reared egg parasitoid, *Anastatus* sp. could be reduced by dividing the exposure boxes up into little cells, thereby decreasing contact between wasps (Huang *et al.*, 1974; Huang, personal communication). A knowledge of LMC theory might have accelerated this discovery.

Theory of progeny and sex allocation might also be useful in predicting the ease with which different species can be reared. In newly-established laboratory cultures of parasitic wasps a rapid and disastrous shift towards the production of males is not uncommon, and may be associated with inbreeding (Waage, 1982b). On the basis of LMC theory, highly sib-mated (often gregarious) parasitoids, which are as a consequence in some way resistant to inbreeding, will also tend to have female-biased sex ratios. On the other hand, highly outbred (often solitary) parasitoids, which would be more susceptible to inbreeding depression, should have sex ratios about 0.5. Thus, the probability of inbreeding depression may be predictable from a knowledge of the field sex ratio of a parasitoid. There is some evidence that wasps with female-biased sex ratios show little inbreeding depression (Wilkes, 1947; Biemont and Boulétreau, 1980), in contrast to wasps with a sex ratio near 0.5 (Simmonds, 1947; Waage, 1982b).

Finally, parasitoid rearers, as optimizers themselves of parasitoid productivity per unit of resource, could do worse than pause to think how parasitoids are solving a similar problem. It may help them realize, for instance, that producing the most parasitoids per unit input may not always maximize the effectiveness of the populations released. Looking far ahead, it may help in the design of artificial hosts for mass production (Li, 1982; Strand and Vinson, 1985) where, given control of both parasitoid and host, the possibilities for manipulating what parasitoids produce should be considerably greater than at present.

Acknowledgements

I would like to thank Charles Godfray and Andy Taylor for helpful discussions on points in this chapter. Many other colleagues provided information and advice. I would particularly like to thank Oscar Alomar, Marianne van Dijken, Guo Ming-fang, Charles Godfray, Rosemary Hails, Andy LeMasurier, Nimmi Pallewatta, Mark Shaw, Rob Smith, Masami Takagi, Andy Taylor, Declan Ward, Kees van Welzen and Paul Wellings for permission to cite unpublished results, and to thank Dr Li Li-ying and her research group for helpful and enjoyable collaboration during part of this research.

REFERENCES

van Achterberg, C. (1977). The function of swarming in *Blacus* spp. (Hymenoptera, Braconidae, Helconinae). *Entomologische Berichtung* **37,** 151–152.

van Alphen, J. J. M., and Nell, H. W. (1982). Superparasitism and host discrimination by *Asobara tabida* Nees (Braconidae: Alysiinae), a larval parasitoid of *Drosophila. Netherlands Journal of Zoology* **32,** 232–360.

van Alphen, J. J. M., and Thunnissen, I. (1983). Host selection and sex allocation by *Pachycrepoideus vindemiae* Rondani (Pteromalidae) as a facultative hyperparasitoid of *Asobara tabida* Nees (Braconidae: Alysiinae) and *Leptopilina heterotoma* (Cynipoidea: Eucolidae). *Netherlands Journal of Zoology* **33,** 497–514.

Askew, R. R. (1968). A survey of leaf miners and their parasites on *Laburnum. Transactions of the Royal Entomological Society of London* **120,** 1–37.

van den Assem, J., Putters, F. A., and Prins, T. C. (1984). Host quality effect on sex ratio of the parasitic wasp, *Anisopteromalus calandrae* (Chalcidoidea, Pteromalidae). *Netherlands Journal of Zoology* **34,** 33–62.

Avilla, J., and Albajes, R. (1984). The influence of female age and host size on the sex ratio of the parasitoid, *Opius concolor. Entomologia Experimentalis et Applicata* **35,** 43–47.

Bakker, K., Eijsackers, H. J. P., van Lenteren, J. C., and Meelis, E. (1972). Some models describing the distribution of eggs of the parasite *Pseudeucoila bochei* (Hymenoptera, Cynipidae) over its hosts, larvae of *Drosophila melanogaster. Oecologia* **10,** 29–57.

Bakker, K., van Alphen, J. J. M., van Batenburg, F. H. D., van der Hoeven N., Nell, H. W., van Strien-van Liempt, W. T. F. H., and Turlings, T. C. (1985). The function of host discrimination and superparasitization in parasitoids. *Oecologia* **67,** 572–576.

Biemont, C., and Boulétreau, M. (1980). Hybridization and inbreeding effects on genome coadaptation in a haplo-diploid Hymenoptera: *Cothonaspis boulardi* (Eucoilidae). *Experientia* **36,** 45–46.

Bryan, G. (1983). Seasonal biological variation in some leaf-miner parasites in the genus *Achrysocharoides* (Hymenoptera, Eulophidae). *Ecological Entomology* **8,** 259–270.

Chacko, M. J. (1969). The phenomenon of superparasitism in *Trichogramma evanescens minutum* Riley. I. *Beiträge zur Entomologie* **19,** 617–635.

Charnov, E. L. (1979). The genetical evolution of patterns of sexuality: Darwinian fitness. *American Naturalist* **113,** 465–480.

Charnov, E. L. (1982). "The Theory of Sex Allocation." Princeton University Press, Princeton.

Charnov, E. L., and Skinner, S. W. (1984). Evolution of host selection and clutch size in parasitoid wasps. *Florida Entomologist* **67,** 5–21.

Charnov, E. L., and Skinner, S. W. (1985). Complementary approaches to the understanding of parasitoid oviposition decisions. *Environmental Entomology* **14,** 383–391.

Charnov, E. L., Los-den Hartogh, R. L., Jones, W. T., and van den Assem, J. (1981). Sex ratio evolution in a variable environment. *Nature* **289,** 27–33.

Clausen, C. P. (1940). "Entomophagous Insects." McGraw-Hill, New York.

Cole, L. R. (1981). A visible sign of a fertilization act during oviposition by an ichneumonid wasp, *Itoplectis maculator. Animal Behaviour* **29,** 299–300.

Colgan, P., and Taylor, P. (1981). Sex ratio in autoparasitic Hymenoptera. *American Naturalist* **117,** 564–566.

Colwell, R. K. (1981). Group selection is implicated in the evolution of female-biased sex ratios. *Nature* **290,** 401–404.

Conde, J. E. and Rabinovich, J. E. (1979). Larval competition between *Telenomus costalimai* (Hymenoptera: Scelionidae) and *Ooencyrtus trinidadensis venatorius* (Hymenoptera: Encyrtidae) after simultaneous oviposition in *Rhodnius prolixus* eggs (Hemiptera: Reduviidae). *Journal of Medical Entomology* **16,** 428–431.

da Costa Lima, A. (1928). Notas sobre a biologia do *Telenomus fariai* Lima, parasito dos ovas de *Triatoma. Memorias do Instituto Oswaldo Cruz* **21,** 201–209.

Cruz, Y. P. (1981). A sterile defender morph in a polyembryonic hymenopterous parasite. *Nature* **294,** 446–447.

Danks, H. V. (1975a). Seasonal life cycle and biology of *Winthemia rufopicta* (Diptera: Tachinidae) as a parasite of *Heliothis* spp. (Lepidoptera: Noctuidae) on tobacco in North Carolina. *Canadian Entomologist* **107,** 639–654.

Danks, H. V. (1975b). Factors determining levels of parasitism by *Winthemia rufopicta* (Diptera: Tachinidae), with particular reference to *Heliothis* spp. (Lepidoptera: Noctuidae) as hosts. *Canadian Entomologist* **107,** 655–685.

DeLoach, C. J., and Rabb, R. L. (1971). Life history of *Winthemia manducae* (Diptera, Tachinidae), a parasite of the tobacco hornworm. *Annals of the Entomological Society of America* **64,** 399–409.

DeLoach, C. J., and Rabb, R. L. (1972). Seasonal abundance and natural mortality of *Winthemia manducae* (Diptera: Tachinidae) and degree of parasitization of its host, the tobacco hornworm. *Annals of the Entomological Society of America* **65,** 779–790.

van Dijken, M., and Waage, J. K. (in press). Self and conspecific superparasitism in *Trichogramma evanescens* Westwood. *Entomologia Experimentalis et Applicata.*

Escalante, G., and Rabinovich, J. E. (1979). Population dynamics of *Telenomus fariai* (Hymenoptera: Scelionidae), a parasite of Chagas' disease vectors. IX. Larval competition and population size regulation under laboratory conditions. *Researches on Population Ecology* **20,** 235–246.

Feijen, H. R., and Schulten, G. G. M. (1981). Egg parasitoids (Hymenoptera, Trichogrammatidae) of *Diopsis macrophthalma* (Diptera: Diopsidae) in Malawi. *Netherlands Journal of Zoology* **31,** 381–417.

Fisher, R. A. (1930). "The Genetical Theory of Natural Selection." Oxford University Press, Oxford.

Fiske, W. F. (1910). Superparasitism; an important factor in the natural control of insects. *Journal of Economic Entomology* **3,** 88–97.

Flanders, S. E. (1939). Environmental control of sex in hymenopterous insects. *Annals of the Entomological Society of America* **32,** 11–26.

Flanders, S. E. (1950). Regulation of ovulation and egg disposal in the parasitic Hymenoptera. *Canadian Entomologist* **82,** 134–140.

Glas, P. C., Smits, P. H., Vlaming, P., and van Lenteren, J. C. (1981). Biological control of lepidopteran pests in cabbage crops by means of inundative releases of *Trichogramma* species (*T. evanescens* Westwood and *T. cacoeciae* March): a combination of field and laboratory experiments. *Mededelingen van de Faculteit Landbouwwetenschappen Rijksuniversiteit Gent* **46,** 487–497.

Godfray, H. C. J. (1986a). Models for clutch size and sex ratio with sibling interaction. *Theoretical Population Biology* **28** (in press).

Godfray, H. C. J. (1986b). Clutch size in a leaf-mining fly (*Pegomya nigritarsis*: Anthomyidae). *Ecological Entomology* **11,** 75–81.

Godfray, H. C. J. (in press). Evolutionary constraints of clutch size production in parasitic wasps. *American Naturalist.*

Grant, B., Burton, S., Contoreggi, C., and Rothstein, M. (1980). Outbreeding via frequency-dependent selection in the parasitoid wasp, *Nasonia* (= *Mormoniella*) *vitripennis* Walker. *Evolution* **39,** 983–992.

Green, R. F. (1982). Optimal foraging and sex ratio in parasitic wasps. *Journal of Theoretical Biology* **95,** 43–48.

Green, R. F., Gordh, G. C., and Hawkins, B. A. (1982). Precise sex ratios in highly inbred parasitic wasps. *American Naturalist* **120,** 653–655.

Haeselbarth, E. (1979). Zur Parasitierung der Puppen von Forleule [*Panolis flammea* (Schiff)], Kiefenspanner [*Bupalus piniarius* (L.)] und Heidelbeerspanner [*Boarmia bistortata* (Goeze)] in bayerische Kiefernwäldern. *Zeitschrift für angewandte Entomologie* **87,** 311–322.

Hamilton, W. D. (1967). Extraordinary sex ratios. *Science* **156,** 477–488.

Hamilton, W. D. (1979). Wingless and fighting males in fig wasps and other insects. *In* "Reproductive Competition and Sexual Selection in Insects" (M. S. Blum and N. A. Blum, eds.), pp. 167–220. Academic Press, New York.

Hartl, D. L. (1971). Some aspects of natural selection in arrhenotokous populations. *American Zoologist* **11,** 309–325.

Hirose, Y., Kimoto, H. and Hiehata, K. (1976). The effect of host aggregation on parasitism by *Trichogramma papilionis* Nagarkatti (Hymenoptera: trichogrammatidae), an egg parasitoid of *Papilio xuthus* Linné (Lepidoptera: Papilionidae). *Applied Entomology and Zoology* **11,** 116–125.

Holmes, H. B. (1972). Genetic evidence for fewer progeny and a higher percent males when *Nasonia vitripennis* oviposits in previously parasitized hosts. *Entomophaga* **17,** 79–88.

Huang Ming-dau, Mai Siu-hui, Wu Wei-Nan, and Poo Chih-lung (1974). The bionomics of *Anastatus* sp. and its utilization for the control of the lichee stink bug, *Tessarotoma papillosa* Drury. *Acta Entomologica Sinica* **17,** 362–375.

Hubbard, S. F., Marris, G. C., Reynolds, A. R., and Rowe, G. W. (in press). Adaptive patterns in the avoidance of superparasitism by solitary insect parasitoids. *Journal of Animal Ecology*.

Hughes, R. N. (1979). Optimal diets under the energy maximization principle: the effects of recognition time and learning. *American Naturalist* **113,** 209–221.

Ikawa, T., and Okabe, H. (1985). Regulation of egg number per host to maximize the reproductive success in the gregarious parasitoid, *Apanteles glomeratus* L. (Hymenoptera: Braconidae). *Applied Entomology and Zoology* **20,** 331–339.

Ikawa, T., and Suzuki, Y. (1982). Ovipositional experience of the gregarious parasitoid, *Apanteles glomeratus* (Hymenoptera: Braconidae), influencing her discrimination of the host larvae, *Pieris rapae crucivora*. *Applied Entomology and Zoology* **17,** 119–126.

Iwasa, Y., Suzuki, Y., and Matsuda, Y. (1984). The oviposition strategy of parasitoids. I. Effect of mortality and limited egg number. *Theoretical Population Biology* **26,** 205–227.

Jackson, D. J. (1966). Observations on the biology of *Caraphractus cinctus* Walker (Hymenoptera: Mymaridae), a parasitoid of the eggs of Dytiscidae (Coleoptera). III. The adult life and sex ratio. *Transactions of the Royal Entomological Society of London* **118,** 23–49.

Jervis, M. A. (1979). Courtship, mating and "swarming" in *Aphelopus melaleucus* (Dalman) (Hymenoptera: Dryinidae). *Entomologist's Gazette* **30,** 191–193.

Jones, W. T. (1982). Sex ratio and host size in a parasitoid wasp. *Behavioural Ecology and Sociobiology* **10,** 207–210.

Juliano, S. A. (1982). Influence of host age on host acceptability and host suitability for a species of *Trichogramma* (Hymenoptera: Trichogrammatidae) attacking aquatic Diptera. *Canadian Entomologist* **114,** 713–720.

King, P. E. (1961). A possible method of sex-ratio determination in the parasitic Hymenopteran, *Nasonia vitripennis*. *Nature* **189,** 330–331.

Kitano, H., and Nakatsuji, N. (1978). Resistance of *Apanteles* eggs to the haemocytic encapsulation by their habitual host, *Pieris*. *Journal of Insect Physiology* **24,** 261–271.

Klomp, H., and Teerink, B. J. (1962). Host selection and number of eggs per oviposition in the egg parasitic *Trichogramma embryophagum* Htg. *Nature* **195,** 1020–1021.

Klomp, H., and Teerink, B. J. (1967). The significance of oviposition rate in the egg parasite,

Trichogramma embryophagum Htg. *Archives Neerlandaises de Zoologie* **17,** 350–375.

Klomp, H., Teerink, B. J., and Wei Chun Ma (1980). Discrimination between parasitized and unparasitized hosts in the egg parasite *Trichogramma embryophagum* (Hym.: Trichogrammatidae): a matter of learning and forgetting. *Netherlands Journal of Zoology* **30,** 254–277.

Kurosu, U. (1985). Male altruism and wing polymorphism in a parasitic wasp. *Journal of Ethology* **3,** 11–19.

Lack, D. (1947). The significance of clutch size. *Ibis* **89,** 309–352.

van Lenteren, J. C. (1976). The development of host discrimination and the prevention of superparasitism in the parasite *Pseudeucoila bochei* Weld. *Netherlands Journal of Zoology* **26,** 1–83.

van Lenteren, J. C. (1981). Host discrimination by parasitoids. *In* "Semiochemicals, Their Role in Pest Control" (D. A. Nordlund, R. L. Jones and W. J. Lewis, eds), pp. 153–179. John Wiley, New York.

van Lenteren, J. C., and Bakker, K. (1975). Discrimination between parasitized and unparasitized hosts in the parasitic wasp *Pseudeucoila bochei*: a matter of learning. *Nature* **254,** 417–419.

Li Li-ying (1982). *Trichogramma* spp. and their utilization in Peoples Republic of China. *In* "Les Trichogrammes", pp. 23–29. C.N.R.S., Paris.

Luck, R. F., and Podoler, H. (1985). Competitive exclusion of *Aphytis lingnanensis* by *A. melinus*: potential role of host size. *Ecology* **66,** 904–913.

Luck, R. F., Podoler, H., and Kfir, R. (1982). Host selection and egg allocation behaviour by *Aphytis melinus* and *A. lingnanensis*: comparison of two facultatively gregarious parasitoids. *Ecological Entomology* **7,** 397–408.

Narayanan, E. S., and Subba Rao, B. R. (1955). Studies in insect parasitism. I-III. The effect of different hosts on the physiology, on the development and behaviour and on the sex ratio of *Microbracon gelechiae* Ashmead. *Beitrage zur Entomologie* **5,** 36–60.

Neser, S. (1973). Biology and behaviour of *Euplectrus* near *laphygmae* Ferriere (Hymenoptera: Eulophidae). *Entomology Memoirs, Department of Agricultural and Technical Services, South Africa* **32,** 1–31.

Owen, R. F. (1983). Sex ratio adjustment in *Asobara persimilis* (Hymenoptera: Braconidae), a parasitoid of *Drosophila*. *Oecologia* **59,** 402–404.

Pak, G. A., and Oatman, E. R. (1982). Biology of *Trichogramma brevicopillum*. *Entomologia Experimentalis et Applicata* **32,** 61–67.

Pallewatta, P. K. T. N. S. (1986). "Factors Affecting Progeny and Sex Allocation by the Egg Parasitoid, *Trichogramma evanescens* Westwood." PhD Thesis, University of London.

Parker, G. A., and Courtney, S. P. (1984). Models of clutch size in insect oviposition. *Theoretical Population Biology* **26,** 27–48.

Pickering, J. (1980). Larval competition and brood sex ratios in the gregarious parasitoid *Pachysomoides stupidus*. *Nature* **283,** 291–292.

Putters, F. A., and van den Assem, J. (1985). Precise sex ratios in a parasitic wasp: the result of counting eggs. *Behavioural Ecology and Sociobiology* **17,** 265–270.

Puttler, B. (1974). *Hypera postica* and *Bathyplectes curculionis*: encapsulation of parasite eggs by host larvae in Missouri and Arkansas. *Environmental Entomology* **3,** 881–882.

Rabinovich, J. E. (1971). Population dynamics of *Telenomus fariai* (Hymenoptera: Scelionidae), a parasite of Chagas' disease vectors. V. Parasite size and vital space. *Revista Biologia Tropical* **19,** 109–120.

Ram, A., and Sharma, A. K. (1977). Selective breeding for improving the fecundity and sex-ratio of *Trichogramma fasciatum* (Perkins) (Hymenoptera: Trichogrammatidae), an egg parasite of a lepidopterous host. *Entomon* **2,** 133–137.

Salt, G. (1961). Competition among insect parasitoids. *In* "Mechanisms in Biological Competition" *Symposium of the Society of Experimental Biology* **15,** 96–119.

Sandlan, K. (1979). Sex ratio regulation in *Coccygomimus turionellae* Linneaus (Hymenoptera: Ichneumonidae) and its ecological implications. *Ecological Entomology* **4,** 365–378.

Sato, Y., and Tanaka, T. (1984). Effect of the number of parasitoid (*Apanteles kariyai*) eggs (Hymenoptera: Braconidae) on the growth of the host (*Leucania separata*) larvae (Lepidoptera: Noctuidae). *Entomophaga* **29,** 21–28.

Schieferdecker, H. (1969). Der Gregärparasitismus von *Trichogramma* (Hymenoptera: Trichogrammatidae). *Beiträge zur Entomologie* **19,** 507–521.

Schmidt, J. M., and Smith, J. J. B. (1985a). Host volume measurement by the parasitoid wasp, *Trichogramma minutum*: the roles of curvature and surface area. *Entomological Experimentalis et Applicata* **39,** 213–221.

Schmidt, J. M., and Smith, J. J. B. (1985b). The mechanism by which the parasitoid wasp *Trichogramma minutum* responds to host clusters. *Entomologia Experimentalis et Applicata* **39,** 287–294.

Shaw, M. R., and Smith, R. H. (unpublished). Local mate competition and sex ratio strategies in *Apanteles* wasps.

Shiga, M., and Nakanishi, A. (1968a). Intraspecific competition in a field population of *Gregopimpla himalayensis* (Hymenoptera: Ichneumonidae) parasitic on *Malacosoma neustria testacea* (Lepidoptera: Lasiocampidae). *Researches on Population Ecology* **10,** 69–86.

Shiga, M., and Nakanishi, A. (1968b). Variation in the sex ratio of *Gregopimpla himalayensis* Cameron (Hymenoptera: Ichneumonidae) parasitic on *Malacosoma neustria testacea* Motschulsky (Lepidoptera: Lasiocampidae), with considerations on the mechanism. *Kontyu* **36,** 369–376.

Simmonds, F. J. (1947). Improvement of the sex-ratio of a parasite by selection. *Canadian Entomologist* **79,** 41–44.

Skinner, S. W. (1985). Clutch size as an optimal foraging problem for insect parasitoids. *Behavioural Ecology and Sociobiology* **17,** 231–238.

Strand, M. R., and Vinson, S. B. (1985). *In vitro* culture of *Trichogramma pretiosum* on an artificial medium. *Entomologia Experimentalis et Applicata* **39,** 203–209.

Suzuki, Y., and Iwasa, Y. (1980). A sex-ratio theory of gregarious parasitoids. *Researches on Population Ecology* **22,** 366–382.

Suzuki, Y., and Hiehata, K. (1985). Mating systems and sex ratios in the egg parasitoids, *Trichogramma dendrolimi* and *T. papilionis* (Hymenoptera: Trichogrammatidae). *Animal Behaviour* **33,** 1223–1227.

Suzuki, Y., Tsuji, H., and Sasakawa, M. (1984). Sex allocation and the effects of superparasitism on secondary sex ratios in the gregarious parasitoid, *Trichogramma chilonis* (Hymenoptera: Trichogrammatidae). *Animal Behaviour* **32,** 478–484.

Tagawa, J., and Kitano, H. (1981). Mating behaviour of the braconid wasp, *Apanteles glomeratus* L. (Hymenoptera: Braconidae) in the field. *Applied Entomology and Zoology* **16,** 345–450.

Takagi, M. (1985). The reproductive strategy of the gregarious parasitoid, *Pteromalus puparum* (Hymenoptera: Pteromalidae). 1. Optimal number of eggs in a single host. *Oecologia* **68,** 1–6.

Taylor, A. D. (unpublished). Density-dependence sex ratios and Charnov's "variation in fitness" model.

Taylor, P. D. (1981). Intra-sex and inter-sex sibling interactions as sex ratio determinants. *Nature* **291,** 64–66.

Tooke, F. G. C. (1955). The eucalyptus snout beetle, *Gonipterus scutellatus* Gyll.: a study of its ecology and control by biological means. *Entomology Memoirs, Department of Agricultural Technical Services, South Africa* **3,** 1–282.

Vater, G. (1971). Ausbreitung und Wandersverhalten parasitischer Hymenopteren. *Biologische Rundschau* **9,** 281–303.

Viktorov, G. A. (1968). The influence of population density on sex ratio in *Trissolcus grandis* Thoms. (Hymenoptera: Scelionidae). *Zoologicheskii Zhurnal* **47,** 1045–039.

Viktorov, G. A., and Kochetova, N. I. (1971). Significance of population density to the control of sex ratio in *Trissolcus grandis* (Hymenoptera: Scelionidae). *Zoologicheskii Zhurnal* **50,** 1753–1755.

Vinson, S. B., and Iwantsch, G. F. (1980). Host suitability for insect parasitoids. *Annual Review of Entomology* **25,** 397–419.

Vu Quang Kong (1985). Influence of oviposition rate, host density and embryonic development on sex ratio in *Trichogramma japonicum* Ashmead (Hymenoptera, Trichogrammatidae). *Entomologicheskoe Obozrenie* **64,** 450–457.

Waage, J. K. (1982a). Sib-mating and sex ratio strategies in scelionid wasps. *Ecological Entomology* **7,** 103–112.

Waage, J. K. (1982b). Sex ratio and population dynamics in natural enemies—some possible interactions. *Annals of Applied Biology* **101,** 159–164.

Waage, J. K., and Godfray, H. C. J. (1985). Reproductive strategies and population ecology of insect parasitoids. *In* "Behavioural Ecology. Ecological Consequences of Adaptive Behaviour" (R. M. Sibly and R. H. Smith, eds), pp. 449–470. Blackwell Scientific Publications, Oxford.

Waage, J. K., and Lane, J. A. (1984). The reproductive strategy of a parasitic wasp. II. Sex allocation and local mate competition in *Trichogramma evanescens. Journal of Animal Ecology* **53,** 417–426.

Waage, J. K., and Ng Sook Ming (1984). The reproductive strategy of a parasitic wasp. I. Optimal progeny and sex allocation in *Trichogramma evanscens. Journal of Animal Ecology* **53,** 401–416.

Waage, J. K., Carl, K., Mills, N. J., and Greathead, D. J. (1984). Rearing entomophagous insects. *In* "Handbook of Insect Rearing. Vol. 1" (P. Singh and R. F. Moore, eds.), pp. 45–66. Elsevier, Amsterdam.

Walker, I. (1967). Effect of population density on the viability and fecundity in *Nasonia vitripennis* Walker (Hymenoptera, Pteromalidae). *Ecology* **48,** 294–301.

Walter, G. A. (1983). Divergent male ontogenies in Aphelinidae (Hymenoptera: Chalcidoidea): a simplified classification and a suggested evolutionary sequence. *Biological Journal of the Linnean Society* **16,** 63–82.

Weiss, A. E., Price, P. W., and Lynch, M. (1983). Selective pressures on clutch size in the gall maker, *Arteomyia carbonifera. Ecology* **64,** 688–695.

Wellings, P. W., Morton, R., and Hart, P. J. (in press). Primary sex ratio and differential progeny survivorship in solitary haplo-diploid parasitoids. *Ecological Entomology* **11**.

van Welzen, K., and Waage, J. K. (unpublished). Adaptive responses to local mate competition by the parasitoid, *Telenomus remus.*

Werren, J. H. (1980). Sex ratio adaptations to local mate competition in a parasitic wasp. *Science* **208,** 1157–1159.

Werren, J. H. (1983). Sex ratio evolution under local mate competition in a parasitic wasp. *Evolution* **37,** 116–124.

Werren, J. H. (1984). A model for sex ratio selection in parasitic wasps: local mate competition and host quality effects. *Netherlands Journal of Zoology* **34,** 81–96.

Wilkes, A. (1947). The effect of selective breeding on the laboratory propagation of insect parasites. *Proceedings of the Royal Society of London* **134,** 227–245.

Wilkes, A. (1963). Environmental causes of variation in the sex ratio of an arrhenotokous insect,

Dahlbominus fuliginosus (Nees) (Hymenoptera: Eulophidae). *Canadian Entomologist* **95,** 182–202.

Williams, G. C. (1979). The question of adaptive sex ratio in outcrossed vertebrates. *Proceedings of the Royal Society of London* **205,** 567–580.

Wylie, H. G. (1965). Discrimination between parasitized and unparasitized house fly pupae by females of *Nasonia vitripennis* (Walker) (Hymenoptera: Pteromalidae). *Canadian Entomologist* **97,** 279–286.

Wylie, H. G. (1966). Some mechanisms that affect the sex ratio of *Nasonia vitripennis* (Walk.) (Hymenoptera: Pteromalidae) reared from super-parasitized housefly pupae. *Canadian Entomologist* **98,** 645–653.

Wylie, H. G. (1967). Some effects of host size on *Nasonia vitripennis* and *Muscidifurax raptor* (Hymenoptera: Pteromalidae). *Canadian Entomologist* **99,** 742–748.

Wylie, H. G. (1973). Control of egg fertilization by *Nasonia vitripennis* (Hymenoptera: Pteromalidae) when laying on parasitized house fly pupae. *Canadian Entomologist* **105,** 709–718.

Wylie, H. G. (1976a). Interference among females of *Nasonia vitripennis* (Hymenoptera: Pteromalidae) and its effects on the sex ratio of their progeny. *Canadian Entomologist* **108,** 655–661.

Wylie, H. G. (1976b). Observations on the life history and sex ratio variability of *Eupteromalus dubius* (Hymenoptera: Pteromalidae), a parasite of cyclorrhaphous Diptera. *Canadian Entomologist* **108,** 1267–1274.

Wylie, H. G. (1977). Observations on *Athrycia cineria* (Diptera: Tachinidae), a parasite of *Mamestra configurata* (Lepidoptera: Noctuidae). *Canadian Entomologist* **109,** 747–754.

Wylie, H. G. (1979). Sex ratio variability of *Muscidifurax zaraptor* (Hymenoptera: Pteromalidae). *Canadian Entomologist* **111,** 105–109.

Ziser, S. W., Wojtowicz, J. A., and Nettles, W. C., Jr. (1977). The effect of the number of maggots per host on length of development, puparial weights and adult emergence of *Eucelatoria* sp. *Annals of the Entomological Society of America* **70,** 733–736.

4

The Physiological Interactions of Parasitoids with their Hosts and their Influence on Reproductive Strategies

M. R. STRAND

I INTRODUCTION

Over the years, a proliferation of terms has developed in the literature in an effort to lend some semblance of organization to the varied life histories of parasitoids. Parasitoids may develop as larvae within their hosts (endoparasitoids) or consume their hosts externally (ectoparasitoids). In turn,

parasitoids may be solitary, producing only one adult per host, or they may be gregarious, producing several adults per host (Clausen, 1940). Regardless of whether a given parasitoid species is endo/ectoparasitic or solitary/gregarious, most species are adapted to developing in a specific host stage such that there are egg, larval, pupal and adult parasitoids. This specificity to a particular host stage is often very precise, and suggests that the developmental status of the host is critical in determining host suitability (Salt, 1935; Vinson and Iwantsch, 1980a). Indeed, a large literature has slowly accrued on the physiological interactions between parasitoids and their hosts (Fisher, 1971; Vinson and Iwantsch, 1980b; Thompson, 1983; Beckage, 1985).

The overwhelming majority of this literature is devoted to larval endoparasitoids and the gross host pathologies associated with their development. In particular, recent studies have concentrated on the endocrine interactions between endoparasitoids and their hosts (Beckage, 1985), the influence of parasitoids on host metabolism (Thompson, 1983), and the function of parasitoid associated viruses, venoms and teratocytes (Salt, 1968) on host development and immunity (Salt, 1970; Stoltz and Vinson, 1979; Vinson and Iwantsch, 1980b).

However, an examination of what is known about the physiological interactions between parasitoids and their hosts reveals three shortcomings. Firstly, the sheer diversity of parasitoids makes generalizations tenuous and in-depth discussions on any given subject difficult. What is known about larval parasitism is restricted to a limited number of species, and our knowledge of egg, pupal and adult parasitism is even poorer. Secondly, the effects of parasitism on hosts are much better understood than the factors responsible for these changes. Thirdly, greater attention is needed in relating physiological events associated with parasitism to other facets of parasitoid biology. The chasms in our knowledge become especially obvious when discussing the forces which shape the evolutionary biology of parasitoids.

To understand how parasitoids function requires a synthesis of all aspects of parasitoid and host biology. Yet, synthesis of information is difficult to achieve simply because workers, by necessity, become specialized. No one disputes the intimate link between population genetics, evolution, ecology and behaviour with these areas of study now unified under the umbrella of behavioural ecology (Krebs and Davies, 1978). The parasitoid literature contains numerous studies in those areas. However, the physiology of parasitoids is largely excluded from discussions on areas such as the assessment of host quality, clutch size and sex allocation, despite their numerous physiological implications.

As is clear from the preceding chapters, current evolutionary theories and models which address questions on sex and progeny allocation by parasitoids are built on assumptions with physiological overtones. For example, models

dealing with local mate competition (LMC) (Hamilton, 1968, 1979), resource quality (Charnov *et al.*, 1981), and optimization of clutch size (Charnov and Skinner, 1984; Waage and Godfray, 1985) refer to host quality factors, parasitoid discrimination and larval competition (see Chapter 3). These models make vital predictions about why parasitoid reproductive strategies should vary with host quality and competition, but they do not, beyond very broad definitions, explain what makes a good quality host, how parasitoids assess quality, or how competition differentially influences reproductive success and progeny survival.

Thus, significant questions concerning parasitoid oviposition decisions and reproductive strategies become segregated into categories of physiological "how" and evolutionary "why" questions (Mayr, 1982; Holldobler, 1983). As discussed by Charnov and Skinner (1984), general predictions on which hosts a parasitoid should accept, how many eggs it should lay, and what sex ratio it should produce are possible without any detailed knowledge of the physiological mechanisms involved. However, these mechanisms are the constraints under which any specific parasitoid-host relationship evolves, suggesting that in order to understand the evolutionary "why" requires an understanding of the physiological "how". Thus, physiological data can provide insight into the oviposition decisions of parasitoids by defining the parameters, limitations and constraints involved. In this chapter, I will examine physiological interactions of parasitoids and hosts as they relate to reproductive strategies. Following a survey of larval parasitism, I will focus on egg parasitism, a less well-known process, to illustrate my points.

II. LARVAL PARASITISM

Hosts often show extreme pathological changes during the development of parasitoids. They are for the most part a manifestation of a complex of factors, some of which are an indirect consequence of feeding by the parasitoid while others are due to factors secreted by the immature parasitoid or injected by the adult at oviposition. The types of host pathologies which occur will be further influenced by the stage of the host when parasitized. That is, the effects observed during larval parasitism are not likely to reflect what occurs during, for example, egg parasitism. However, because most of the literature deals with larval parasitism, many of the host pathologies commonly associated with parasitoid development are drawn from this pool of information. Several good reviews already exist on the physiological interactions of endoparasitoids with their hosts (Vinson and Iwantsch, 1980a; Thompson, 1983; Beckage, 1985). Even within the limited number of

species which have been studied, a myriad of examples underscore the variety of parasitoid factors and host defences which play a role in successful larval parasitism. Nonetheless, a brief summary of a few of the more important physiological adaptations will serve as a useful contrast to parasitoids of other life stages, hopefully illustrating how parasitoid-host interactions often reflect the characteristics of the life stage parasitized.

Despite the diversity of larval parasitoids certain salient features of their hosts have had a pronounced effect on their developmental adaptations. The most important include firstly, the presence of a competent immune system and secondly, active host feeding which typically results in marked changes in host size due to moulting, variations in metabolic activity, fluctuating hormone titres, and mobility. To contend with these features larval parasitoids have diverged into ecto- and endoparasitoids (see Chapter 8).

From a mechanistic standpoint, the physiological adaptations of endoparasitoids are more intriguing, yet ectoparasitism is an effective strategy for circumventing many of the problems associated with attacking larvae. They are neither exposed to internal host defences nor to host hormones other than through oral ingestion. On the other hand, ectoparasitoids are potentially more vulnerable to physical damage or removal from the host. Not surprisingly, ectoparasitism has evolved most often in those parasitoids which attack concealed hosts. One apparent physiological adaptation to developing externally is the production by many ectoparasitoids of venoms which either partially or totally paralyse the host and inhibit further development (Piek *et al.*, 1974; Beard, 1978).

The most detailed work on the mode of action of ectoparasitoid venoms has been with the genus *Bracon*. The active components of *Bracon hebetor* Say venom consist of two highly labile proteins with molecular weights of 57 and 42 K (Spanjer *et al.*, 1977). *B. hebetor* venom acts presynaptically at somatic neuromuscular junctions, but seemingly does not affect the excitability of nerve or muscle (Piek and Owen, 1982). In addition, paralysed hosts fail to moult or undergo metamorphosis. The prevention of moulting tends both to maintain the host as well as to prevent shedding of the parasitoid immatures with the exuvium. For example, larvae which are paralysed by *B. hebetor* never moult even though they can remain alive for several weeks.

In contrast to ectoparasitoids, endoparasitoids are exposed to host cellular defences, hormones and metabolic events. In order to develop in such an environment they are often highly specialized and promote a variety of host pathologies. Of vital importance is the parasitoid's ability to avoid the host's immune system. Insects seemingly do not possess the ability to produce immunoglobulins against foreign antigens, but are capable of mounting a cellular immune response (Nappi, 1975; Stoltz and Vinson, 1979; Dunn, 1986). Host haemocytes form an enveloping, multicellular capsule against parasitoids, as well as other large targets. Although a few parasitoids avoid encapsulation by ovipositing into sites which are not exposed to host haemocytes, most species oviposit directly into the haemocoel where they are subject to encapsulation. Obviously, parasitoids are able to evade encapsulation in their habitual, permissive, hosts while the failure to develop successfully in other nonpermissive hosts is often attributable to the host immune system (Vinson and Iwantsch, 1980a).

Larval endoparasitoids contend with host immunity either by possessing in the egg or larval stage surface properties which resist encapsulation, or by inhibiting the immune system in some manner so that it is incapable of encapsulating the developing parasitoid. There are very few data to substantiate the first of these mechanisms, even though several observations suggest that the surface properties of some parasitoids are resistant to encapsulation (Salt, 1973; Lynn and Vinson, 1977). For example, both *Heliothis virescens* (F.) and *Heliothis zea* (Boddie) are readily parasitized by *Cardiochiles nigriceps* Viereck, but *C. nigriceps* fails to develop in *H. zea* because the eggs are encapsulated. *C. nigriceps* eggs possess an outer flocculent layer which is removed in *H. zea*, possibly by enzymes present in the haemolymph, but which persists until eclosion in *H. virescens*. It is only after the outer layer has been removed in *H. zea* that *C. nigriceps* eggs are encapsulated, suggesting that the outer layer possesses properties which resist encapsulation (Davies, personal communication).

Some data also indicate that either separately or in tandem with parasitoid surface properties, parasitoid survival is influenced by the direct inhibition of the host immune system. In particular, attention has been focused on the possible influence of symbiotic viruses possessed by certain braconids and ichneumonids (Stoltz and Vinson, 1979; Edson *et al.*, 1981). These viruses replicate in calyx cells of the parasitoid reproductive tract and the virions stored in the calyx lumen. At oviposition a quantity of virus is injected into the host along with the egg. Once in the host the virus enters a variety of host tissues and in the case of *Campoletis sonorensis* (Cameron) virus specific mRNA transcripts are detectable within 2 h of parasitoid oviposition (Fleming *et al.*, 1983). For some parasitoids the virus must be present in the host, otherwise the eggs or larvae are encapsulated. However, it was not clear until recently how the virus might influence the host immune system.

For parasitoids such as *Cardiochiles nigriceps* the egg is protected in *H. virescens* by its outer coating, while the larva is protected, presumably, by the virus (Stoltz and Vinson, 1979). However, other abiotic targets placed in *Cardiochiles nigriceps* parasitized larvae remain susceptible to encapsulation, which has led to the assumption that parasitoid suppression of host immune systems is likely to be specific only to the developing parasitoid. Yet, in several recent studies evasion of immunity has involved the selective elimination of haemocytes which are known to be involved in encapsulation not only of the parasitoid but of other targets as well (Rizki and Rizki, 1984; Stoltz and Guzo, 1986; Davies *et al.*, in press). For example, injection of *Campoletis sonorensis* polydnavirus (Csv) eliminates the majority of plasmatocytes from circulation, and may affect the ability of Csv-infected larvae to encapsulate either the parasitoid or other foreign targets (Davies *et al.*, in press). Whether Csv destroys plasmatocytes or merely eliminates them from circulation is unclear, but virus viability is necessary for the effect to occur and *Campoletis sonorensis* is encapsulated in the absence of virus. Obviously, evasion of immunity by parasitoids is an extremely complicated subject with protective egg coatings and viruses being only two possible means of evasion.

From an ecological standpoint an interesting consequence of the necessity for endoparasitoids to evade the host immune systems is in the understanding of specificity. Firstly, one would expect that in general larval endoparasitoids would be more host-specific than ectoparasitoids or, as will be discussed shortly, egg parasitoids simply because of problems associated with host immunity. The fact that 82% of surveyed North American ichneumonids, the majority of which are larval endoparasitoids, have three or fewer hosts would seem to support this view (Townes and Townes, 1951). Furthermore, we might predict that the host ranges of larval endoparasitoids may be partially dictated by the mechanisms used in avoiding encapsulation. Some species may have a broader host range because of having adopted a more general means of compromising host immunity, while other species may be more specific due to very specialized mechanisms. Currently, there are insufficient data to support or reject this view; however, workers are becoming increasingly aware of the importance of developmental constraints in shaping the course of evolution (Maynard Smith *et al.*, 1985).

As important as contending with the host immune system is the necessity for larval endoparasitoids to adapt to host growth and development. Since insect larvae moult regularly and pupate, often within a relatively short period of time and to the potential detriment of a parasitoid, it is not surprising that numerous developmental interactions have been reported with larval hosts. Some of the more obvious act through the host endocrine system and involve developmental synchrony between endoparasitoids and their hosts (Lawerence, 1986) as well as the alterations which often occur in host growth, behaviour, and reproduction.

The endocrine interactions which occur between parasitoids and their hosts have been reviewed recently by Beckage (1985). The most commonly reported anomalies in host development include: (1) the inhibition of host metamorphosis by either elevating juvenile hormone (JH) titres (Beckage and Riddiford, 1983) or inhibiting ecdysteroid release (Webb and Dahlman, 1986); (2) precocious host metamorphosis due possibly to a decline in JH synthesis and/or increase in JH esterase activity (Jones *et al.*, 1981); (3) induction of supernumerary larval instars due to a failure of the host's JH titre to decline (Jones *et al.*, 1982) and (4) disruption of lipid, carbohydrate and protein metabolism through perturbations of JH or other hormone titres (Beckage, 1985).

In some cases these hormonal perturbations may be due to stress or nutritional imbalances, but for others it is possible that the parasitoid itself is directly responsible through the injection of the aforementioned viruses and venoms, or secretion of factors from the developing parasitoid immature or associated teratocytes. Hosts which have been pseudoparasitized (i.e. oviposited into by the adult but which fail to produce any parasitoid offspring) often show the same developmental pathologies as hosts which contain a viable parasitoid, suggesting the possibility that factors other than parasitoid feeding are responsible for some of the symptoms associated with parasitism (Vinson and Iwatnsch, 1980b; Beckage, 1985). Even within the larval parasitoids much more study is needed to understand the factors associated with successful development. The mechanisms employed by parasitoids attacking particular life stages may follow similar trends but vary in specifics; however, of equal importance is the need to understand that what is observed in, for example, larval parasitism need not pertain to parasitoids of other life stages, i.e. egg parasitoids.

III. EGG PARASITISM: AN OVERVIEW

A. Eggs as Hosts

Almost nothing is known about the interactions of egg parasitoids with their hosts, despite the fact that three major families of Hymenoptera are composed exclusively of egg parasitoids (Trichogrammatidae, Mymaridae and Scelionidae), 13 other families of Hymenoptera are composed partially of egg parasitoids (Clausen, 1940), and the importance of these insects in biological control projects throughout the world (van Lenteren and Pak, 1984; see Chapter 10). The paucity of information concerning egg parasitoid-host interactions can be attributed largely to the very small size of these insects and their hosts which makes their study difficult, especially at a physiological level.

Three features distinguish eggs from other arthropod life stages as hosts for parasitoids. Firstly, eggs lack a cellular defence response to foreign intrusion (Salt, 1970; Askew, 1971). Although data are scarce, the general trend seems to be that the younger the stage, the fewer the circulating haemocytes in the haemocoel (Ratcliffe and Rowley, 1979; Sminia, 1981; Dikkeboom *et al.*, 1984). Haemocytes are mesodermally derived and appear during late gastrulation and the initiation of organogeny, and so are present during only a portion of embryogenesis (Anderson, 1972). Thus, they are never very numerous, and in all likelihood are insufficiently differentiated to function as a competent immune system during this period. Instead, the majority of egg defences are directed toward rupture of the chorion and external wounding which only prevent or retard predators and parasitoids from gaining access (Hinton, 1981).

The second feature is the developmental changes associated with insect embryogenesis. Eggs progress from a single cell to a pharate first instar over a varying length of time. For most Hemimetabola, embryogenesis is relatively slow, lasting, under near standard laboratory conditions, from 4 days in some Heteroptera and Psocoptera to a month or more for various Orthoptera, Odonata, and Embioptera (Anderson, 1972). In contrast, holometabolous embryogenesis can be extremely rapid, especially for certain species of Coleoptera, Lepidoptera, and Diptera which often hatch in 3 days or less (Anderson, 1972). Thus, in a developmental and temporal sense, egg parasitoids contend with a variable and transitory host stage.

The third feature is the uncertain source and role of hormones in development. The role of ecdysteroids or juvenile hormone in embryogenesis remains largely unknown even though fluctuating titres of both have been found in *Bombyx mori* (L.) (Mizuno *et al.*, 1981; Gharib and de Reggi, 1983), *Manduca sexta* (L.) (Bergot *et al.*, 1981), *Locusta migratoria* (L.) (Lagueux *et al.*, 1981), and *Schistocerca gregaria* (Forskål) (Scalia and Morgan, 1982). It is generally accepted that ecdysone and/or 20-hydroxyecdysone triggers cuticle deposition in the pharate first instar just as it does in larvae (Gharib and de Reggi, 1983). Similarly, the deposition of serosal cuticle very early in embryogenesis is also associated with high levels of 20-hydroxyecdysone in *B. mori* (Gharib *et al.*, 1983). Since deposition of the serosal cuticle occurs prior to the differentiation of endocrine glands, ecdysteroids may be synthesized in the female and stored in eggs as an inactive conjugate.

Hormones probably play an important role in egg diapause even though *B. mori* remains the only species to be studied to any extent. In *B. mori*, a maternal "diapause hormone" is synthesized and released by the female's suboesophageal ganglion when the moth is exposed to high temperature and long-day photoperiods which arrest oviposited eggs in the germ band stage (Downer and Laufer, 1983). Conversely, moths exposed to standard temperatures and short-day photoperiods oviposit eggs which develop without interruption. Despite the uncertainties egg parasitoids are presumably exposed to host hormones and could be influenced by them.

B. The Interactions of Egg Parasitoids with their Hosts

Before one can begin to address how host-parasitoid interactions influence reproductive strategies and competition, it is necessary to characterize how egg parasitoids develop successfully. Since eggs lack a developed immune system, the most serious problems facing egg parasitoids are those associated with host development.

To use eggs as hosts, parasitoids appear to have evolved two broad strategies which employ both behavioural and physiological adaptations. The behavioural adaptations are directed toward somehow assuring that foraging wasps preferentially encounter hosts in the proper stage of development. Such an adaptation is seen in phoretic egg parasitoids (Clausen, 1976). Phoresy is most common among scelionids which parasitize the eggs of Orthoptera, Heteroptera, and Lepidoptera which lay large, often protected, clutches. Commonly cited advantages of phoresy are that it allows the wasp to parasitize poorly accessible eggs, gain access to multiple clutches from the same carrier and it facilitates host location. However, in developmental terms, phoresy may be critical for progeny development because it ensures that the parasitoid encounters newly oviposited eggs.

Nonphoretic egg parasitoids may encounter recently oviposited eggs through different behavioural adaptations. Although long range attractants have not been well investigated, some egg parasitoids appear capable of orienting over long distances through the use of chemical stimuli such as sex pheromones (Lewis *et al.*, 1972; Buleza and Mikheev, 1979; Noldus and van Lenteren, 1985). Others use contact kairomones from such diverse sources as host mandibular and labial gland secretions (Clausen, 1940), faecal material (Boucek and Askew, 1968; Hinton, 1981), or accessory gland secretions used in attaching eggs to substrates (Strand and Vinson, 1982, 1983a). All of these chemical stimuli could provide behavioural information about host age.

Egg parasitoids also perceive the physical configuration of hosts (Klomp and Teerink, 1962; Schmidt and Smith, 1985) and appear capable of assessing host age by measuring curvature and surface area. Strand and Vinson (1983b) found that *Telenomus heliothidis* Ashmead preferentially oviposits in younger *Heliothis virescens* eggs by detecting alterations in host shape with age. *H. virescens* eggs are nearly spherical at oviposition, but progressively become more conical as the embryo develops.

Acting either separately or concomitantly with these behavioural adaptations are adaptations which alter host embryogenesis physiologically. These could be due firstly to indirect stress effects of parasitoid feeding or secondly to the parasitoid female injecting, or her progeny secreting, factors which directly alter embryogenesis. Many early workers (Clausen, 1940; Salt,

1968) stated that indirect stress via larval feeding damage was the primary means by which egg parasitoids disrupt host development. However, at least for egg parasitoids of rapidly developing hosts, feeding damage would seem to be an unsatisfactory strategy, primarily because of the brief period when the host would be susceptible to parasitism. One could easily envisage a non-diapausing host with an embryonic development period of 60 h. Then a parasitoid with an embryonic development period of say 35 h would be unable to parasitize successfully eggs any older than 20–25 h, simply because there would be insufficient time for its larvae to begin feeding prior to hatching of the host. Consequently, the period for which this host would be vulnerable to attack is very brief. Although hypothetical, the above scenario illustrates that it may be advantageous for species to evolve mechanisms which directly retard host embryogenesis.

Active alteration of host physiology by parasitoids has been referred to by Vinson and Iwantsch (1980b) as host regulation. Yet, demonstrating host regulation in the terms they intended—i.e. that observed host pathologies are due to factors directly attributable to the parasitoid, while absolutely excluding the possibility that such pathologies are due to indirect stress—is often difficult (Thompson, 1983).

IV. HOST EMBRYOGENESIS AND PHYSIOLOGICAL ARRESTMENT OF DEVELOPMENT

A. Egg Parasitoids and their Host Associations

If the rate of host embryogenesis is indeed an important selective force in determining whether or not an egg parasitoid evolves the ability to arrest host development, one would predict that such pressure would be strongest on those parasitoids which attack rapidly developing hosts. As the duration of host embryogenesis increases, the necessity to arrest development would decline since the host would be vulnerable to parasitism for increasing periods of time. Lastly, parasitoids of hosts with extremely long development periods or phoretic parasitoids would be least likely to evolve mechanisms for arresting host development.

By examining life history data, it has been possible to gain an estimate of the host arrestment capabilities of a species by comparing the duration of host embryogenesis with the proportion of host embryogenesis which is susceptible to parasitism. Data on non-diapausing hosts whose development was measured under standard or near standard conditions were used, and the maximum host age which could be successfully parasitized was arbitrarily set at a 60% or greater rate of progeny survival. Such an estimate defines a lower limit of host suitability since it is well documented that parasitoids often show a preference for a particular age or size of host, but include less suitable hosts in their diet when preferred hosts are scarce (Salt, 1935; Charnov *et al.*, 1981).

Figure 1 illustrates that a significant negative correlation exists between the duration of host development and the percentage of this period which is

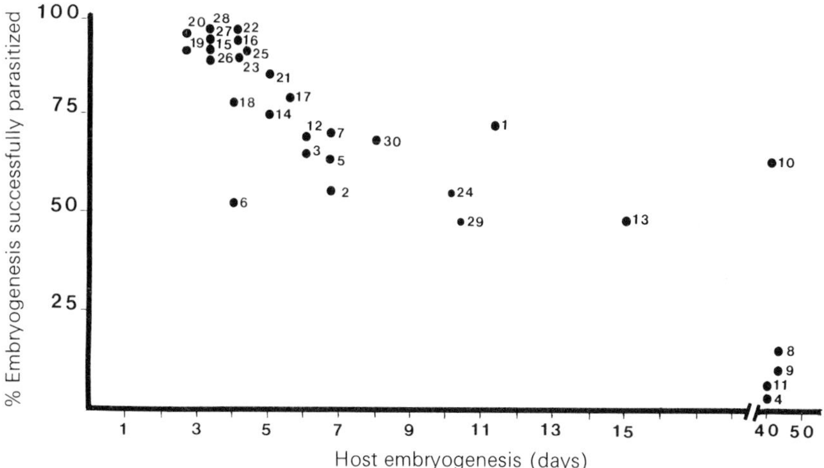

Fig. 1. Percentage of host embryogenesis which is successfully parasitized in 30 parasitoid-host relationships. (1) *Anagrus incarnatus* Haliday (M: Chantarasa-ard and Hirashima, 1984); (2) *Anaphes ovijentatus* (Crawford & Leonardi) (M: Stoner and Surber, 1971); (3) *Anaphes anomocerus* Girault (M: Rajakulendran, personal communication); (4) *Sceliocerda viatrix* (Brues) (S: Channa Basavanna, 1953); (5) *Patasson lameerei* Debauche (M: Leibee *et al.*, 1979); (6) *Platytelenomus busseolae* (Gahan) (=*hylas* Nixon) (S: Hafez *et al.*, 1977); (7) *Prestwichia aquatica* Lubbock (T: Henriksen, 1922); (8) *Scelio calopteni* Riley (S: Pickford, 1964); (9) *Scelio javanica* Roepke (S: Vinokurov, 1927); (10) *Scelio pembertoni* Timberlake (S: Pemberton, 1933); (11) *Sceliocerdo viatrix* (Brues) (S: Brues, 1917); (12) *Telenomus alsophilae* Viereck (S: Fedde, 1977); (13) *Telenomus fariai* Lima (S: Rabinovich, 1970); (14) *Telenomus gifuensis* Ashmead (S: Hidaka, 1958); (15) *Telenomus heliothidis* Ashmead (S: Strand *et al.*, 1983); (16) *Telenomus remus* Nixon (S: Gerling and Orion, 1973; Strand, unpublished); (17) *Telenomus ulyetti* Nixon (S: Jones, 1937); (18) *Trichogramma brevicapillum* Pinto & Platner (T: Pak and Oatman, 1982); (19) *Trichogramma embryophagum* Hartig (T: Klomp and Teerink, 1978); (20) *Trichogramma evanescens* Westwood (T: Lewis and Redlinger, 1969); (21) *Trichogramma evanscens* Westwood (T: Gatenby, 1917); (22) *Trichogramma lutea* Girault (T: Jones, 1937); (23) *Trichogramma maidis* Pintureau and Voegele (T: Hawlitzky and Boulay, 1982); (24) *Trichogramma minutum* Riley (T: Jennings and Houseweart, (1983); (25) *Trichogramma minutum* Riley (T: Peterson, 1930); (26) *Trichogramma nubilalum* Ertle and Davis (T: Davis and Burbutis, 1974); (27) *Trichogrammatoidea nana* Zehntner (T: Tothill *et al.*, 1930); (28) *Trichogramma pretiosum* (Riley) (T: Strand, 1985); (29) *Trissolcus basalis* (Wollaston) (S: Jubb and Watson, 1971); (30) *Trissolcus rungsi* (Voegelé) (S: Voegele, 1970). (M = Mymaridae, S = Scelionidae, T = Trichogrammatidae).

susceptible to parasitism by 27 egg parasitoid species (Spearman rank correlation, $P < 0.001$). As predicted, more rapidly developing hosts tend to be susceptible to parasitism over a greater percentage of their development. Several other notable trends are suggested in Fig. 1.

While the major egg parasitoid families each contain species which parasitize hosts in several orders, the dominant host group(s) for trichogrammatids is Lepidoptera, for mymarids is Homoptera, and for scelionids are Lepidoptera, Hemiptera, and Orthoptera (Clausen, 1940; Askew, 1971). Clausen (1940) states that many trichogrammatids are capable of developing in hosts during most of their development while scelionids are more variable, with some species capable of parasitizing well developed eggs while others require younger hosts. Thus, as shown in Fig. 2, of the hosts which are successfully parasitized over 75% of their developmental period, 80% are rapidly developing Lepidoptera parasitized by trichogrammatids (60%) and scelionids (30%). For hosts which are successfully parasitized over 50–75% of their developmental period, 82% are Heteroptera parasitized by scelionids or mymarids. Lastly, hosts which are successfully parasitized during less than 50% of their development are all Orthoptera parasitized by scelionids.

These data broadly support Clausen's generalization, but may also reflect a bias in sampling. Thus, trichogrammatids may appear to show greater flexibility in parasitizing well developed hosts because the genus *Trichogramma* and their lepidopteran hosts are the dominant group for which such data are available. It is possible that if more studies are conducted on other genera which parasitize hosts in other orders, more variability may be apparent. In contrast, scelionids may appear more variable only superficially, because of the greater variety of orders which they commonly parasitize and which have been studied. From Figs 1 and 2 it is clear that the scelionids which parasitize Lepidoptera tend to be as well adapted to parasitizing rapidly developing hosts as trichogrammatids, but those species which parasitize more slowly developing hosts (Hemiptera, Orthoptera) are restricted to earlier stages of host development. Also in agreement with earlier predictions, two of the four scelionids which parasitize only very young orthopteran eggs are phoretic. Lastly, the mymarids listed in Fig. 1 parasitize Homoptera, and develop successfully in approximately the same proportion of host embryogenesis as scelionids parasitizing Hemiptera.

B. Host Eggs and their Influence on Specialization

As suggested by Price (1980) the potential for specialization among parasites is high due to the discontinuous nature of the environment, host diversity, host specialization and the transience of resources. The probability for

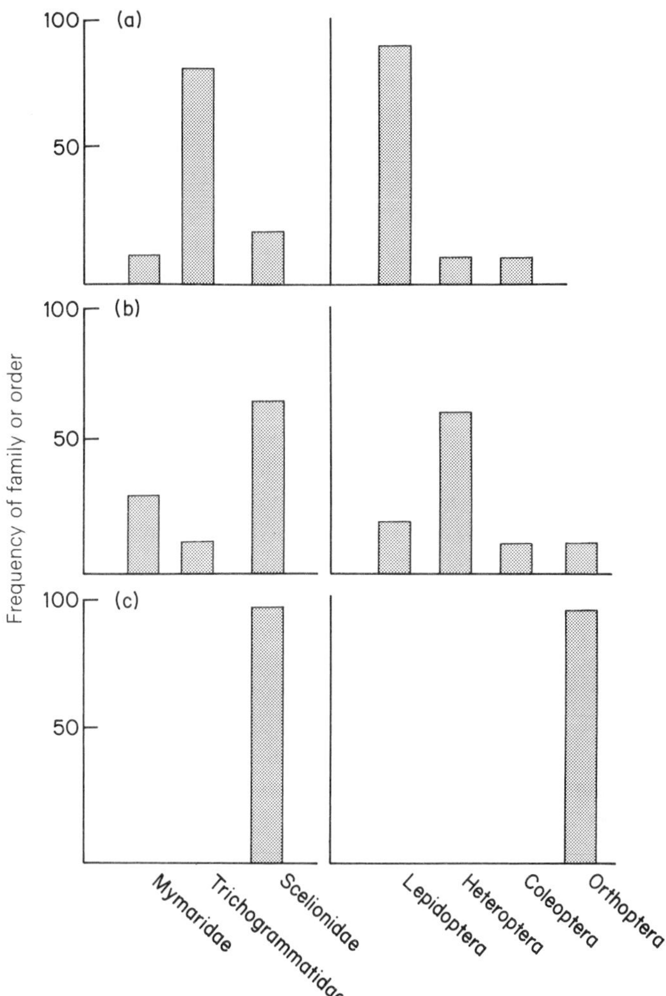

Fig. 2. The relationship between host order, parasitoid family, and the percentage of host embryogenesis susceptable to parasitism: (a) hosts which are susceptible to parasitism over more than 75% of their development; (b) from 50–75% of their development and (c) less than 50% of their development.

colonization of a given host is further enhanced by host size, host population size and host geographic range. Indeed, there is no greater testimony to the potential for adaptive radiation by parasites than the parasitic Hymenoptera for which there are an estimated 200,000 species (Askew, 1971). Townes (1969) estimates that there are at least 60,000 species of ichneumonids worldwide and as noted, the majority of North American species are highly specialized, being restricted to three hosts or fewer (Townes and Townes, 1951).

In contrast, many egg parasitoids such as *Trichogramma* spp. are notoriously polyphagous (Pinto *et al.*, 1978; Burks, 1979). Since chemical (Noldus and van Lenteren, 1985) and physical stimuli (Salt, 1935) have been found to influence *Trichogramma* host selection, local specialization, host preferences and geographic races may exist. Nonetheless, the recognized species within the genus remain capable of parasitizing a vast array of species in the laboratory. It is possible that some of the physiological characteristics of the eggs which trichogrammatids parasitize may have influenced its adaptive radiation and their polyphagous habits.

The high degree of specialization in parasites is often attributed to the chemical and physiological heterogeneity of hosts (Strong *et al.*, 1984). The biochemical, immunological and behavioural defences of many plants and animals probably result in parasite specialization and narrowing of host range as suggested for larval endoparasitoids. However, as Feeny (1976) suggests for plants, unpredictable rare species may invest relatively little reproductive effort in defence, because of the ability to escape from herbivores in time and space. By analogy, many insects appear to invest very little in the defence of eggs. Orders with relatively slow rates of embryogenesis often provide external defence of eggs by secretion of oothecae or similar structures, but many orders of insects provide no external defence except, on occasion, concealment. This, combined with the lack of an immune system or other pervasive internal defence suggests that eggs are in a sense relatively unspecialized, and in consequence may be a transient but physiologically very uniform resource which promotes polyphagy in parasitoids such as *Trichogramma*.

Although the eggs of some species may be quite numerous for brief periods, especially in agroecosystems, the eggs of many other species are probably rare in time and space, which make them a difficult resource to exploit. Many eggs may in fact be too rare to support a specialist, and so restrict adaptive radiation (Janzen, 1975). The near complete absence of obligate hyperparasitoids of egg parasitoids lends further support to this idea (Strand and Vinson, 1984). Although host size is often cited as a limitation to the number of trophic levels which can exist, the size of many egg parasitoids falls within the size range of other parasitoids which are hyperparasitized (Askew, 1971). Since hyperparasitoids themselves tend to be generalists (Janzen, 1975), their absence combined with the generalist tendencies of parasitoids such as *Trichogramma* spp. suggests that eggs may be a difficult resource on which to specialize.

Not all egg parasitoid groups are as species poor as the genus *Tricho-*

gramma. For example, the scelionid genus *Telenomus* contains an estimated 200–300 nearctic species (Johnson, 1984) which are at least broadly host-specific to given orders and families, but which show considerable variation in the number of hosts parasitized (Bin and Johnson, 1982; Johnson, 1984). Why *Telenomus* has specialized to a greater degree and in turn adapted a somewhat more specific host range than *Trichogramma* is unknown, but some of the aformentioned factors may have played a role.

Regardless of the differences between these genera, some species in both share common hosts and presumably must contend with similar problems. Knowing the age of host which a given parasitoid can successfully parasitize does not reveal the mechanisms by which this is accomplished. Unfortunately, very few studies in the literature examine egg parasitoid development. However, recent work by the author on *Telenomus heliothidis* and *Trichogramma pretiosum* (Riley) parasitism of *Heliothis virescens* offers a useful comparison of some of the factors involved.

C. *Telenomus heliothidis*

Telenomus heliothidis is a solitary egg parasitoid of several species of *Heliothis* (Strand and Vinson, 1983a). *H. virescens* eggs at 27°C complete gastrulation and segmentation in 14–18 h, dorsal closure in 36 h, initiate cuticle deposition by 55 h, and hatch in 72 h. In *H. virescens* eggs 12-h old when parasitized, *T. heliothidis* hatches in 18 h, moults to a second instar in 60 h, to a third instar in 72 h, pupates in 144 h, and emerges as an adult in 312 h (13 days) (Strand *et al.*, 1986). Of particular note is that coincident with hatching is the liberation of 25–35 teratocytes (Salt, 1968). These unusual cells, thought until recently to be present only in braconids (Gerling and Orion, 1973; Strand *et al.*, 1985), are derived from the parasitoid's embryonic serosa and continue to grow until the host is consumed.

Under the premise that larval feeding damage is responsible for successful development, one would predict that the parasitoid could not parasitize eggs any older than *ca.* 48–50 h, because of the time necessary for parasitoid embryogenesis. Yet, *T. heliothidis* is able to parasitize successfully *H. virescens* eggs up to 63-h old (Strand *et al.*, 1983). Thus, for temporal reasons alone, feeding damage does not adequately account for successful parasitism.

In addition, host pathology during the early stages of parasitism does not support a feeding damage hypothesis. *H. virescens* embryos, 36-h old when parasitized, begin to lose some structural integrity and ectodermal cells become pycnotic by 12 h postparasitism (Fig. 3a and b). Extensive degeneration of host tissue occurs between 20 and 24 h postparasitism, and by 30 h postparasitism virtually no host cells remain unlysed (Fig. 3c and d). The

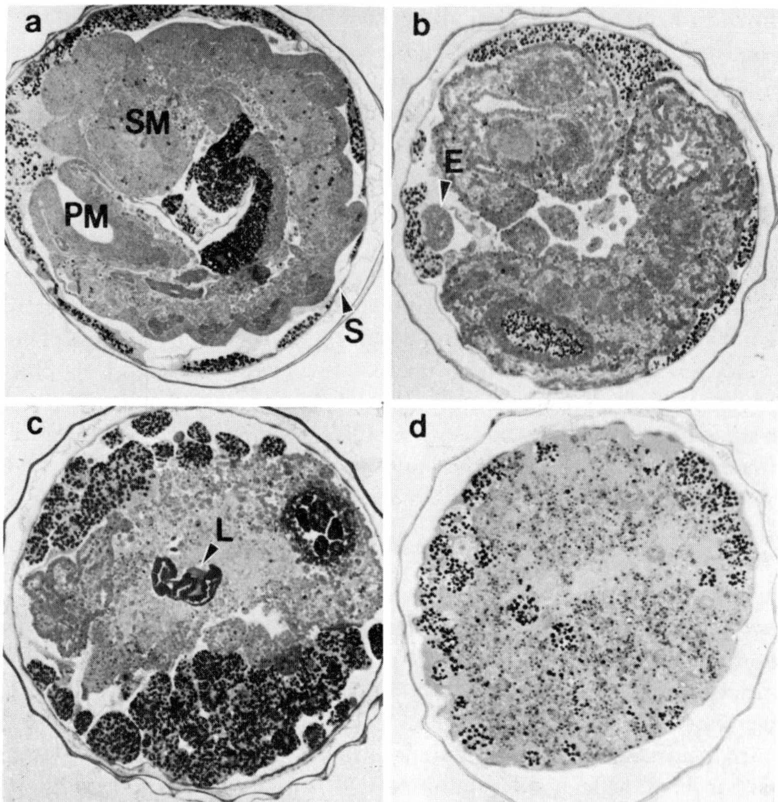

Fig. 3. Pathology of *Heliothis virescens* (F.) egg parasitized by *Telenomus heliothidis* Ashmead. (a) 36-h-old *H. virescens* egg parasitized for 1 h with the host embryo's segments, stomodaeum (SM), proctodaeum (PM), and the embryonic serosa (S) present; (b) 12 h postparasitism with the *T. heliothidis* embryo (E) lying externally to the host embryo; (c) 20 h postparasitism with the first instar parasitoid (L) visible along with a great loss of host structural integrity; (d) 36 h postparasitism with no host tissue remaining intact. All sections 650 ×.

only portion of the host which is not digested is the serosal endocuticle. All of these events occur prior to or within 12 h of parasitoid first instar eclosion. Obviously, events prior to eclosion cannot be due to larval feeding, but the necrosis observed by 30 h postparasitism is also not due to larval feeding. Observations of larval behaviour *in vitro* (Strand *et al.*, 1986), show that the first instar feeds very little and that the second is the dominant feeding stage. Yet, by the time the second instar moult occurs, the host is already necrotic. Parasitism of older hosts is basically the same as described with the exception that highly developed *H. virescens* embryos in which cuticle deposition has begun do not decompose as quickly.

Although these data point away from the feeding damage hypothesis, they do not indicate how 48-h old and older *H. virescens* eggs are prevented from hatching and whether the adult or progeny are responsible for the pathological events observed. *T. heliothidis* oviposition may be interrupted so as to prevent egg deposition but without significantly affecting normal behaviour. By interrupting oviposition at 5-s intervals and monitoring the hosts for emergence of *H. virescens* larvae, it was found that when oviposition was interrupted at *ca.* 25 s, the hosts did not contain a parasitoid although they failed to hatch (Strand *et al.*, 1983). These data strongly suggest that *T. heliothidis* injects an "arrestment factor" into the host immediately prior to egg release. Naturally, hosts in which oviposition is interrupted between 30 and 35 s contain a parasitoid and also fail to hatch. Of particular interest, however, is that pseudoparasitized hosts do not exhibit the same pathological symptoms as parasitized hosts. Embryonic meso- and ectodermal cells become pycnotic at *ca.* 12 h post pseudoparasitism, but the host does not subsequently become necrotic (Strand *et al.*, 1986). Extracts and implants of various regions of the *T. heliothidis* reproductive tract indicate that the distal common oviduct is the source of the arrestment factor (Strand *et al.*, 1983). In this species it has become modified into an exocrine organ which appears to secrete material into the lumen of the oviduct that is possibly injected into the host at oviposition (Strand *et al.*, 1986). It is possible that the arrestment factor also has a role in host necrosis by facilitating the action of other factors, but the data indicate that its primary function is in preventing host development and hatching.

Since the arrestment factor produced by the mother is probably not responsible for the observed necrosis in parasitized hosts, this leaves either the first instar larva or teratocytes as possible sources of this effect. There is no evidence that the first instar larva has an active role in host decomposition since it has no exocrine organs with which to produce a cytolytic factor (Strand *et al.*, 1986). Some workers (Balduf, 1926; Clausen, 1940; Sahad, 1982) have suggested that the movement of scelionid and mymarid larvae is sufficient for dissociation of host tissues. However, in relation to the size of the host, the newly eclosed first instar parasitoid is extremely small (*ca.* 70 μm in length) and fragile, and judging by observations of its behaviour *in vitro*, it seems insufficiently active to disrupt host tissue to the degree observed. Second and third instars possess labial glands and malpighian tubules which potentially could be secretory in function, but host necrosis is completed well before moulting of the first instar to a second.

Three lines of evidence support the contention that teratocytes are responsible for host decomposition. Firstly, teratocyte formation and the initiation of necrosis coincide. They begin to form *ca.* 17 h postparasitism and are released at first instar eclosion, and by 24 h necrosis is very extensive. In addition, if *T. heliothidis* are reared at varying temperatures, hatching can be varied by several hours. Yet, the initiation of host necrosis is synchronous with first instar eclosion and teratocyte release regardless of when it occurs.

Secondly, ultrastructural evidence indicates that the teratocytes are highly secretory during the period when host necrosis occurs (Strand *et al.*, 1985; Strand *et al.*, 1986). Vesicular bodies are very numerous in the serosa during formation of the teratocytes. Exocytosis of the vesicular bodies from the teratocytes continues until 36 h postparasitism, at which time secretory activity progressively declines coincident with the completion of host necrosis. Based upon structure, the vesicular bodies resemble lysosomes (Kurosumi and Fujita, 1974). Leucine amino peptidase, esterase and phosphatase activity is present in media in which *T. heliothidis* teratocytes are cultured, further supporting a secretory function and the presence of lysosomes. Conversely, no such secretory activity is associated with the first instar. Why teratocytes and the parasitoid are not digested is unknown, but the presence of cuticle may be important.

The last and strongest piece of evidence which implicates teratocytes in host decomposition is that when they are introduced into hosts in the absence of both the arrestment factor and parasitoid, host necrosis still occurs (Strand *et al.*, 1986). Necrosis is more rapid if hosts are first pseudoparasitized and held for 18 h before introducing teratocytes. Similarly, digestion of culture medium occurs during *in vitro* rearing of *T. heliothidis* only when teratocytes are present (Strand and Vinson, 1985; Strand, unpublished). Thus, parasitism by *T. heliothidis* is a multifaceted process in which several factors interact to produce the observed pathological responses.

D. *Trichogramma pretiosum*

Trichogramma pretiosum is a facultatively gregarious parasitoid of numerous species of Lepidoptera, including *H. virescens* into which it usually oviposits two eggs. In *H. virescens* eggs, 12-h old when parasitized, *T. pretiosum* eggs hatch in 26 h, the larvae pupate in 108 h, and emerge as adults in 216 h (9 days). Unlike the findings of Voegele *et al.* (1974) for *Trichogramma brasiliensies* Ashmead, I have found no evidence that *T. pretiosum* possesses teratocytes. Two polar bodies are liberated per *T. pretiosum* egg at eclosion, but they degenerate within 8 h. Since only one teratocyte was reported in association with *T. brasiliensis*, it is likely that this cell is in fact a polar body.

Using the same logic as for *Telenomus heliothidis*, *Trichogramma pretiosum* should not be able to parasitize successfully *H. virescens* eggs any older than 42–44 h. Yet, as first shown in Fig. 1, it parasitizes at least 60% of eggs up to 68-h old even though they are within 2 h of hatching at this time (Strand, unpublished). Host pathology does not support a feeding damage hypothesis for *T. pretiosum* parasitism. It usually oviposits in the vitellophages in *H. virescens* eggs up to about 36 h but oviposits into the embryo in older hosts. Embryos 36-h old when parasitized become dissociated and show signs of degeneration 6–12 h postparasitism (Fig. 4a) when structural integrity is lost and tissues begin to degenerate. By 12 h postparasitism, plasma membranes begin to degenerate and cells lyse (Fig. 4b). Such degeneration is progressive with host tissue becoming totally dissociated by 22 h (Fig. 4c) and necrotic by 26 h, which coincides with hatching of the parasitoid eggs (Fig. 4d). The rate of tissue degeneration in eggs over 60-h old is slower since it does not become

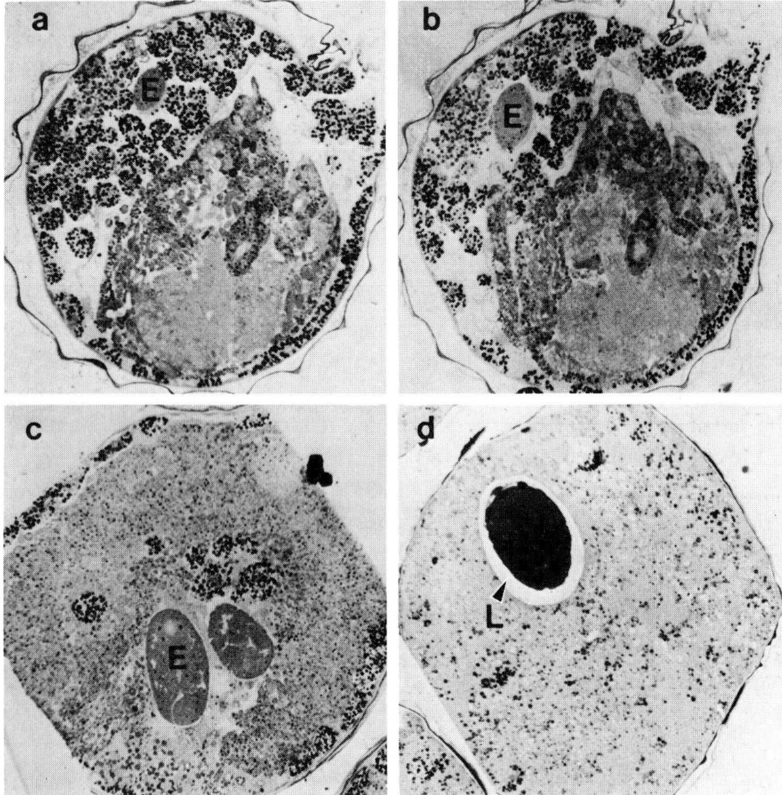

Fig. 4. Host pathology of 36-h-old *Heliothis virescens* (F.) eggs parasitized by *Trichogramma pretiosum* (Riley): (a) 6 h postparasitism with the *T. pretiosum* egg (E) developing externally of the host embryo; (b) 12 h postparasitism; (c) 22 h postparasitism with host tissue dissociated extensively and (d) 26 h postparasitism with newly eclosed larva (L) feeding.

totally necrotic until *ca.* 36–42 h postparasitism. Much as in the case of *Telenomus heliothidis*, the presence of a cuticle in the pharate first instar *H. virescens* embryo seems to slow the rate of degeneration, possibly due to poorer penetration of the factors responsible. The only host components which remain undigested are the endocuticle of the serosa and the host pharate larval cuticle if present.

Because the host is digested before hatching of the eggs, *Trichogramma pretiosum* larvae are able to feed extremely rapidly. In *H. virescens* eggs containing two *T. pretiosum*, the larvae completely consume the digested contents of the egg within 10 h of hatching. In consequence, the larvae increase in size from *ca.* 300^3 μm at hatching to 2700^3 μm in 10 h, and occupy the entire volume of the host chorion. The rate of feeding in 60-h and older *H. virescens* eggs is somewhat slower since they are not consumed until *ca.* 36 h after parasitoid hatching. This appears, once again, to be due to the fact that the host embryo possesses a cuticle which interferes with the rate of necrosis as well as the ability of the hatching larvae to feed (Strand, unpublished).

Whether the adult female injects, or the progeny secrete a factor which is responsible for the observed host pathology can be approached in the same manner as before. *T. pretiosum* host acceptance behaviour toward *H. virescens* may be divided into five stages: examination (E), drilling (D), first oviposition (O_1), second oviposition (O_2), and marking (M). Oviposition takes *ca.* 34 s and is characterized as the period during which the ovipositor is fully inserted into the host until it is withdrawn and marking is initiated. Oviposition is divided into two phases because of *T. pretiosum*'s tendency to lay two eggs in *H. virescens*. Precise movements during oviposition differ for male and female eggs (Suzuki *et al.*, 1984; Strand, unpublished; see Chapter 3), but for both sexes, the egg is deposited during the later part of an "abdominal trembling" phase, and interruption of oviposition before this point produces hosts without eggs which are pseudoparasitized.

Pseudoparasitized hosts exhibit the same symptoms as parasitized hosts. Degeneration of host tissue is complete by 24 h, indicating that an adult factor is primarily responsible for both the cessation of host development and necrosis. Research into the source of the adult factor indicates the source to be the *T. pretiosum* poison gland (Strand, unpublished). Because of the source and the pathology of hosts exposed to it, the factor is hereafter referred to as a venom. The larvae appear to play no active role in host pathology.

It is not known whether the wasp injects the venom during each egg-laying sequence or only during the first. Experiments performed with *H. virescens* show that necrosis occurs after interruption of the first egg-laying sequence, suggesting that a sufficient amount of the venom is placed in the host at this time to cause host necrosis. Yet one would predict that since *T.*

pretiosum usually lays two eggs in *H. virescens*, interruption of oviposition before release of the first egg would inhibit the injection of venom before release of the second egg. Since oviposition patterns for male and female eggs are the same regardless of whether a male or female egg is laid first, or how many eggs are laid, it is likely that the venom is injected prior to each oviposition sequence. Since it is well established that *Trichogramma* spp. adjust the number of eggs laid to the size of the host (Klomp and Teerink, 1962) it is possible that a set dosage of the venom is injected with each egg such that the amount of venom is also correlated with the size of host.

E. Conclusions

These two case histories show that at least some egg parasitoids are capable of rapidly halting host development and parasitoid feeding plays little or no role in the observed host pathologies. The phylogenetic distance between scelionids and trichogrammatids (Hennig, 1981) is indicative of strong evolutionary convergence. The sequence of pathological events in *H. virescens* parasitized by these species as well as the factors responsible for these events vary, but the overall result is quite similar.

The developmental strategies of *Telenomus heliothidis* and *Trichogramma pretiosum* differ markedly from the trends observed in endoparasitoids of host larvae (Thompson, 1983). Most notably, the egg parasitoids do not maintain the host during any part of their development, nor do they feed on living tissue. In effect, there is no metabolic integration between parasitoid and host. In fact, it may be a general rule that little or no integration occurs with idiobiotic parasitoids (see Chapter 8) which attack non-moulting hosts such as eggs, pupae or paralysed larvae.

Alterations in host endocrine function and development are common in larval host-parasitoid interactions (Beckage, 1985); however, the host's endocrine system does not play a role in *Telenomus heliothidis* or *Trichogramma pretiosum* parasitism. The severity of host tissue destruction probably eliminates any active production of hormones by the host. It is possible that maternally produced hormones sequestered in the host egg could influence the developing parasitoid, but no such interaction has been found. The fact that both species can be cultured to adulthood *in vitro* in the absence of any host material also suggests that host hormones do not play a role in parasitoid development (Strand and Vinson, 1985). The serosal endocuticle is the only portion of the host which is maintained during development. This structure appears to be important in water regulation and probably prevents desiccation of the progeny prior to their pupation. Both parasitoids are able to parasitize infertile eggs less than 12-h old, but fail to develop in older ones. Infertile eggs lack an endocuticle, because it is secreted by the serosal cells which form early in embryogenesis. In young, infertile eggs the parasitoids produce a pupal case which protects the parasitoid from desiccation. However, in older infertile eggs the parasitoid desiccates prior to becoming a prepupa (Strand, unpublished). Fedde (1977) also found that *Telenomus alsophilae* Viereck does not survive well in older infertile eggs, perhaps for the same reasons.

Both arrestment of host embryogenesis and subsequent tissue necrosis occur in other scelionids and trichogrammatids (Jones, 1937; Voegele, 1970; Gerling and Orion, 1973; Hawlitzky and Boulay, 1982) as well as mymarids, encyrtids and eulophids (van Alphen, 1980; Chantarasa-ard and Hirashima, 1984; Brown, 1984).

However, no other studies indicate the source of the factors responsible for the changes which occur in host eggs. Voegele (1970) and Gerling and Orion (1973) suggest that scelionid teratocytes play a role in host necrosis and Marston and Ertle (1969) and Hawlitzky and Boulay (1982) suggest that *Trichogramma* spp. inject substances into hosts which prevent hatching. The effects *Telenomus heliothidis* and *Trichogramma pretiosum* have on *H. virescens* may be common among egg parasitoids, but much more study is necessary before any conclusions are possible.

F. Diapause

So far, egg parasitism has been discussed for hosts which do not overwinter or diapause. Yet, many insects do overwinter or diapause as eggs, and these eggs are sometimes parasitized (Askew, 1971). As discussed earlier, it is possible that endocrine factors which influence host diapause could also influence the parasitoid. Very few studies have addressed this question, but life history data and inference by the author suggest that most egg parasitoids overwinter rather than enter obligatory diapause, and are not influenced by their hosts. However, a few data suggest that some egg parasitoids do enter obligatory diapause, and in at least one case, host hormones may play a significant role.

Usually, overwintering parasitoids have a protracted development, but are able to emerge quickly under the proper environmental conditions. For example, *Trichogramma minutum* Riley parasitizes *Peridroma saucia* (Hübner) eggs during autumn and winter months (Parker and Pinnell, 1971). It does not emerge when the temperature is below 13°C, but does emerge during brief periods of warm weather. Similarly, in other hosts (Peterson, 1930) it remains as a larva or pupa during periods of low temperature, but rapidly completes development when temperatures rise. Other egg parasitoids overwinter, with the only variation being the stage in which the parasitoid quiesces. Thus, *Scelio calopteni* Riley and *Anagrus optabilis* (Perkins) overwinter as first instars (Pickford, 1964; Sahad, 1984) while *Gonatocerus* sp. overwinters as a pupa (Sahad, 1982). It is likely that the stage of overwintering reflects both the time at which the host is parasitized and the temperature, since development is able to proceed normally if the parasitoids are moved to warmer conditions.

In contrast, Jackson (1963) demonstrated that the mymarid *Caraphractus cinctus* Walker diapauses in *Agabus* eggs. It enters diapause as a prepupa when host eggs are maintained at a photoperiod of less than 15 h light, but develops directly when they are maintained at a photoperiod of greater than 16.5 h light. The chorion is completely transparent, which allows even low intensity light to reach the developing parasitoid. Thus, *C. cinctus* diapause is dependent upon an external stimulus rather than on any host influence.

Trichogramma cacoeciae Marchal diapause may be determined by the status of its host, the tortricid *Archips rosanus* (L.) (= *Cacoecia rosana*) (Marchal, 1936). *C. rosana* is univoltine, laying eggs in late July which diapause until March when they hatch; the immature develop, and the adults emerge and oviposit the following July. *T. cacoeciae* is bivoltine in *C. rosana* eggs. The first generation parasitizes freshly laid eggs in July. The parasitoid eggs hatch, the larvae consume the host, and diapause as terminal instar larvae until the subsequent March. They then pupate, emerge and oviposit in unparasitized *C. rosana* eggs laid the previous July which by this time have broken diapause. The second generation of parasitoids develops without interruption, and parasitizes freshly laid *C. rosana* eggs in July.

From these observations, it is possible that temperature or photoperiod could have induced diapause in first generation *T. cacoeciae* as described for *Caraphractus cinctus*. However, *T. cacoeciae* passes through five to six generations per year without diapausing when it parasitizes *Mamestra brassicae* (L.), which does not diapause (Marchal, 1936). Thus, *T. cacoeciae* seems to enter diapause when the larva consumes a diapausing egg. There is a striking similarity between *T. cacoeciae* diapause and the influence of diapause hormone in *Bombyx mori* which provides the first evidence that host endocrine factors can play a role in egg parasitism. Egg diapause hormones exist in other species (Downer and Laufer, 1983), but further research is needed to determine whether diapause by the parasitoid is actually due to host hormones.

V. PARASITOID DEVELOPMENT AND ITS INFLUENCE ON DISCRIMINATION

Many parasitoids are capable of intraspecifically discriminating between parasitized and unparasitized hosts (van Lenteren, 1976). Egg parasitoids discriminate by detection of either external marking pheromones or some form of internal marker (Salt, 1940; Rabb and Bradley, 1970; Bosque and Rabinovich, 1979; van Alphen, 1980; Klomp *et al.*, 1980). Scelionids and trichogrammatids apply external markers by wiping the extruded ovipositor across the host after oviposition. Some workers suggest that the external marker is due to the ovipositor "scratching" the surface of the host (Hokyo *et al.*, 1966; Safavi, 1968), but it is now generally accepted that the marker is chemical. Parasitoids detect the presence of the external marker with their antennae (Strand, unpublished) with the marker usually being active immediately following deposition. Previous oviposition experience has a strong influence on the parasitoid's response to the external mark (van Lenteren, 1976); however, even frequently ovipositing females sometimes cannot detect the external mark after certain periods of time. For example, the marks of *Trichogramma embryophagum* Hartig, *Trichogramma pretiosum* and *Telenomus heliothidis* persist for 12–18 h (Klomp *et al.*, 1980; Strand, unpublished) and that of *Telenomus fariai* Lima for 3 days (Bosque and Rabinovich, 1979).

Neither the exact source nor the composition of egg parasitoid external marks has been characterized, but the *Telenomus heliothidis* external mark is produced by the accessary gland (Strand, 1985). Rabb and Bradley (1970) and Bosque and Rabinovich (1979) respectively report that the external marks of *Telenomus remus* Nixon and *Telenomus fariai* are water soluble; however, the external mark of *Telenomus heliothidis* is soluble in more apolar solvents such as ether and acetone (Strand, unpublished). The fact that the mark is often not persistent suggests that the chemical(s) are either unstable or slightly volatile.

The factors responsible for internal marking are much less understood even though many egg parasitoids appear capable of internal discrimination. In fact, internal marking is the only means of discrimination available to some egg parasitoids of aquatic hosts (Jackson, 1958). Internal discrimination is frequently latent, becoming discernable to the parasitoid several hours after oviposition, yet persisting throughout development (Klomp *et al.*, 1980; Strand *et al.*, 1986).

While intraspecific marking is pervasive, there are very few reports of interspecific marking (Vet *et al.*, 1984). Ables *et al.* (1981) report that *Telenomus heliothidis* and *Trichogramma pretiosum* do not interspecifically discriminate, and while this is true for the external mark, I have found that both species possess a latent internal mark which is intra- and interspecific

(Strand, unpublished). In response to internal markers, *Telenomus heliothidis* rejects all hosts 21 h after parasitism by a conspecific at 27°C, but rejects all hosts after 18 h if the hosts are incubated at 29°C (Fig. 5). In both cases *Telenomus heliothidis* egg hatching must occur 2–3 h before a second female is able to detect parasitism by internal marker. This is coincidentally the period in which host necrosis occurs. Pseudoparasitized eggs do not become necrotic and are not discriminated against (Fig. 5). In response to internal markers, *Telenomus heliothidis* rejects most eggs parasitized by *Trichogramma pretiosum* 12 h after parasitism, possibly because of the more rapid rate of host necrosis. If host necrosis is responsible for internal discrimination by *Telenomus heliothidis* one would also expect that hosts pseudoparasitized by *Trichogramma pretiosum* would be rejected since the venom is responsible for necrosis. In fact this is what occurs. Lastly, *Telenomus heliothidis* rejects all eggs which are mechanically disrupted. Host embryonic cells are easily lysed by sonication, resulting in hosts which are pathologically similar to parasitized eggs, and which are rejected by *Telenomus heliothidis*.

Internal discrimination by *Trichogramma pretiosum* is very similar, in that the parasitoid discriminates against necrotic hosts. *Trichogramma pretiosum* is able to detect conspecific parasitism *ca.* 14 h after oviposition at 27°C; however, *Telenomus heliothidis* parasitism is not detected until *ca.* 24 h. Thus, it may be slightly less sensitive to internal changes than *Telenomus heliothidis*. Although *Trichogramma pretiosum* which have never oviposited or which

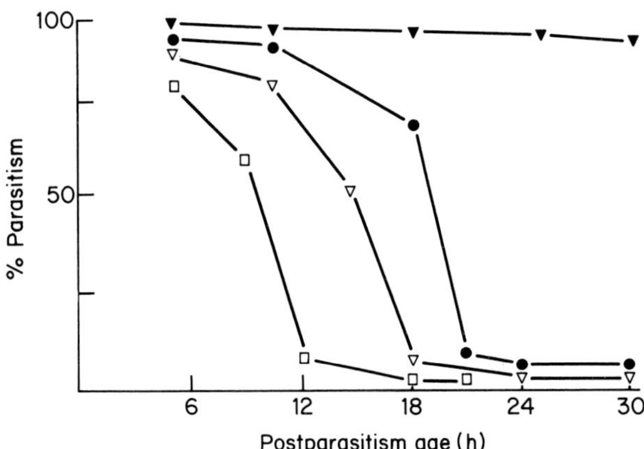

Fig. 5. Percentage superparasitism by *Telenomus heliothidis* Ashmead of *Heliothis virescens* (F.) eggs of increasing age when: pseudoparasitized (solid triangles); parasitized and maintained at 27°C (solid circles); parasitized and maintained at 29°C (open triangles) and parasitized by *Trichogramma pretiosum* (Riley) (open squares).

oviposit infrequently usually reject hosts in which the internal mark is present, they are more prone to superparasitize than *Telenomus heliothidis*. The same observation was also made for *Trichogramma embryophagum* by Klomp *et al.* (1980).

VI. PARASITOID DEVELOPMENT AND ITS INFLUENCE ON INTRINSIC COMPETITION

Since super- or multiparasitism does occur under certain conditions, it is important to understand the fate of the progeny. Supernumerary larvae of solitary endoparasitoids are eliminated by physical attack of physiological suppression via toxins, anoxia, or nutritional deprivation (Salt, 1961; Vinson and Iwantsch, 1980b).

First instar scelionids are teleaform as first instars, possessing large mandibles, which several authors have suggested are used in physically eliminating competitors (Schell, 1943; Hokyo *et al.*, 1966; Gerling, 1972). Yet the fact that some larvae appear to be physically adapted to combat does not necessarily mean that conspecifics are eliminated only in this way. For example, *Telenomus heliothidis* usually eliminates conspecifics by physiological suppression using the same mechanisms which are involved in host necrosis and internal marking (Strand and Vinson, 1984; Strand, 1985). Through dissection of parasitized hosts and observations during *in vitro* culture (Strand *et al.*, 1986), competition can be very carefully observed. First, the relative age of the competitors is very important. When ovipositions are within 2 h of one another, both eggs hatch and one of the larvae is eliminated by physical attack. If the interval is more than 2 h, the second female's egg develops normally until the first female's egg hatches. Then the second egg becomes necrotic along with the rest of the host. Apparently, cuticle deposition has initiated in the second embryo prior to hatching of the first embryo, thus protecting it from necrosis. If newly formed teratocytes are introduced into a host containing a *Telenomus heliothidis* egg, it is killed. Yet, as expected, if a female oviposits into a pseudoparasitized host, the egg hatches and develops normally. These experiments suggest that the elimination of younger competitors is due to teratocytes and that the living first instar and adult arrestment factor play no role.

Several unexplained observations by other workers seem to support this conclusion. Spencer (1926), Thompson and Parker (1930), Mackauer (1959) and Hokyo *et al.* (1966) all suggest that competitors are eliminated by cytolytic factors which arise with the hatching of the older egg, and Mackauer (1959) discounted the proposition that suppression of supernumeraries was due to anoxia or nutrient depletion. Interestingly, the parasitoids used in these studies produce teratocytes, and the hosts become necrotic after the parasitoid hatches. This role for cytolytic factors has been criticized because it assumes the factors act on competitors but not on the producer (Salt, 1961). Yet, because the production of cytolytic factors is synchronized with the hatching of *Telenomus heliothidis*, these observations are quite plausible.

Since gregarious larvae frequently contact one another under normal development (Salt, 1961), it is reasonable to assume that physical attack will not occur when they are crowded. Since the amount of food ingested by larvae is important for survival, one would expect that mortality will increase with both the number of eggs laid per host and the hatching time between larvae. This is precisely what occurs for *Trichogramma embryophagum* and *Trichogramma pretiosum* (Klomp *et al.*, 1980; Strand and Vinson, 1985). When hosts are superparasitized simultaneously, the number of emerging adults increases slightly with the number of eggs laid, but the increase in mortality is much greater (Fig. 6). In both species most of the eggs hatch, but some of the larvae die from starvation early in development. What determines which larvae survive is unclear, but small differences in the time when the eggs hatch is important.

Considerable insight on how hatching time influences supernumerary survival can be gained if larval behaviour is observed directly. Although preliminary, some recent experiments using *Trichogramma pretiosum* are worth mentioning. Methods were developed which allow for the successful *in vitro* culture of *Trichogramma pretiosum*, resulting in nearly 100% rates of pupation and 75% adult emergence (Strand and Vinson, 1985). Such

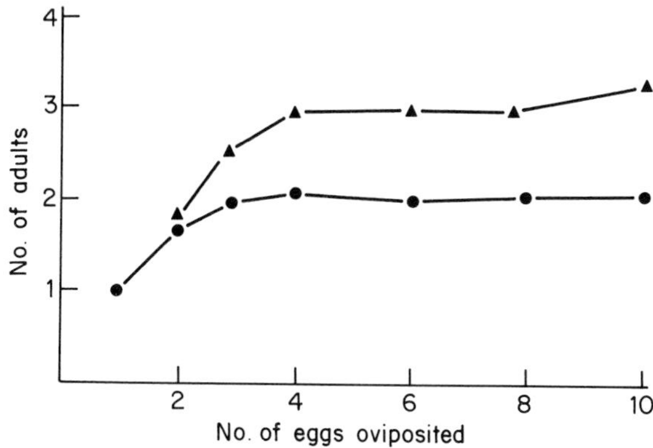

Fig. 6. Number of adults emerging from superparasitized hosts with increasing numbers of eggs for *Trichogramma embryophagum* Hartig (solid circles) and *Trichogramma pretiosum* (Riley) (solid triangles).

techniques provide a means of manipulating and monitoring levels of superparasitism. In one experiment, a volume of medium equal in volume to two *H. virescens* eggs was provided for a cohort of five *Trichogramma pretiosum* eggs. A second cohort of five or one *Trichogramma pretiosum* was then added to the medium at 2–3-h intervals up to 18 h. The results indicated that all supernumerary eggs hatched if placed in medium up to 9 h after the first cohort. However, very few supernumerary eggs hatched after 12 h, because the primary larvae consume the medium before the supernumerary eggs complete development. The number of supernumerary larvae which hatch and produce pupae also diminishes with time. Approximately 25% of supernumeraries survived to pupation if the eggs were placed in the medium within 2 h of the first cohort. This reduced survival at 5 h to less than 10%, and zero survival occurred after 8 h. Yet, as expected, if only one supernumerary was added, 60% survived if placed in medium within 2 h of the first cohort, 28% survived at 5 h, and 5% survived at 8 h. More experiments are necessary but these data also indicate that male progeny survive better than female, because of their more rapid feeding and ability to pupate at a smaller larval size (Strand, unpublished).

Interspecific competition associated with multiparasitism has been described as often as intraspecific competition and superparasitism (Vinson and Iwantsch, 1980b). Since interspecific discrimination is less common, the possibility for multiparasitism may be potentially greater, and may influence the structure of parasitoid guilds (Force, 1974). The same mechanisms involved in intraspecific competition have been reported for interspecific competitions and need not be reviewed here. Instead, a brief account of multiparasitism by *Telenomus heliothidis* and *Trichogramma pretiosum* illustrate the major points.

A consideration of interspecific competition between *Telenomus heliothidis* and *Trichogramma pretiosum* raises several questions, including how do their rates of development influence the interaction, do the cytolytic factors associated with *Telenomus heliothidis* teratocytes influence *Trichogramma pretiosum*, and does the lack of fighting by *Trichogramma pretiosum* adversely affect its ability to compete against *Telenomus heliothidis*? At 27°C *Trichogramma pretiosum* usually emerges if allowed to oviposit 6–12 h before *Telenomus heliothidis* (Fig. 7). *Telenomus heliothidis* usually emerges if allowed to oviposit from 72 h before to 3 h after parasitism by *Trichogramma pretiosum*. Curiously, *Trichogramma pretiosum* emerges from hosts first parasitized by *Telenomus heliothidis* 84–96 h previously, but loses to it if the first oviposition precedes the second by 108 h or more (Fig. 7).

These data are not surprising in the light of what has been described. Firstly, *Telenomus heliothidis* does not oviposit in hosts parasitized more than 12 h previously by *Trichogramma pretiosum*, because of the onset of

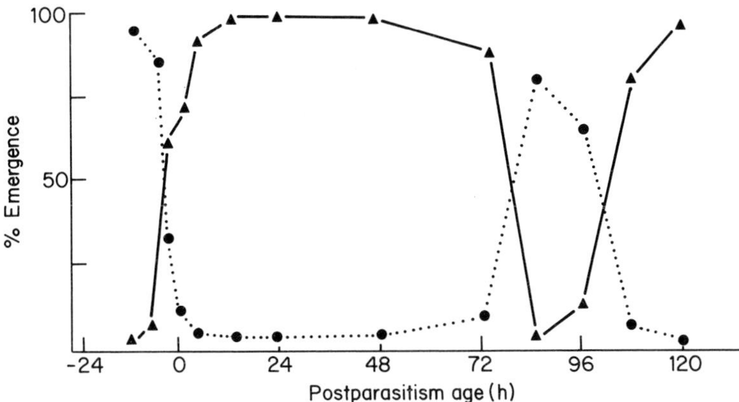

Fig. 7. Percentage emergence of *Telenomus heliothidis* Ashmead (solid triangles) and *Trichogramma pretiosum* (Riley) (solid circles) from multiparasitized hosts with varying times between ovipositions. Negative values indicate *Trichogramma pretiosum* oviposition preceded *Telenomus heliothidis* and positive values indicate *Telenomus heliothidis* oviposition preceded *Trichogramma pretiosum*.

internal discrimination. However, *Trichogramma pretiosum* will oviposit in eggs parasitized by *Telenomus heliothidis*, because inexperienced females disregard the mark. Secondly, at 27°C *Trichogramma pretiosum* eggs hatch in *ca.* 25 h and *Telenomus heliothidis* eggs hatch in *ca.* 18 h, but the former consume the host by *ca.* 30 h postoviposition and the latter in *ca.* 80 h. As a result, *Trichogramma pretiosum* egg hatching precedes *Telenomus heliothidis* only if given a *ca.* 8-h head start. From Fig. 6 it is clear that *Trichogramma pretiosum* emerges when oviposition precedes *Telenomus heliothidis* by 6 h. Dissection and *in vitro* experiments indicate that the eggs of both species hatch, but the *Trichogramma pretiosum* larvae consume the host before *Telenomus heliothidis* even moults to a second instar. Neither larva is able to physically harm the other, nor does the *Trichogramma pretiosum* venom affect *Telenomus heliothidis*. Thus, *Trichogramma pretiosum* outcompetes *Telenomus heliothidis* in the same manner as it outcompetes conspecifics; that is, through rapid ingestion of the host.

Telenomus heliothidis is the victor if its egg hatches first, because the *Trichogramma pretiosum* eggs become necrotic and never hatch. The *Trichogramma pretiosum* larvae may be unaffected because their cuticle protects them in the same way it appears to protect *Telenomus heliothidis*. *Trichogramma pretiosum* eggs continue to die throughout *Telenomus heliothidis* larval development if laid into the host milieu, but survive and adults emerge if the female oviposits into the *Telenomus heliothidis* third instar (Strand and Vinson, 1984) which fills the entire volume of the host egg, making this trichogrammatid facultatively hyperparasitic. It usually oviposits into the gut where the eggs develop successfully. The *Telenomus heliothidis* third instar becomes necrotic very quickly after hyperparasitism, due to the adult venom. *Trichogramma pretiosum* is apparently unable to oviposit into *Telenomus heliothidis* pupae.

The most striking feature of these results is how integrated marking has become with development and the mechanisms involved in intrinsic competition. Of particular interest is the relationship between external and internal marking. Superficially, there would seem to be little need for such duplication in that both marks indicate that a host is parasitized. Yet, under natural conditions many parasitoids probably experience relatively low levels of competition within patches, since mobility is limited and hosts are often dispersed; which leads to the question of why marking evolved at all? External marking may have evolved as a means by which a female can rapidly discriminate and avoid duplication of her own reproductive effort. As discussed earlier, small differences in oviposition time can put a superparasitizing female's progeny at a large competitive disadvantage, particularly for solitary species. Combined with the fact that many egg parasitoids attack hosts which lay egg masses which are likely to be re-encountered soon after parasitism, this suggests the need for an efficient means of discrimination. That other conspecifics also respond in the laboratory is not surprising, but this may not have as much relevance in the field. This suggestion is consistent with the ephemeral nature of the mark, for if it was important in warding off conspecifics, it would seem likely that these marks would be more enduring. On the other hand, the internal mark persists throughout the course of parasitoid development and so manifests itself long after the ovipositing female has left a patch. If the parasitized host is re-encountered then there is a strong likelihood that it would be by another female. The fact that the internal mark is less frequently disregarded by the wasps discussed here, regardless of oviposition experience, further suggests that their functions are different. Indeed, parasitoid marking strategies and the behavioural responses they elicit may reflect the parasitoid's own estimate of host quality, and provide important information in understanding parasitoid oviposition decisions.

VII. PARASITOID DEVELOPMENTAL PHYSIOLOGY AND ITS INFLUENCE ON OVIPOSITION DECISIONS

Undoubtedly, numerous factors act together to define the stage and variety of host that parasitoids are capable of exploiting (Wood and Graniti, 1976). In general, the more closely knit the life histories of the parasitoid and host, the more host-specific the parasitoid will be (Price, 1980). Factors such as parasitoid mobility, the spatial and chemical heterogeneity of the host (Thompson, 1982) and host rarity (Janzen, 1975) collectively determine the degree of parasitoid specialization. The confused states of the phylogeny of the major egg parasitoid families and the lack of knowledge about the life history of most parasitoid species currently hinder efforts to discern trends in parasitoid specificity and adaptive radiation (Burks, 1979; Hennig, 1981; Johnson, 1984). Nonetheless, some of the features alluded to in this chapter may play a role in shaping parasitoid host preferences and reproductive strategies.

A key physiological feature which influences parasitoid reproductive strategies is host quality (see Chapters 2 and 3). Certain host stages will be more suitable than others for a given parasitoid with parasitized hosts or hosts which deviate in size or age from the optimum providing less or a lower quality resource. Reduced quality often causes a decline in progeny survival or adult size, which in turn affects other fitness measures such as fecundity, longevity, or mobility (Charnov *et al.*, 1981; Waage and Godfray, 1985). Similarly, there can be Allee effects due to the inability of some endoparasitoids to contend physiologically with excess resource (Strand and Vinson, 1985). Sex ratio shifts may also be expected (Charnov, 1979; Charnov *et al.*, 1981; Waage, 1982) since it is possible that the fitness of one sex (usually female) is more adversely affected by host size than the other. For host species which do not vary much in size (eggs and pupae) one might expect similar clutch and sex ratio shifts to occur with changes in host age as the amount and/or accessibility of food available for parasitoid progeny declines. Such reductions have been observed for *Trichogramma minutum* (Pak and Oatman, 1982), and *Trichogramma pretiosum* (Strand and Vinson, 1984), while increases in sex ratio have been observed in *Anagrus incarnatus* Haliday (Chantarasa-ard and Hirashima, 1984) and *Telenomus heliothidis* (Strand, 1985). How a parasitoid might detect changes in age is uncertain. Changes in host shape and surface area with age may be one method by which parasitoids could make progeny adjustments (Strand and Vinson, 1983a; Schmidt and Smith, 1985), but detection of internal changes in the host may also be important.

Obviously, quality differences exist between parasitized and unparasitized hosts, yet under certain circumstances, numerous species readily superparasitize. Whether a parasitoid should superparasitize depends on the value of a parasitized host as a resource. For most parasitoids the value of the parasitized host will depend on the time spent between ovipositions by the first and second female and for gregarious species, the clutch size of the first female, since both parameters influence within-brood mortality. Recent optimal diet models (Iwasa *et al.*, 1984; Charnov and Skinner, 1984) predict that superparasitism will become more frequent when rates of host encounter are low or when rates of encounter with parasitized hosts are high. Several studies empirically support these predictions (Jackson, 1958; Chacko, 1969; van Lenteren, 1976; Bosque and Rabinovich, 1979; Werren, 1980), but a still largely unanswered question is how the parasitoid assesses changes in host quality.

If parasitoid markers in some way reflected the status of a host, then marking would be a plausible mechanism by which quality differences could be detected. In the light of earlier discussion, the external/internal marking systems of some telenomines and trichogrammatids seem well adapted for assessing host quality after parasitism. Consider, for example, marking by telenomines such as *Telenomus heliothidis*. As solitary parasitoids only one individual emerges per host, and, at least for *T. heliothidis*, host quality declines precipitously after parasitism. Superparasitism is restricted to females which have not oviposited previously or which have not oviposited for long periods, and occurs only when the external mark is present and the internal mark absent. While the external mark is present, a superparasitizing female's progeny have a small probability of surviving, but by the time the internal mark manifests itself, the progeny have no chance. The rather strong aversion to superparasitism shown by *T. heliothidis* may simply reflect the selective forces which work against the survival of superparasitic offspring of solitary species.

Contrast this with the situation seen for gregarious species such as *Trichogramma pretiosum*. It also possesses an external/internal marking system, but host quality does not decline as rapidly with time. The onset of host necrosis associated with the *Trichogramma pretiosum* internal mark is more gradual than for *Telenomus heliothidis*, and *Trichogramma pretiosum* responds differently. Probabilities for survival of a superparasitizing female's offspring are greater for the gregarious parasitoid since the larvae do not fight, physiological suppression usually does not occur, and the number of parasitoids a host can support is somewhat variable. Furthermore, depending on the number of hosts available it might be advantageous for a gregarious female to superparasitize herself in order to adjust brood size, provided the time between ovipositions does not give the first clutch too much of a developmental advantage. Both *Trichogramma evanescens* Westwood and *Trichogramma embryophagum* have been observed to oviposit in a host, resume searching, and if no other hosts are found, return and oviposit again before leaving (Klomp *et al*., 1980; Waage and Godfray, 1985; see Chapter 3). For both species the external mark is present and the internal mark absent immediately after oviposition, possibly indicating to the female that the host was only recently parasitized, most likely by herself. Thus, both external and internal marking could provide parasitoids with the information necessary to adjust clutch size by indicating for how long the host has been parasitized.

To say that telenomines are better discriminators than trichogrammatids is misleading in that the decision to superparasitize is based on potential fitness gains for the superparasitizing female. It is reasonable to suspect that species with low brood mortality will produce larger clutches (i.e. be gregarious; Waage and Godfray, 1985) and at the same time be more prone to superparasitize. Thus, although the marking systems of *Trichogramma pretiosum* and *Telenomus heliothidis* are very similar, their very different behavioural responses are in accord with such predictions.

Currently, there is only a rudimentary understanding of the physiological interactions between parasitoids and hosts. This lack of knowledge is especially evident for the egg parasitoids discussed in this chapter in that they are probably the poorest subjects for physiological investigations due to their small size. Yet, the purpose of this discussion was not so much to describe the variety of mechanisms involved in parasitoid physiology but to illustrate the integration which exists between many of the components involved in host selection. Several factors involved in ensuring successful development of a parasitoid in a given host also play a role in other aspects of parasitoid biology. Combining physiological data with other information on parasitoid reproductive strategies can help to understand parasitoid host preferences in the field as well as improve rearing efficiency in the laboratory.

Acknowledgements

I would like to thank Prof. S. B. Vinson for support during my graduate tenure. In addition I would like to thank Drs H. Davies and J. K. Waage and Mr B. Dover for their comments and ideas.

REFERENCES

Ables, J. R., Vinson, S. B., and Ellis, J. S. (1981). Host discrimination by *Chelonus insularis* (Hym: Braconidae), *Telenomus heliothidis* (Hym: Scelionidae), and *Trichogramma pretiosum* (Hym: Trichogrammatidae). *Entomophaga* **26,** 149–156.

van Alphen, J. J. M. (1980). Aspects of the foraging behaviour of *Tetrastichus asparagi* Crawford and *Tetrastichus* spec. (Eulophidae), gregarious egg parasitoids of the asparagus beetles *Crioceris asparagi* L. and *C. duodecimpunctata* L. (Chrysomelidae). *Netherlands Journal of Zoology* **30,** 307–325.

Anderson, D. T. (1972). The development of holometabolous insects. *In* "Developmental

Systems: Insects" Vol. 1 (S. J. Counce and C. H. W. Waddington, eds.), pp. 165–242. Academic Press, London.

Askew, L. L. (1971). "Parasitic Insects." Elsevier, New York.

Balduf, W. U. (1926). *Telenomus cosmopeplae* Gahan, an egg parasite of *Cosmopepla bimaculata* Thomas. *Journal of Economic Entomology* **19,** 829–841.

Beard, R. L. (1978). Venoms of Braconidae. *In* "Handbuch der experimentellen Pharmakologie, Volume 48 (Arthropod Venoms)" (S. Bettini, ed.), pp. 773–800. Springer-Verlag, Berlin.

Beckage, N. E. (1985). Endocrine interactions between endoparasitic insects and their hosts. *Annual Review of Entomology* **30,** 371–413.

Beckage, N. E., and Riddiford, L. M. (1983). Lepidopteran anti-juvenile hormones: effects on development of *Apanteles congregatus* in *Manduca sexta. Journal of Insect Physiology* **29,** 633–637.

Bergot, B. J., Baker, F. C., Cerf, G., Janieson, G., and Schooley, D. A. (1981). Qualitative and quantitative aspects of juvenile hormone titres in developing embryos of several insect species: discovery of a new JH-like substance extracted from eggs of *Manduca sexta. In* "Juvenile Hormone Biochemistry" (G. H. Pratt and G. T. Brookes, eds.), pp. 33–45. Elsevier, New York.

Bin, F., and Johnson, N. F. (1982). Potential of Telenominae in biocontrol with egg parasitoids (Hym., Scelionidae). *In* "Les Trichogrammes, Antibes (France)," pp. 275–287. Institut National de la Recherche Agronomique, Paris.

Bosque, C., and Rabinovich, J. E. (1979). Population dynamics of *Telenomus fariai* (Hymenoptera: Scelionidae), a parasite of Chagas' disease vectors. VII. Oviposition behavior and host discrimination. *Canadian Entomologist* **111,** 171–180.

Boucek, Z., and Askew, R. R. (1968). "Index of Entomophagous Insects. Palearctic Eulophidae (excl. Tetrastichinae) (Hym. Chalcidoidae)." Le François, New York.

Brown, M. W. (1984). Literature review of *Ooencyrtus kuvanae* (Hym.: Encyrtidae), an egg parasite of *Lymantria dispar* (Lep.: Lemantriidae). *Entomophaga* **29,** 249–265.

Brues, C. T. (1971). Notes on the adult habits of some hymenopterous egg parasites of Orthoptera and Mantoidea. *Psyche* **24,** 195–196.

Buleza, V. B., and Mikheev, A. V. (1979). On the interactions of *Trissolcus grandis* and *T. simoni*, egg parasites of *Eurygaster integriceps. Zoologicheskii Zhurnal* **58,** 54–60.

Burks, B. D. (1979). Trichogrammatidae. *In* "Catalog of Hymenoptera in America North of Mexico", pp. 1033–1042. Smithsonian Institution Press, Washington D.C.

Chacko, M. J. (1969). The phenomenon of superparasitism in *Trichogramma evanscens minutum* Riley. I. *Beiträge zur Entomologie* **19,** 617–635.

Channa Basavanna, G. P. (1953). Phoresy exhibited by *Lepidoscelio viatrix* Brues (Scelionidae, Hymenoptera). *Indian Journal of Entomology* **15,** 264–266.

Chantarasa-ard, S., and Hirashima, Y. (1984). Host range and host suitability of *Anagrus incarnatus* Haliday (Hymenoptera: Mymaridae), an egg parasitoid of delphacid planthoppers. *Applied Entomological Zoology* **19,** 491–497.

Charnov, E. L. (1979). The genetical evolution of patterns of sexuality: Darwinian fitness. *American Naturalist* **113,** 465–480.

Charnov, E. L., and Skinner, S. W. (1984). Evolution of host selection and clutch size in parasitoid wasps. *Florida Entomologist* **67,** 5–21.

Charnov, E. L., Los-den Hartogh, R. L., Jones, W. T., and van den Assem, J. (1981). Sex ratio evolution in a variable environment. *Nature* **289,** 27–33.

Clausen, C. P. (1940). "Entomophagous Insects." Hafner, New York.

Clausen, C. P. (1976). Phoresy among entomophagous insects. *Annual Review of Entomology* **21,** 343–368.

Davies, D., Strand, M. R., and Vinson, S. B. (in press). Changes in differential haemocyte count and *in vitro* behaviour of plasmatocytes from host *Heliothis virescens* caused by *Campoletis sonorensis* polydna virus. *Journal of Insect Physiology.*

Davis, C. P., and Burbutis, P. P. (1974). The effect of age selective rearing on the biological quality of females of *Trichogramma nubilale. Annals of the Entomological Society of America* **67,** 765–766.

Dikkeboom, R., van den Knaap, Wil, P. W., Meuleman, E. A., and Sminia, T. (1984). Differences between blood cells of juvenile and adult specimens of the pond snail *Lymnaea stagnalis. Cell and Tissue Research* **238,** 43–47.

Downer, G. H., Laufer, H. (1983). "Endocrinology of Insects." Alan R. Liss, New York.

Dunn, P. E. (1986). Biochemical aspects of insect immunology. *Annual Review of Entomology* **31,** 321–340.

Edson, K. M., Vinson, S. B., Stoltz, D. B., and Summers, M. D. (1981). Virus in a parasitoid wasp: suppression of the cellular immune response in the parasitoid's host. *Science* **211,** 582–583.

Fedde, G. F. (1977). Laboratory study of egg parasitization capabilities of *Telenomus alsophilae. Environmental Entomology* **6,** 773–776.

Feeny, P. (1976). Plant apparency and chemical defense. *Recent Advances in Phytochemistry* **10,** 1–40.

Fisher, R. C. (1971). Aspects of the physiology of endoparasitic Hymenoptera. *Biological Reviews* **46,** 243–278.

Fleming, J. G. W., Blissard, G. W., Summers, M. D., and Vinson, S. B. (1983). Expression of *Campoletis sonorensis* virus in the parasitized host, *Heliothis virescens. Journal of Virology* **48,** 74–78.

Force, D. C. (1974). Ecology of insect host-parasitoid communities. *Science* **184,** 624–632.

Gatenby, J. B. (1917). The embryonic development of *Trichogramma evanescens*, Westw., monembryonic egg of *Donacia simplex*, Fab. *Quarterly Journal of Microscopic Science* **62,** 149–177.

Gerling, D. (1972). The developmental biology of *Telenomus remus* Nixon (Hym.: Scelionidae). *Bulletin of Entomological Research* **61,** 385–388.

Gerling, D., and Orion, T. (1973). The giant cells produced by *Telenomus remus. Journal of Invertebrate Pathology* **21,** 164–171.

Gharib, B., and de Reggi, M. (1983). Changes in ecdysteroid and juvenile hormone levels in developing eggs of *Bombyx mori. Journal of Insect Physiology* **29,** 871–876.

Gharib, B., de Reggi, M., Connat, J., and Chaix, J. (1983). Ecdysteroid and juvenile hormone changes in *Bombyx mori* eggs, related to the initiation of diapause. *Federation of European Biochemical Societies* **160,** 119–123.

Hafez, M., El-Kifl, A. H., and Fayad, Y. H. (1977). On the bionomics of *Platytelenomus hylas* Nixon, an egg parasite of *Sesamia cretica* Led. in Egypt. *Bulletin of the Entomological Society of Egypt* **61,** 161–178.

Hamilton, W. D. (1968). Extraordinary sex ratios. *Science* **156,** 477–488.

Hamilton, W. D. (1979). Wingless and fighting males in fig wasps and other insects. *In* "Reproductive Competition and Sexual Selection in Insects" (M. S. Blum and N. A. Blum, eds.), pp. 167–220. Academic Press, New York.

Hawlitzky, N., and Boulay, C. (1982). Régimes alimentaires et développement chez *Trichogramma maidis* Pintureau et Voegele (Hym. Trichogrammatidae) dans l'oeuf d'*Anagasta keuhniella* Zeller (Lep. Pyralidae). *In* "Les Trichogrammes, Antibes (France)", pp. 100–106. Institut National de la Recherche Agronomique, Paris.

Hennig (1981). "Insect Phylogeny" (A. C. Pont, ed.). John Wiley, New York.

Henriksen (1922). Notes upon some aquatic Hymenoptera. *Annales de Biologie Lacustre* **11,** 19–37.

Hidaka, T. (1958). Biological investigation on *Telenomus gifuensis* Ashmead (Hym.: Scelionidae), an egg parasite of *Scotinophora lurida* Burmeister (Hem.: Pentatomidae) in Japan. *Acta Hymenopterologica* **1,** 75–93.

Hinton, H. E. (1981). "Biology of Insect Eggs." Pergamon, Oxford.

Hokyo, N., Shiga, M., and Nakashji, F. (1966). The effect of intra- and interspecific conditioning of host eggs on the ovipositional behavior of two scelionid egg parasites of the southern green stink bug, *Nezara viridula* L. *Japanese Journal of Ecology* **16,** 67–71.

Holldobler, B. (1983). Evolution of insect communication. *In* "Insect Communication" (T. Lewis, ed.), pp. 349–379. Academic Press, London.

Iwasa, Y., Suzuki, Y., and Matsuda, H. (1984). Theory of oviposition of strategy of parasitoids. I. Effect of mortality and limited egg number. *Theoretical Population Biology* **26,** 205–227.

Jackson, D. (1958). Observations on the biology of *Caraphractus cinctus* Walker (Hymenoptera: Mymaridae), a parasitoid of the eggs of Dytiscidae. I. Methods of rearing and numbers bred on different hosts' eggs. *Transactions of the Royal Entomological Society of London* **110,** 533–554.

Jackson, D. (1963). Diapause in *Caraphractus cinctus* Walker (Hymenoptera: Mymaridae), a parasitoid of the eggs of Dytiscidae (Coleoptera). *Parasitology* **53,** 225–251.

Janzen, D. H. (1975). Interactions of seeds and their insect predators/parasitoids in a tropical deciduous forest. *In* "Evolutionary Strategies of Parasitic Insects and Mites" (P. W. Price, ed.), pp. 154–186. Plenum, New York.

Jennings, D. T., and Houseweart, M. W. (1983). Parasitism of spruce budworm (Lepidoptera: Tortricidae) eggs by *Trichogramma minutum* and absence of overwintering parasitoids. *Environmental Entomology* **12,** 535–540.

Johnson, N. F. (1984). Systematics of nearctic *Telenomus*: classification and revisions of the *Podisi* and *Phymatae* species groups (Hymenoptera: Scelionidae). *Bulletin of the Ohio Biological Survey, Knull Series Volume 6.*

Jones, D., Jones, G., and Hammock, B. D. (1981). Developmental and behavioral responses of larval *Trichoplusia ni* to parasitization by an imported braconid parasite *Chelonus* sp. *Physiological Entomology* **6,** 387–394.

Jones, D., Jones, G. Van Steenwyk, R. A., and Hammock, B. D. (1982). Effect of the parasite *Copidosoma truncatellum* on development of its host *Trichoplusia ni. Annals of the Entomological Society of America* **75,** 7–11.

Jones, E. P. (1937). The egg parasites of the cotton boll worm, *Heliothis armigera*, Hubn. (*obsoleta*, Fabr.), in southern Rhodesia. *Publications of the British South Africa Company* **6,** 37–105.

Jubb, G. L. Jr., and Watson, T. F. (1971). Development of the egg parasite *Telenomus utahensis* in two pentatomid hosts in relation to temperature and host age. *Annals of the Entomological Society of America* **64,** 202–205.

Klomp, H., and Teerink, B. J. (1962). Host selection and number of eggs per oviposition in the egg parasite *Trichogramma embryophagum* Htg. *Nature* **195,** 1020–1021.

Klomp, H., and Teerink, B. J. (1978). The elimination of supernumerary larvae of the gregarious egg-parasitoid *Trichogramma embryophagum* (Hym.: Trichogrammatidae) in eggs of the host *Ephestia kuehniella* (Lep.: Pyralidae). *Entomophaga* **23,** 153–159.

Klomp, H., Teerink, B. J., and Wei Chun Ma (1980). Discrimination between parasitized and unparasitized hosts in the egg parasite *Trichogramma embryophagum* (Hym.: Trichogrammatidae): a matter of learning and forgetting. *Netherlands Journal of Zoology* **30,** 254–277.

Krebs, J. R., and Davies, N. B. (1978). "Behavioural Ecology." Blackwell Scientific Publications, London.

Kurosumi, K., and Fujita, H. (1974). "An Atlas of Electron Micrographs: Functional Morphology of Endocrine Glands." Igaku Shoin, Tokyo.

Lagueux, M., Sall, C., and Hoffman, J. A. (1981). Ecdysteroids during embryogenesis in *Locusta migratoria*. *American Zoologist* **21,** 715–726.

Lawerence, P. O. (1986). Host-parasite interactions: an overview. *Journal of Insect Physiology* **32,** 295–298.

Leibee, C. L., Pass, B. C., and Yeargan, K. V. (1979). Developmental rates of *Patasson lameerei* (Hym.: Mymaridae) and the effect of host egg age on parasitism. *Entomophaga* **24,** 345–348.

van Lenteren, J. C. (1976). The development of host discrimination and the prevention of superparasitism in the parasitic wasp *Pseudeucoila bochei* Weld. (Hym.: Cynipidae). *Netherlands Journal of Zoology* **26,** 1–83.

van Lenteren, J. C., and Pak, G. (1984). Behavioral variations between *Trichogramma* spp. strains, a technique for candidate-strain evaluation. *Proceedings of the 17th International Congress of Entomology, Hamburg* (abstract) 800.

Lewis, W. J., and Redlinger, L. M. (1969). Suitability of eggs of the almond moth, *Cadra cautella*, of various ages of parasitism by *Trichogramma evanescens*. *Annals of the Entomological Society of America* **62,** 1482–1485.

Lewis, W. J., Jones, R. L., and Sparks, A. N. (1972). A host-seeking stimulant for the egg parasite, *Trichogramma evanescens*. Its source and demonstration of its laboratory and field activity. *Annals of the Entomological Society of America* **65,** 1087–1089.

Lynn, D. C., and Vinson, S. B. (1977). Effects of temperature, host age, and hormones upon the encapsulation of *Cardiochiles nigriceps* eggs by *Heliothis* spp. *Journal of Invertebrate Pathology* **29,** 50–55.

Mackauer, M. (1959). Histologische Untersuchungen an parasitierten Blattenlausen. *Zeitschrift für Parasitenkunde* **19,** 322–352.

Marchal, P. (1936). Recherches sur la biologie et de développement des Hymenoptères: les Trichogrammes. *Annales des Épiphyties* **22,** 447–550.

Marston, N., and Ertle, L. R. (1969). Host age and parasitism by *Trichogramma minutum* (Hymenoptera: Trichogrammatidae). *Annals of the Entomological Society of America* **62,** 1476–1481.

Maynard Smith, J., Burian, R., Kauffman, S., Alberch, P. Campbell, J., Goodwin, B., Lande, R., Raup, D., and Wolpert, L. (1985). Developmental constraints and evolution. *Quarterly Review of Biology* **60,** 265–287.

Mayr, E. (1982). "The Growth of Biological Thought." Harvard University Press, Cambridge, Massachusetts.

Mizuno, T., Watanabe, K., and Ohnishi, E. (1981). Developmental changes of ecdysteroids in the eggs of the silkworm, *Bombyx mori*. *Insect Biochemistry* **2,** 155–159.

Nappi, A. J. (1975). Parasite encapsulation in insects. *In* "Invertebrate Immunity" (K. Maramorosch and R. E. Shope, eds.) pp. 293–326. Academic Press, New York.

Noldus, L. P. J. J., and van Lenteren, J. C. (1985). Kairomones for the egg parasite *Trichogramma evanescens* Westwood: I. Effect of volatile substances released by two of its hosts, *Pieris brassicae* L. *Journal of Chemical Ecology* **11,** 781–792.

Pak, G. A., and Oatman, E. R. (1982). Biology of *Trichogramma brevicapillum*. *Entomologia Experimentalis et Applicata* **32,** 61–67.

Parker, F. D., and Pinnell, R. E. (1971). Overwintering of some *Trichogramma* spp. in Missouri. *Journal of Economic Entomology* **64,** 80–81.

Pemberton, C. E. (1933). Introduction to Hawaii of Malayan parasites (Scelionidae) of the Chinese grasshopper *Oxya chinensis* (Thun.) with life history notes. *Proceedings of the Hawaiian Entomological Society* **8,** 253–264.

Peterson, A. (1930). A biological study of *Trichogramma minutum* Riley as an egg parasite of the oriental fruit moth. *United States Department of Agriculture Technical Bulletin* **215.**

Pickford, R. (1964). Life history and behaviour of *Scelio calopteni* Riley (Hymenoptera: Scelionidae), a parasite of grasshopper eggs. *Canadian Entomologist* **96,** 1167–1172.

Piek, T., and Owen, M. D. (1982). "Hymenoptera Venom Systems." Academic Press, London.

Piek, T., Spanjer, W., Njio, K. D., Veenendaal, R. L., and Mantel, P. (1974). Paralysis caused by the venom of the wasp, *Microbracon gelechiae. Journal of Insect Physiology* **20,** 2307–2319.

Pinto, J. D., Platner, G. R., and Oatman, E. R. (1978). Clarification of the identity of several common species of North American *Trichogramma* (Hymenoptera: Trichogrammatidae). *Annals of the Entomological Society of America* **71,** 169–180.

Price, P. W. (1980). "Evolutionary Biology of Parasites." Princeton University Press, Princeton, New Jersey.

Rabb, R. L., and Bradley, J. R. (1970). Marking host eggs by *Telenomus sphingis. Annals of the Entomological Society of America* **63,** 1053–1056.

Rabinovich, J. E. (1970). Population dynamics of *Telenomus fariai* (Hymenoptera: Scelionidae), a parasite of Chagas' disease vectors. II. Effect of host-egg age. *Journal of Medical Entomology* **7,** 477–481.

Ratcliffe, N. A., and Rowley, A. F. (1979). Role of hemocytes in defense against biological agents. *In* "Insect Hemocytes" (A. P. Gupta, ed.), pp. 331–341. Cambridge University Press, Cambridge.

Rizki, R. M., and Rizki, T. M. (1984). Selective destruction of a host blood cell type by a parasitoid wasp. *Proceedings of the National Academy of Science USA* **81,** 6154–6158.

Safavi, M. (1968). Etude biologique et écologique des Hymenoptères parasites des oeufs punaises des céréales. *Entomophaga* **13,** 381–495.

Sahad, K. A. (1982). Biology and morphology of *Gonatocerus* sp. (Hymenoptera, Mymaridae), an egg parasitoid of the green rice leafhopper, *Nephotettix cincticeps* Uhler (Homoptera, Deltocephalidae) I. Biology. *Kontyu, Tokyo* **50,** 246–260.

Sahad, K. A. (1984). Biology of *Anagrus optabilis* (Perkins) (Hymenoptera, Mymaridae), an egg parasitoid of delphacid planthoppers. *Esakia* **22,** 129–144.

Salt, G. (1935). Experimental studies in insect parasitism. III. Host selection. *Proceedings of the Royal Entomological Society of London* **117,** 414–435.

Salt, G. (1940). Experimental studies in insect parasitism. VII. The effects of different hosts on the parasite *Trichogramma evanescens* Westwood. *Proceedings of the Royal Entomological Society of London* **15A,** 81–95.

Salt, G. (1961). Competition among insect parasitoids. *Symposium of the Society of Experimental Biology* **15,** 96–119.

Salt, G. (1968). The resistance of insect parasitoids to the defence reactions of their hosts. *Biological Reviews* **43,** 200–23.

Salt, G. (1970). "The Cellular Defence Reaction of Insects." Cambridge University Press, Cambridge.

Salt, G. (1973). Experimental studies in insect parasitism. XVI. The mechanism of the resistance of *Nemeritis* to defence reactions. *Proceedings of the Royal Society of London* **183,** 337–350.

Scalia, S., and Morgan, E. D. (1982). A reinvestigation on the ecdysteroids during embryogenesis in the desert locust *Schistocerca gregaria. Journal of Insect Physiology* **28,** 647–654.

Schell, S. C. (1943). The biology of *Hadronotus ajax* Girault (Hymenoptera-Scelionidae), a parasite in the eggs of squash-bug (*Anasa tristis* DeGreer). *Annals of the Entomological Society of America* **36,** 625–635.

Schmidt, J. M., and Smith, J. J. B. (1985). Host volume measurement by the parasitoid wasp *Trichogramma munutum*: the role of curvature and surface area. *Entomologia Experimentalis et Applicata* **39,** 213–221.

Sminia, T. (1981). Gastropods. *In* "Invertebrate Blood Cells" (N. A. Ratcliffe and A. F. Rowley, eds.), pp. 191–232. Academic Press, London.

Spanjer, W., Grosu, L., and Piek, T. (1977). Two different paralyzing preparations obtained from a homogenate of the wasp *Microbracon hebetor* (Say). *Toxicon* **15,** 413–421.

Spencer, H. (1926). Biology of parasites and hyperparasites of aphids. *Annals of the Entomological Society of America* **19,** 119–157.

Stoltz, D. B., and Guzo, D. (1986). Apparent haemocytic transformations associated with parasitoid-induced inhibition of immunity in *Malacosoma disstria* larvae. *Journal of Insect Physiology* **32,** 377–388.

Stoltz, D. B., and Vinson, S. B. (1979). Viruses and parasitism in insects. *Advances in Virus Research* **24,** 125–170.

Stoner, A., and Surber, D. E. (1971). Notes on the biology and rearing *Anaphes ovijentatus*, a new parasite of *Lyuus hesperus* in Arizona. *Journal of Economic Entomology* **64,** 501–502.

Strand, M. R., and Vinson, S. B. (1982). Source and characterization of an egg recognition kairomone of *Telenomus helithidis*, a parasitoid of *Heliothis virescens*. *Physiological Entomology* **7,** 83–90.

Strand, M. R., and Vinson, S. B. (1983a). Factors affecting host recognition and acceptance in the egg parasitoid *Telenomus heliothidis* (Hymenoptera: Scelionidae). *Environmental Entomology* **12,** 1114–1119.

Strand, M. R., and Vinson, S. B. (1983b). Analysis of an egg recognition kairomone of *Telenomus heliothidis* (Hymenoptera: Scelionidae). *Journal of Chemical Ecology* **9,** 423–432.

Strand, M. R., and Vinson, S. B. (1984). Facultative hyperparasitism by the egg parasitoid *Trichogramma pretiosum* (Hymenoptera: Trichogrammatidae). *Annals of the Entomological Society of America* **77,** 679–686.

Strand, M. R., and Vinson, S. B. (1985). *In vitro* culture of *Trichogramma pretiosum* on an artificial medium. *Entomologia Experimentalis et Applicata* **39,** 203–209.

Strand, M. R., Ratner, S., and Vinson, S. B. (1983). Maternally induced host regulation by the egg parasitoid *Telenomus heliothidis*. *Physiological Entomology* **8,** 469–475.

Strand, M. R., Quarles, J. M., Meola, S. M., and Vinson, S. B. (1985). Cultivation of teratocytes of the egg parasitoid *Telenomus heliothidis* (Hymenoptera: Scelionidae). *In Vitro Cellular and Developmental Biology* **21,** 361–367.

Strand, M. R., Meola, S. M., and Vinson, S. B. (1986). Correlating pathological symptoms in *Heliothis virescens* eggs with development of the parasitoid *Telenomus heliothidis*. *Journal of Insect Physiology* **32,** 389–402.

Strong, D. R., Lawton, J. H., and Southwood, T. R. E. (1984). "Insects on Plants Community Patterns and Mechanisms." Harvard University Press, Cambridge, Massachusetts.

Suzuki, Y., Tsuji, H., and Sasakawa, M. (1984). Sex allocation and effects of superparasitism on secondary sex ratios in the gregarious parasitoid, *Trichogramma chilonis* (Hymenoptera: Trichogrammatidae). *Animal Behaviour* **32,** 478–484.

Thompson, J. N. (1982). "Interaction and Coevolution". John Wiley, New York.

Thompson, S. N. (1983). Biochemical and physiological effects of metazoan endoparasites on their host species. *Comparative Biochemistry and Physiology* **74,** 183–211.

Thompson, W. R., and Parker, H. J. (1930). Morphology and biology of *Eulimneria crassifemur*, an important parasite of the corn-borer. *Journal of Agricultural Research* **40,** 321–345.

Tothill, D. G., Taylor, T. H. C., and Paine, R. W. (1930). "The Coconut Moth in Fiji." Imperial Bureau of Entomology, London.

Townes, H. (1969). The genera of Ichneumonidae. Part I. *Memoirs of the American Entomological Institute* **11,** 1–300.

Townes, H., and Townes, M. (1951). Family Ichneumonidae. *In* "Hymenoptera of America North of Mexico" (C. F. W. Muesebeck, K. V. Krombein, and H. K. Townes, eds.), pp. 184–409. U.S. Department of Agriculture Monograph 2.

Vet, L. E. M., Meyer, M., Bakker, K., and van Alphen, J. J. M. (1984). Intra- and interspecific host discrimination in *Asobara* (Hymenoptera) larval endoparasitoids of Drosophilidae: comparison between closely related and less closely related species. *Animal Behaviour* **32,** 871–874.

Vinokurov, G. M. (1927). Grasshoppers and areas of their outbreaks in eastern Siberia. *Bulletin of the Irkutsk Plant Protection Station* **1,** 3–52.

Vinson, S. B., and Iwantsch, G. F. (1980b). Host regulation by insect parasitoids. *Quarterly Review of Biology* **55,** 143–165.

Voegele, M. (1970). "Les Aélia du Maroc et Leurs Parasites Oophages." Thése, Faculté des Sciences D'Orsay.

Voegele, M., Brun, P., and Daumal, J. (1974). Modalités de la prise de possession et de l'élimination de d'hôte chez le parasite embryonnaire *Trichogramma brasiliensois*. *Annals de la Societé Entomologique de France* **10,** 757–762.

Waage, J. K. (1982). Sib-mating and sex ratio strategies in scelionid wasps. *Ecological Entomology* **7,** 103–112.

Waage, J. K., and Godfray, H. C. J. (1985). Reproductive strategies and population ecology of insect parasitoids. *In* "Behavioural Ecology, Ecological Consequences of Adaptive Behaviour" (R. M. Sibly and R. H. Smith, eds.), pp. 449–470. Blackwell Scientific Publications, London.

Webb, B. A., and Dahlman, D. L. (in press). Ecdysteroid influence on the development of the *Heliothis virescens* and its endoparasite *Microplitis croceipes*. *Journal of Insect Physiology*.

Werren, J. H. (1980). Sex ratio adaptation to local mate competition in a parasitic wasp. *Science* **208,** 1157–1159.

Wood, R. K. S., and Graniti, A. (1976). "Specificity in Plant Diseases." Plenum, London.

5

Mating Behaviour in Parasitic Wasps

J. VAN DEN ASSEM

I. INTRODUCTION

This paper is intended to be an ethologist's contribution. Ethologists are interested in the role of behaviour in animal life: the role of mating behaviour is the topic of interest here.

We observe that on a behavioural level animals are adapted to a particular niche, in which individuals have to maintain themselves and attempt to reproduce profitably. "Adapted" means that new generations need not start from scratch; rather, they carry information about the environment in their genes, and, in consequence, have at their disposal more or less adequate hardware—their morphological and physiological properties—and corresponding software—their behavioural programmes. Such programmes are expected to maximize reproductive success, either directly or indirectly. The programmes are run at appropriate times because a second source of information about the environment is available—sensory input. Keeping this in mind, the study of behaviour boils down to the study of information processing and its subsequent output, the observed overt behaviour.

There are three aspects to behaviour which are inseparable: (1) its proximate causation; (2) its effect, function and survival value and (3) its origin and long-term changes during evolution (Tinbergen, 1963). These aspects will be discussed with reference to mating behaviour, emphasizing problems of causality: how sequences of mating get started, how they come to an end, and what happens in between. I do not discuss at length the functional aspects and reproductive strategies because extensive reviews have been published recently (Krebs and Davies, 1978, especially the second part; Alcock *et al.*, 1978; Thornhill and Alcock, 1983). I have very little to add. Instead, in the last part of this paper I speculate about some questions on the evolution of courtship in parasitic wasps.

Few genuinely ethological studies of mating behaviour in parasitic wasps have been published so far. To be sure, many descriptions of mating procedures are scattered in the literature, but most are fragmentary and anecdotal (review by Gordh and DeBach, 1978). Quantitative work is rare; apart from our own group only Barrass (1960a,b, 1961, 1976), and Gordh and DeBach (1976, 1978) have made extensive use of quantitative data. (In this respect the situation in several other groups of insects is radically different. To mention just a few: ethology and communication of social bees and ants, Hölldobler, 1978, 1984; ethology and genetics of behaviour of flies (*Drosophila* spp.), Hall, 1979; Hall *et al.*, 1980; and the neurophysiological substrates of behaviour of grasshoppers and crickets, Loher and Huber, 1966; Riede, 1983.)

Mating behaviour (i.e. courtship and copulation) is obviously an aspect of sexual reproduction (Chapman, 1983). In parasitic wasps it implies separate reproductive strategies for males and females, producers of sperm and eggs, respectively. The quintessence of sexual reproduction is the recombination of genetic material in the offspring. In general, fertilization (fusion of a male and a female gamete) is required for production of a viable offspring. In parasitic wasps this applies to daughters only since sons result from unfertilized eggs. Insemination precedes fertilization: male products are introduced into a female's genital tract by means of sexual contact: copulation, and gametes are stored in a spermatheca until recruited for fertilization (King and Radcliffe, 1969). As a rule, potentially inseminable females do not simply allow copulation; rather, males have to induce readiness to copulate by means of some sort of effort. Usually, it takes the form of a display of some kind, called courtship (Bastock, 1967).

Of course, courtship and copulation are by no means restricted to parasitic wasps, and the nature of their performances is not a characteristic of their way of life. Frequently, however, due to the parasitic way of life, the mating pool is very limited in space and time. This has implications for the way in which contributors can maximize reproductive success. For males, it is of paramount importance to be present ahead of female emergence since inseminable females usually become available only for relatively short periods.

The mating pool may not just be limited by the number of participants: quite often only sibs take part. In such circumstances male sibs may compete for mates, which would impair the reproductive success of the foundress (the sibs' mother). Such deleterious effects can be minimized by moving the offspring sex ratio away from parity, making males the minority sex (Hamilton, 1967; see Chapter 3). In some cases they represent an extreme minority (Hamilton, 1979; Werren, 1980). Is the production of males still profitable in such circumstances? A mating pool with only one or two males will lead to a severe reduction of genetic variability in the offspring, especially in parasitic wasps, where a male's sperms are identical. Indeed, production of males may cease altogether: such thelytokous species are by no means rare. Where there are no males, there is no longer courtship or copulation. Yet, even under such conditions, all is not apparently lost. Thelytokous females may be manipulated in such a way that they lay haploid eggs from which viable males ensue (Legner, 1985). In one of our species (*Muscidifurax uniraptor* Kogan & Legner) such males exhibit a normal courtship repertoire, judging by that of related species. The necessary genetic coding must have been transferred and kept intact over many generations of females.

However, thelytoky does not seem to be a success (Maynard Smith, 1978). In parasitic wasps, gonochoristic species are the large majority and provide a sound basis for the comparative study of mating behaviour. The conclusion is that, biologically speaking, copulation seems to be an important occupation for both males and females. The next question is, do they enhance their reproductive prospects by copulating more than once?

Males and females seem to differ in this respect. Obviously, males may increase their reproductive success with every successful insemination. However, other males copulating with the same female on later occasions might interfere. It comes as no surprise that the ability to "switch off" females after copulation is widespread in insects. Of course, any male is free to introduce, while copulating, more substances than just gametes (Leopold, 1976) and several have been found to render females unresponsive to further courtship

(Craig, 1967). This will reduce the probability of another copulation, and hence, sperm competition.

In theory, the reproductive success of females may also increase with multiple copulations. This would be the case if they were able to assess and compare the genetic quality of potential partners, or if a copulation involved an energetic gain of some kind (Borgia, 1979; Thornhill and Alcock, 1983), or both. A male of high quality should be allowed to replenish a spermatheca or replace its contents with his own sperm. In parasitic wasps, sperm of a second male can be utilized in certain circumstances; complete replacement does not occur. However, it seems very improbable that females will grant a copulation because of direct quality assessment. They do not seem to have much to choose from in a mating pool with one or two males. They might refuse to mate, but the hazards of finding another more suitable partner elsewhere may be overwhelming. In other cases, where there is severe competition between males for inseminable females (Hamilton, 1979; van den Assem *et al.*, 1980a) it is probably correct to state that a male which is able to court at all must already have proven his male qualities in competition. Therefore, the female's ability to choose may be absent, and indeed, virgin females will usually copulate with the first conspecific male encountered.

Energetic gains associated with insemination are unknown in parasitic wasps. Females may profit from genes only. The conclusion is that unlike males, they do not gain from extra copulations, unless they exhaust their sperm supply (Cousin, 1933).

In general, male parasitic wasps are the more active sex in courtship. On encounter a male will manoeuvre in such a way that he comes to take up a stereotypical position relative to the female (Fig. 1), either on top of her or on the substrate. This accomplished, a display will start. In principle, displays consist of a repetition of fixed patterns of movements in which several limbs may participate. These repetitions occur continuously or at intervals, separated by short or long pauses (van den Assem, 1975). All displays have characteristic temporal structures.

The female's role in courtship is not as conspicuous. A female need not emit stimuli periodically, as if responding to the male, to keep him going, nor does she influence the characteristic form of his movements or the precise timing of the repetitions. Once initiated, his display proceeds in a highly stereotypical fashion, on a living partner just as well as on a dead female or on a dummy (at least in many species). Clearly, simple chain reactions cannot be involved. On the contrary, the temporal structure of male displays seems to result in the first place from endogenous processes. This is not to say that external stimuli, originating from the female, are not continuously perceived by the male. In fact, this is what he does and I will return to this point below (Section II. B, C).

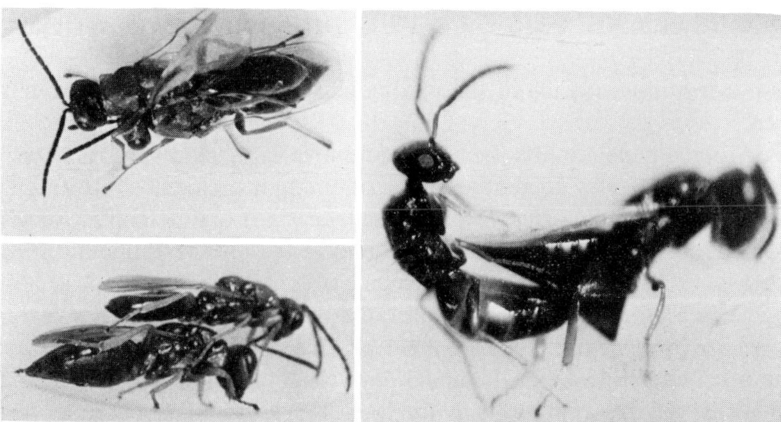

Fig. 1. Mating behaviour of parasitic wasps. Left: a courting couple of *Stenomalina liparae* (Giraud), Pteromalidae, from above (top), and from the side (bottom). Elements of courtship in this species include conspicuous movements with the antennae and head nodding. The male's fore tarsi are placed on the female's head. Right: copulating couple of *Muscidifurax zaraptor* Kogan & Legner, Pteromalidae. The female has raised her abdomen to expose the genital orifice, and has drawn the antennal flagellae to her head.

There is an immense variety of species of parasitic wasps: so much so that, with the present knowledge, generalizations can hardly be made. Instead, a few general questions must be asked which are applicable to almost any species. The answers will be derived from work on a single species: the pteromalid wasp *Nasonia vitripennis* (Walker), a gregarious parasitoid of the pupae of many flies, which has been used in a wide field of scientific endeavour (Whiting, 1967; Holmes, 1976). It is widespread and frequently found in nests of song birds, parasitizing *Protocalliphora* spp., flies whose maggots are parasitic on nestlings (Werren, 1983; Abraham, 1984). *Nasonia* males emerge ahead of the females; one of them positions himself near an escape hole from which (virgin) females will later emerge (King *et al.*, 1969). This male (the dominant) chases the others but they stay in the vicinity and may interfere repeatedly. Upon emergence, females are mounted immediately; they will copulate with the first male to court them. There is a short post-copulatory display. Next, the successful male will mark the substrate with the tip of his abdomen. Markings appear attractive to virgin females (and to other males) (van den Assem *et al.*, 1980b). To give an indication of the time scales involved, a complete mating sequence takes about 35 s (25°C): 10 s for courtship until readiness, 15 s for copulation, and about 10 s for the post-copulatory period.

II. CAUSAL ANALYSIS OF COURTSHIP AND COPULATION

First, a few more details of the mating routines of parasitic wasps will be given.

A *Nasonia* male mounts the female and invariably places his fore tarsi on her head. His display is composed of drumming movements (with fore tarsi on the female's eyes), wing vibrations, several kinds of antennal movements, nodding movements with the head, and extrusion of the mouth parts (Barrass, 1960a). The nodding movements (Fig. 2) have our special attention for several reasons: they can easily be quantified, they can be manipulated by the person carrying out the experiment, and they seem to reflect more clearly than other components the dynamics of internal processes which underly the display; I will return to this point later on. Head nods come in series; numbers per series vary but never in a random way. Nods follow each other closely, series are separated by longer pauses. The upward stroke of every nod coincides with mouth-part extrusion. First nods of every series are more elaborate and of a longer duration than the following ones. Each series except the first one is preceded by an antennal sweep.

A. Some Questions

An unmistakable function of male courtship is to induce readiness to copulate in a female. Only display by a conspecific partner will have this effect. Apparently a male produces species-characteristic stimuli with his performance. Which stimuli provoke sexual receptivity in the female, and how much of it is needed? Are they continuously produced during a display? At what point will receptivity occur? With the onset of female receptivity a male stops displaying and switches to copulatory behaviour. What stimuli

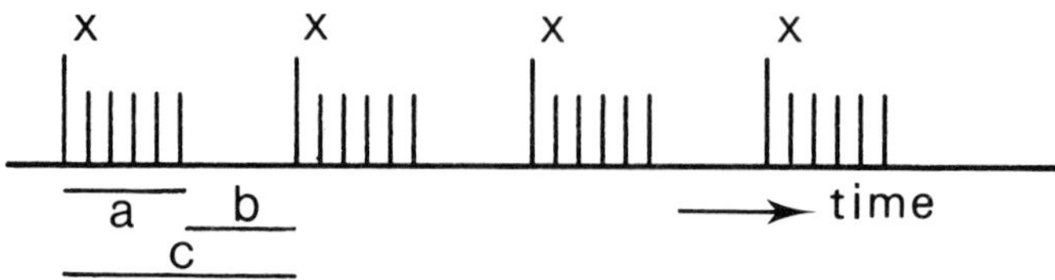

Fig. 2. Schematic representation of the succession of head noddings during a sequence of courtship of *Nasonia vitripennis* (Walker). Vertical lines represent single nods. Nods are clustered in time and occur in series (a). Single nods are separated by short pauses. Series are separated by longer pauses, called intervals (b). First nods of any series differ from the following ones: they are more elaborate and last longer. The period from first nod to first nod (c) is called a courtship cycle. Essentially, a display consists of a repetition of cycles. X marks the points in time where a virgin female may signal sexual receptivity.

are effective here and what underlying mechanism is responsible for this change? In the continuing absence of female readiness a male will give up at a certain point. Why are certain females unreceptive? What mechanism causes a male to dismount, and at what point? For both types of sequences (with and without following copulation) we can ask when a male will be able to court again, and what mechanisms are responsible for the allocation of time to be invested in these successive displays.

B. Courtship Stimuli in Males

1. External Stimuli Which Release Courtship

No male will court any object; certain specific stimuli are always required to get him started. Stimuli which release mounting, which guide him to the courtship position, which are necessary to start a display, which keep it going, and those which bring it to an end, can be investigated with dummies (our unpublished data; Yoshida and Hidaka, 1979) (Fig. 3). Most chalcidoids are small insects, *N. vitripennis* averages 2–3 mm in length, and such

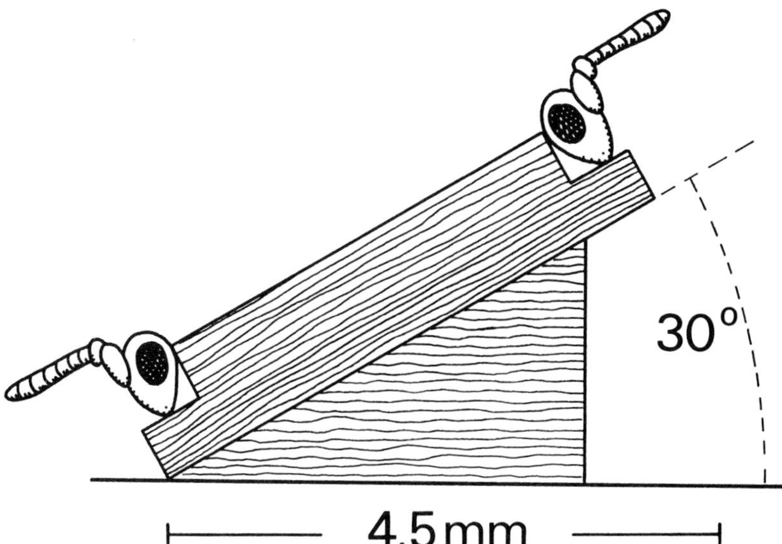

Fig. 3. Example of a dummy used to investigate the role of external stimuli in releasing courtship behaviour in males of *Nasonia vitripennis* (Walker), and guiding them to the proper courtship position. With dummies as illustrated we investigated a male's front/rear choices following mounting, and the role of gravity (see also Yoshida and Hidaka, 1979).

sizes put constraints on experimentation, but a bit of manual dexterity will go a long way. Experiments with dummies have made it clear that chemical, tactile and visual stimuli are involved in the release and continuation of courtship (van den Assem and Jachmann, 1982). Specific contact pheromones play a role in mounting behaviour (Obara and Kitano, 1974). Many males will court dead females but treatment with a solvent makes such females unattractive. Neutral objects, such as a small piece of plywood or filter paper can be made attractive by application of the dissolved substances (Tagawa and Hidaka, 1982; Takahashi and Sugai, 1982). Other cues play a role in becoming oriented in a proper way. The direction in which the wing is implanted seems to provide a male with such cues. The male's search for the correct courtship position stops when he perceives "something projecting from the main body", usually the female's antennae. On an antennaless female or dummy he will turn around and orient himself to the protruding tip of the ovipositor.

During courtship males seem to perceive chemical stimuli continuously: they will, for example, court on a dummy composed of a female abdomen and a male head, but dismounting comes earlier than with an all-female dummy. There are specific differences in the nature of the substances involved, which are very stable in some species, where dead females remain attractive over very long periods, even months, while in others they deteriorate rapidly: dead *N. vitripennis* females, for example, lose much of their attractiveness within a couple of hours.

2. *Stimuli Produced by Courting Males*

Apparently, specific combinations of stimuli originate from displays about which we know little detail. Performance of certain display movements is no proof that these movements as such are required to induce sexual receptivity in a female. Nodding, for example, is not necessary in that sense. An *N. vitripennis* male's head can be immobilized by glueing it to his thorax, but his courtship success is not affected by this operation. Similarly, his wing vibrations are not necessary: males with their wings removed were just as successful as intact males. Clearly, this result cannot be generalized: Kitano (1975) obtained very different results with the braconid *Apanteles glomeratus* L.: removal of the wings reduced courtship success to almost nil. Obviously, wings may serve different functions in different species.

In *N. vitripennis* the effectiveness of at least three kinds of stimuli has been demonstrated. The first one is a pheromone which is released with the extrusion of the mouth parts during head nodding. It is probably synthesized in the mandibular glands, a source of chemical stimuli in a wide array of insects (Butler, 1964; Blum, 1974; Bergström, 1979; Hölldobler, 1978; van Honk *et al.*, 1978). Sealing a male's mouth-parts with a droplet of glue (we used alpha-cyanoacrylate rapid bonding adhesive) (Fig. 4) does not affect his readiness to court but prevents him from inducing receptivity. Drawing air, with a syringe, from a culture vial in which numerous males are courting and blowing this air on a female that is being courted by a sealed male, will immediately provoke sexual receptivity. Males which have their mandibular palpi protruding from the glue will then back up and copulate; those with their palpi entirely covered will not react, and continue their courtship. Drawing air from other vials, which contained only males or only females, had no such effects (van den Assem *et al.*, 1980b).

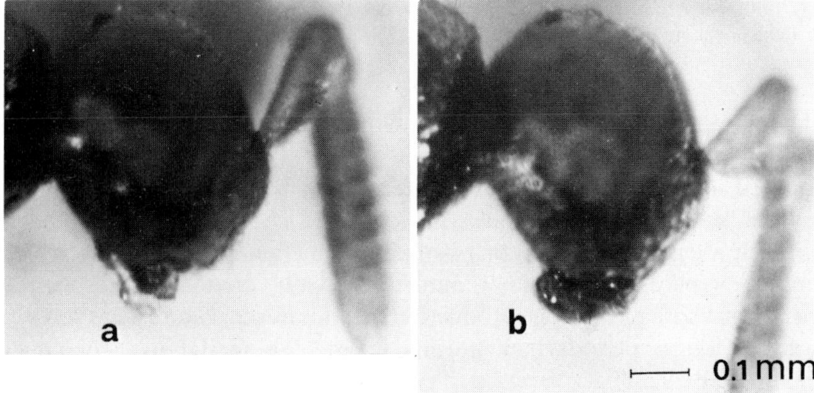

Fig. 4. Heads of (non-courting) *Nasonia vitripennis* (Walker) males. (a) Intact insect; the right-side maxillary palp and other mouth-parts are visible. (b) Situation after "sealing": all mouth-parts are completely covered with a droplet of glue.

Extrusion of the mouth-parts, discharge of the pheromone and nodding seem to result from pressure changes inside the male's head capsule. Abdomectomized males will court vigorously but the ability to nod and extrude the mouth-parts is lost. Consequently, they never induce receptivity. However, sealing the injury with a droplet of glue will restore nodding and discharging, as if, by sealing, the system can be re-pressurized. Sealing was no longer effective if 15 min or more elapsed between cutting and glueing, perhaps due to a loss of body fluids.

Males of almost all species of parasitic wasps thus far investigated produce sounds during their displays (Miller and Tsao, 1974; van den Assem and Putters, 1980; Fig. 5). These sounds are species-characteristic and serve a biological function. Males can be silenced by applying a drop of glue to the top of the thorax. The wing musculature seems to be the source of sound production and the chitinous wall of the thorax is probably serving as a resonator. This function is disrupted by the glue.

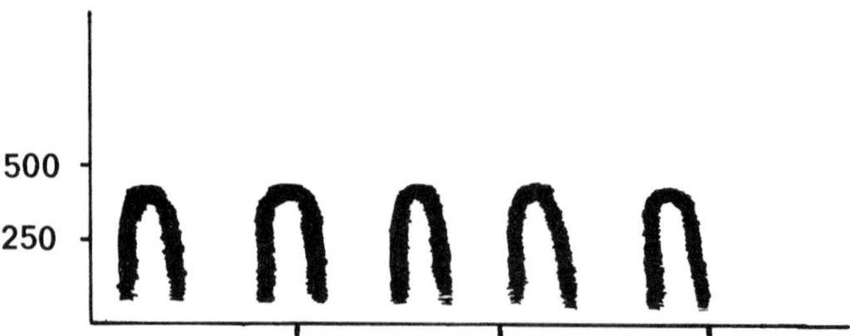

Fig. 5. Sonagram tracings of sounds produced by courting *Nasonia vitripennis* (Walker) males. Only the fundamentals are illustrated. Frequencies in Hz, along the ordinate; the markings along the abscissa denote successive seconds.

Dumb *N. vitripennis* males appeared handicapped in two ways, compared with controls. First, a female which was mounted by a dumb male usually produced a freezing reaction, as if she had not been aware of the other's presence before. This postpones the onset of courtship. Normal males approach a female with vibrating wings, making continuous sounds, which seems to prepare her for being mounted. Secondly, dumb males, especially older ones whose pheromone production deteriorates to some extent, have to court for longer periods than normal males of a similar age, to induce receptivity. Playing back their taped courtship sounds improved their efforts (van den Assem and Putters, 1980).

As well as chemicals and sounds, tactile stimuli are involved in inducing receptivity. Legs, with their bristles or brushes, which are moved to and fro are such a source, but are not considered here.

C. Female Receptivity and Mating

1. Onset of Sexual Receptivity in Females

Virgin females will show readiness to copulate during the first display by a conspecific male. Onset of receptivity is never a random event. This holds for all species which have been observed. With displays composed of a repetition of fixed patterns of movements, or cycles, receptivity may occur at any cycle, always related to the same display movement. In *Nasonia vitripennis*, this happens to be the first nod in a head nod series (see Fig. 2). Apparently, an essential combination of stimuli is reached at this moment. Since this may occur with any cycle (van den Assem and Visser, 1976) it is impossible to predict its occurrence with high precision. Other species are different. Males of most species of *Melittobia* (Eulophidae), for example, seem to produce an essential combination of key stimuli only once per display: female receptivity occurs at only one point, at the end, with what has been termed the finale (van den Assem, 1975). I will return to this point later on (Section V. C.).

With the onset of sexual receptivity there is a change in the female's posture, which is probably a necessary consequence of the readiness to copulate: a female has to raise her abdomen to expose the genital orifice to make copulation possible. A courting male may directly perceive this change, and it could provide him with a tactile stimulus which blocks further courtship and makes him switch to copulatory behaviour. Usually the change is abrupt. Both male and female profit from a prompt reaction because it makes a copulation by an intruder or "sneaker" more unlikely (Section III. C.). Moreover, the state of overt receptivity cannot be maintained, for unknown reasons, over a prolonged period.

However, in many chalcidoids, including *N. vitripennis*, males court from a frontal position and they are too small to perceive directly the female's abdomen rising. Other signals seem to be called for. Such signals differ between groups. In *N. vitripennis* (and all other Pteromalinae) movements with the antennae serve as a signal: females lower the antennae and draw them tightly in to their head at the onset of overt receptivity (see Fig. 1, right). These movements prove sufficient to stop a male's display and make him back up. The signalling function of a movement can be proven irrevocably by using a dummy which mimics the movement (Tinbergen, 1963), in this case a dummy female with movable antennae (van den Assem and Jachmann, 1982; Fig. 6). It works perfectly. The person experimenting pushes a button to lower the dummy's antennae and the male reacts by backing up.

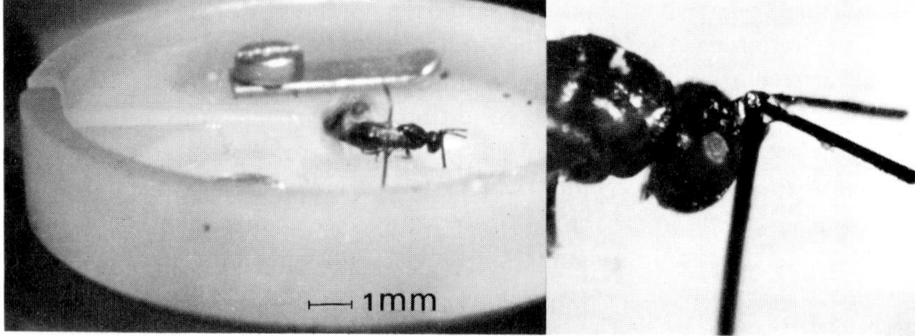

Fig. 6. Left: perspex observation cell, showing a dummy female with movable fake antennae *in situ*. The dummy is fastened to the substrate with a hair, to allow changes. Right: details of the dummy's head and the fitting of the fake antennae. The vertical rod passes through a hole in the bottom of the observation cell, allowing an up and down movement of the entire antennal construction.

2. *Amount of Courtship Necessary to Induce Receptivity*

Do all females signal receptivity following stimulation with similar quantities of display behaviour? We may well assume that they vary in this respect. Indeed, this is demonstrated in species in which displays consist of a succession of cycles. Using standardized males, some females reacted more promptly than others, and individuals differed consistently when, by experimental manipulation, a second or even third signal was provoked. This seems to indicate that cumulative processes are at work to bring about receptivity. Exceeding a threshold value—a different one for different individuals—results in signalling. The increase in probability of signalling during an ongoing display seems to point to the same conclusion. In other species, e.g. *Melittobia* of the *acasta, assemi* or *hawaiiensis* groups, the temporal organization of a male's display is such that the question of variable female susceptibility cannot be answered (van den Assem, 1975). Induction of receptivity may be due to cumulative processes but a (variable?) threshold has always been crossed before the male comes to produce his finale with the key-stimulus configuration (Section V. C.).

On the other hand, the quality of male display is not constant. Ageing is an important factor: older *Nasonia vitripennis* males produce less mandibular pheromone than younger males, judging by performances on standard females. We have no reliable data on systematic differences between males of similar age groups.

3. *Switching off Willingness to Copulate*

Virgin females will signal receptivity during the first display by a conspecific male. The normal sequel to signalling is copulation and insemination. Usually an *N. vitripennis* female will signal a second time during the short bout of post-copulatory courtship which follows immediately. However, a male will never react to this second signal, although it comes with a first head nod. Obviously, the male's internal condition has changed: he has become unresponsive to an external stimulus which effectively released copulatory behaviour shortly before. Similarly, females are no longer in a reactive mood following signalling and copulation. In many species of parasitic wasps the ability to signal is limited—sometimes it is even a once-in-a-lifetime affair.

In several groups of insects, female unresponsiveness results from being inseminated, and is due either to the presence of sperm in the spermatheca or to additional substances introduced by the male during copulation (Manning, 1962; Merle, 1968). Cousin (1933) hypothesized that the presence of sperm in the spermatheca precludes further receptivity in female *N. vitripennis*. This factor could indeed account for an almost instantaneous switch-off:

King (1960) and Wilkes (1965) found sperm to be present within 1 min following termination of copulation. However, other factors must be involved in parasitic wasps, because females were also unreceptive following signalling even when copulation has been prevented: e.g. when the male is brushed off by the experimenter before he could make contact. Brushing off in the absence of a signal has no such effects. Thus, it is the achievement of the state of overt receptivity and/or the production of the receptivity signal itself that causes the change of female condition.

In many species of wasps the female's condition seems to change abruptly: immediately following copulation no further signalling could be provoked (e.g. in the pteromalids *Spalangia cameroni* Perkins, *Asaphes vulgaris* Walker, or *Anogmus strobilorum* Thompson). In others it is a more gradual process which proceeds with time (van den Assem and Visser, 1976), when more and more courtship is required to provoke a second or third signal. The nature of the processes which underly these changes is unknown, nor do we know whether they are peripherally located, or more centrally. In *N. vitripennis* the changes affect reactions to several, probably unrelated, chemical signals. The markings, which males apply to the substrate, appear to be attractive to virgin females. This attractiveness disappears instantaneously with signalling, even when copulation was prevented (see also Gonzalez *et al.*, 1985).

Switching off a female's responsiveness following insemination is to a male's advantage because it reduces the probability of a second copulation, and hence, sperm competition. To be sure, a second male was never found to inseminate successfully if he copulated soon after the first male and this first male still had a full ejaculate to offer (Section III. A.). However, a second male's sperm can be utilized when first and second copulations are 24 h or more apart, even when the female has not been egg-laying during that period. On such occasions considerable quantities of sperm may be added, judging by the ratio of daughters sired by first and second males in subsequent batches of eggs. Since these ratios remained rather constant over time, the conclusion is that ejaculates come to be mixed inside the spermatheca, at least in most cases. With a minority of females the opposite was found. The period of time in between two inseminations may have been the decisive factor here.

In species where female unresponsiveness sets in gradually, and sperm can be added with a second copulation, it may be to the first male's advantage to prevent a second copulation during this period, and stay with the female instead of searching for a new mate immediately. Cases of mate-guarding (Parker, 1974) have been recorded for parasitic wasps. *Aphelinus* species have a very prolonged post-coital phase of courtship (Gordh and DeBach, 1978) during which the male fends off other males which try to make contact with the recently mated female. Prolonged post-copulatory displays of *Asaphes* species (Pteromalidae) may serve a similar function. Viggiani and Battaglia (1983) described post-copulatory displays of 30 min or more in *Encarsia asterobemisiae* Viggiani & Mazzone. The probability of an encounter with another male will be an important factor for the development of post-copulatory displays: if females tend to leave the mating patch as soon as copulation is terminated, which seems to be the rule in many species, then performance of a post-copulatory display would probably mean a waste of time for a male, even if females were not yet fully unreceptive at the time of leaving. In the majority of parasitic wasps post-copulatory displays are absent.

D. Periodic Endogenous Changes in Courting Males

In experiments with dummy *Nasonia vitripennis* females, fitted with movable antennae for faking antennal signalling (Section II. C.), we found delays to occur before a male backed up. A delay appears to depend upon the precise point in a display where the antennal signal is produced (van den Assem and Vernel, 1979; Jachmann, 1983). If it was with the first head nod, delays were minimal but they increased the greater the time lapse since the first nod. This phenomenon is illustrated in Fig. 7. Experiments were arranged in such a way that the antennal signal coincided either with a male's first head nod, a later nod, an antennal sweep, or was delivered somewhere during the interval between nodding series. Once produced, the signal stayed on, i.e., the antennae remained in the low position, which is similar to how a living receptive female behaves. No differences were found between successive cycles and all data have been pooled. In Fig. 7, the timing of the signal relative to the display is entered on the abscissa, and the delay, in seconds, on the ordinate. Figure 7 represents the time course of the male's delay over an entire cycle of courtship.

One conclusion is that a male does not delay his backing-up reaction because he did not perceive the signal when it was given. This supposition seems plausible for delays with signals given over an interval. If this were true, then one would expect a male to back up at the onset of the next cycle, with the antennal sweep, or with the first nod: points in time which can be exactly predicted. This was not the case; males backed up much later. Therefore, another conclusion is that a male's readiness to react to a receptivity signal is not continuous at the same level. Rather, a male's readiness fluctuates markedly and changes periodically. Yet another conclusion is that, once the female's signal is given, the periodic, endogenous changes do not proceed unaffected. Signal production has, at any moment, a profound effect on a male's internal condition which, in turn, implies that a

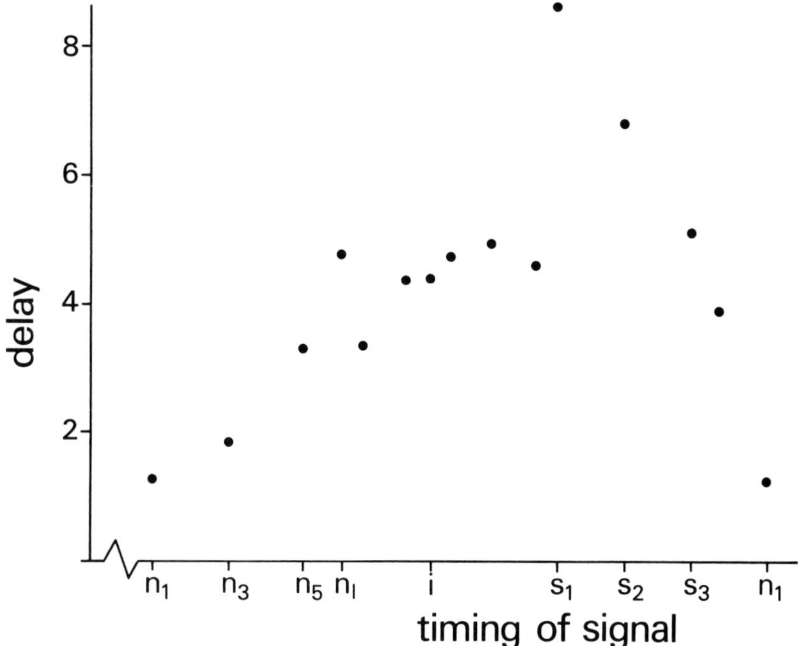

Fig. 7. Relationship between the points in the display where a signal is produced and the delay before backing up. Ordinate: delay time in seconds; the signal is always produced at time 0. Abscissa: specific points in the display that were tested: n_1, n_3, n_5, n_l: first, third, fifth and last nod of a series, respectively: i, interval; s_1, s_2, s_3: onset, halfway and end of antennal sweep. From the last nod of a series until the end of the sweep the delay has a more or less constant value, but signalling during the sweep adds the duration of the performance to the delay. Delay time builds up during nodding. Each point in the figure represents the average of about 200 observations. (Further explanation in text. Section II. D.)

male is monitoring a female's behaviour continuously, although he may not react to it immediately in an overt way.

III. FACTORS AFFECTING MALE MATING SUCCESS

A. Male Mating Potential

In normal circumstances courtship and copulation are preliminaries to insemination. However, performance of both is in no way dependent upon the ability to inseminate. Any male is able to serve a number of females, but his capacity is limited; usually it is related to body size. In *Nasonia vitripennis* it ranges between about 15 females for small individuals to about twice as many for large individuals. These numbers may be achieved only rarely under natural conditions, if ever, though reliable data on this point are non-existent. A rough estimate of the amount of sperm transferred with successive copulations is obtained from counts of daughters produced over a lifetime by the series of females with which a test male had copulated. Males may copulate with hundreds of females offered at the shortest possible intervals (Barrass, 1961) but a full batch of sperm is transferred with only a few copulations. Progressively fewer daughters resulted from successive inseminations. The rate of depletion was again correlated to male body size. Approximately similar amounts of sperms were present in first ejaculates of males of different sizes. Females served first produced an average of 500 daughters over a lifetime; females served 15 copulations later about half as many (large males) or nil (small males). Following a rest period of one or a few days an incomplete recovery of insemination potential was apparent.

Sperm depletion is no barrier to further mating (Dautert-Willemzik, 1931). An effect of this is that females are switched off, inseminated or not, and consequently are no longer available to other males, at least not for some time. Is this an example of spiteful behaviour? To be sure, a male achieves his reproductive success by inseminating females, but also by preventing rivals from doing the same. The ability of *Nasonia vitripennis* males to copulate with hundreds of females when given a chance, is perhaps without biological significance, but in other species, where males are far more rapidly depleted, the effects may be not entirely artificial (Nadel and Luck, 1985; Gordh and DeBach, 1976; Jones, 1982).

Males pass on genes through daughters. An excess production of sons by mated females is therefore detrimental to male reproductive success. This can happen in two ways. The first concerns a temporary factor. Immediately following insemination the fertilization apparatus does not seem to be operational, for reasons unknown. All eggs laid in a period of up to 24 h (the duration varies between individual females and is sometimes much shorter) will pass unfertilized. In normal circumstances virgin females are inseminated upon emergence, at a time when few mature eggs are present. Because fecundity only reaches a peak level a few days later, after the female has located a host and fed upon it (Velthuis *et al.*, 1965), the practical consequence of this effect will be minimal. However, where re-insemination has occurred a female may produce a considerable number of eggs during the period of inactivation, thus adding to the disadvantageous position of the first male: not only will the second insemination lead to sperm competition but inactivation of the fertilization apparatus, as a result of the second insemination, results in extra losses (van den Assem and Feuth-de Bruin, 1977).

The second exception is caused by a so-called sex-converting factor of which several have been found to exist in natural populations of *Nasonia vitripennis*; its effects are permanent. It is an extra chromosomal factor, which is paternally inherited (Werren *et al.*, 1981). A female mated to a carrier of the factor will produce offspring practically 100% male, with the majority of her sons being new carriers. Experiments indicate that the factor utilizes sperm to travel from one generation to the next. An "infected" sperm penetrates an egg in the normal way but there is no subsequent fusion of genetic material, hence the sex conversion: what should have been a (diploid) daughter becomes a (haploid) son instead. We were unable to find any difference in the mating behaviour of carrier males and healthy ones (Werren and van den Assem, unpublished). Skinner (1982) reported a maternally inherited factor of the conversion type which skews the sex ratio toward females.

B. Courting Unwilling Females: Giving-up Times

Males of several species discriminate between virgin (potentially receptive) and mated (usually unreceptive) females [I have observed this in, for example, the pteromalid *Cecidostiba semifascia* (Walker)], but this feature is not universal. *Nasonia vitripennis* males will readily court mated females; a small fraction of these will eventually signal receptivity and copulation may lead to successful insemination after all. Displays on genuinely unreceptive females are by definition never cut short by signalling. Because a female's behaviour does not change noticeably during a display, one must assume that the male's giving-up results from endogenous processes in the first place. We have analysed these endogenous processes, using two different approaches: firstly, quantification of courtship outputs on a very fine time-scale; and secondly, quantification of a male's output in displays which follow one another at variable, manipulable intervals.

It is only partly true that a male's display is composed of a repetition of identical elements, or cycles because systematic, quantitative changes in the output pattern occur along the way. Displays have characteristic time courses: the number of nods in the first series is high, there are fewer in the second series and a gradual increase as the display progresses. Moreover, the duration of successive intervals in between series increases gradually. A male will dismount during nodding or at the end of a series, but never during an interval. This precludes the possibility that a display ends randomly. Apparently, during or with nodding some threshold value is exceeded which makes a male give up. My colleagues Jachmann and Putters (in press) have developed a model which gives a satisfactory description of a male's output.

Two factors are required to account for the dynamics of any *Nasonia vitripennis* display. Courtship as a result of a two-factor interaction is interesting because it is suggestive of an alternation of internal states. I will return to this point in Section V. D.

Theoretical considerations (Parker, 1978) predict that giving-up events result from optimization procedures, aimed at maximizing the rate of copulations. Hence, a male should not give up too soon, so as not to miss an opportunity to inseminate, nor too late, so as not to waste time which would be better allocated to courting other females. How to designate "soon" and "late" depends upon the external stimulus situation. In addition, our data show that a male is unable to maintain a receptivity-induceable level of pheromone discharge over the entire length of prolonged displays. Short-term depletion occurs, and to continue to court in such conditions seems to be counterproductive.

This brings me to the second approach: for successions of courtship displays with variable intervals of non-courtship in between. These procedures provide insight into proximate causative mechanisms, but they also provide data on functional aspects, i.e. on a male's mating success under variable circumstances.

An inexperienced male's first output is always a long display, composed of many cycles, when he happens to encounter an unwilling female (Barrass, 1961; van den Assem *et al.*, 1984). In the displays which follow the moment of giving-up comes much earlier. If unwilling females are frequently encountered, then a male comes to court them with minimum displays, composed of one or two cycles. The durations of the intervening non-courtship periods completely determine the male's display production as if, at the moment of dismounting (at which a male's readiness to court has reached a zero level) recovery of his tendency to court is initiated (van den Assem *et al.*, 1984). This recovery happens to be a non-linear process. Long intervals, in the order of 24 h or so, lead to complete recovery: a male will then court as if he had never before encountered a female. All quantitative changes of the males' courtship output which occur after variable periods of non-courtship are predicted by the two-factor model (Jachmann and Putters, in press).

A variety of mechanisms could be responsible for the decrease in a male's tendency to court: for example, sensory adaptation, fatigue, exhaustion of a specific internal factor, inhibition. The results of a somewhat unorthodox experiment point to the latter possibility. Males which had encountered many females at short intervals could be instantaneously induced to court at the maximum output level by exposure to an extremely low temperature for a short time (we used a freezer with −30°C). Immediately following this treatment males courted as if they were producing their very first display. However, when there was a 10-min pause between exposure and test there was no longer any difference between treatment and controls: output had dropped to a low level again. This result supports the hypothesis, originally proposed by Barrass (1976), that it is the performance of the display itself which has a cumulatively inhibiting effect on the performance of the next display (van den Assem *et al.*, 1984). Inhibition diminishes, i.e. recovery takes place, during periods of non-courtship. Apparently, the inhibitory process had been temporarily removed by the cold treatment but it appeared re-established shortly afterwards. Once inhibition is effective again, the processes in control of the strength of inhibition appear not to have been affected by cold treatment. Strength is at any moment solely determined by time passed since the male last dismounted.

C. Patterns of Female Emergence and Positions Preferred by Males

Are the mechanisms underlying giving-up tuned in such a way that maximum mating success is ensured? To answer this question, data are required on the rate at which females become available for insemination, on the ratio of virgin and mated females which are encountered, and in what order.

In gregarious species, potential partners are present from the beginning, so searching for mates does not seem to pose much of a problem. However, to have any success at all, males should emerge ahead of the bulk of females, at least in species where females are available over relatively short periods. Usually males await the emergence of females outside a host cocoon, preferably near an escape hole through which they will leave the host. Is this the most profitable position? It depends on three factors: (1) the time required to court, copulate and "switch off" a female effectively, (2) the predictability of the pattern of female emergence and (3) the costs of maintaining itself in that position. Satellite males, positioned nearby, try continuously to oust a resident from his holding, and aggressive charges are necessary to ward them off, which is time-consuming and involves the risk of being injured, at least in some species, including *Nasonia vitripennis*.

Durations of displays and copulations depend upon the ambient temperature. On average, it takes an *N. vitripennis* male about 1 min to go through a full mating. Inseminable females leave a host at unpredictable times, and at such a rate that in ideal circumstances (when no competitors are around) about 10% of them will escape unattended. These females become fully available to males of the second rank: the satellites. In practice much more than 10% escape because competitors may interact frequently during mating, and also because females may walk out during aggressive interactions in defence of the preferred position. In general, the more satellites present, the more time has to be spent on defence. When there are about six of them a position near an escape hole is no longer profitable. Observations show that territorial defence breaks down at this density, and is replaced by another strategy: scrambling for mates (van den Assem *et al.*, 1980a).

At least one more strategy may be profitably pursued by a male of a species with a frontal courtship position. He may clasp the female's abdomen while she is being courted by another male and copulate with her as soon as receptivity occurs, thanks to the courting male's efforts. This male will find his place taken when he backs up. Sometimes a dominant male will lose a copulation to a rival which emerges from the host in the wake of a female and mounts or clasps immediately.

Do males optimize their efforts so as to score a maximum success? We do not know because the situation is far from simple in most species. Only with more observations made in field studies will our knowledge deepen, but these studies are not so easy to obtain. For *Nasonia vitripennis* one positive statement can be made: the causal mechanism underlying giving-up has properties required for optimizing outputs: in a succession of displays on mated females the duration of the display is strictly related to the rate of encounter (Putters *et al.*, in press).

IV. SEARCHING FOR MATES

Solitary parasitoids pose additional problems with respect to acquisition of mates. Species which have been called quasi-gregarious, in which hosts are clumped together (e.g. parasitoids of *Drosophila* spp. larvae or pupae), will not differ from genuinely gregarious parasitoids, but those with hosts well spaced-out do. In such species, searching for mates seems to be an important life history feature. A random search can be ruled out; instead, signalling seems to be called for. In principle, either sex may signal (Alexander and Borgia, 1979; Greenfield, 1981). In at least a number of solitary parasitoids males seem to be the signallers and females the searchers. Males have been observed to swarm: crawling about on a substrate (usually part of the vegetation) in large numbers and keeping close (Southwood, 1957; Jervis, 1979; Graham, personal communication; Nadel, personal communication). Swarms may be present at the same site over a period of several days; one or more species—not necessarily closely related species—may take part. Nadel (personal communication) ascertained that virgin females were attracted to swarms, landed in the melée, and were mated.

Perhaps swarms emit chemical signals that work over relatively long distances. Males would signal communally this way and thus increase signal efficiency. In *Nasonia vitripennis* (a gregarious species), males mark the substrate at intervals (especially following a bout of courtship or copulation). These markings are remarkably persistent (once settled a male may stay for a long time, in the order of several days) and they appear to be attractive to other males and to virgin females (but not to mated females). Swarms may result from comparable procedures (though nobody has recorded actual marking behaviour in this context). The persistence of swarms at the same site over several days supports this theory. One might speculate that mixed swarms are composed of species which use similar long-distance signals. Species recognition follows at close quarters by means of specific contact-pheromones (see also Section II. B.).

Chemical signals which work over relatively long distances have been identified and widely reported for several species of wasps, mostly ichneumonids and braconids (Lewis *et al.*, 1971; Eller *et al.*, 1984). In fact, they can be used in monitoring parasitoid species (see Chapter 11).

V. THE EVOLUTION OF COURTSHIP BEHAVIOUR IN CHALCIDOID WASPS

The last section of this paper deals with some aspects of the evolution of courtship behaviour of parasitoid wasps. My aim is not so much to reconstruct the phylogeny of certain displays (as attempted by Alexander (1964) for arthropod mating behaviour), but to identify what Hölldobler and Wilson (1983) termed phylogenetic grades: to uncover successively more advanced combinations of traits (or, perhaps, combinations of more specialized traits) by examining the displays of as many species as possible. In this way it is hoped to expose consistent trends which must support hypothetical evolutionary reconstructions. The Chalcidoidea are ideally suited for such comparative studies because of their enormous diversity. As stated above, anecdotal observations on one or more aspects of mating abound in the literature. Only a few authors have dealt with one group or another in a systematic way, so as to expose trends (Dahms, 1973, 1984; Goodpasture, 1975; Gordh and DeBach, 1978; Bryan, 1980; Grissell and Goodpasture, 1981; van den Assem *et al.*, 1982; Orr and Borden, 1983).

A. Phylogenetic Grades

Courtship displays of *Nasonia vitripennis* and other Pteromalinae have many features in common, such as the kind of appendages taking part and the general characteristics of the movements, the temporal structure, the male's frontal position (with the fore tarsi placed on the female's head) and the way females signal receptivity. Yet, details of the performances are different, in fact so much so that they can conveniently be used as taxonomic characters. Another feature, common to many species, is the size difference between the sexes, males being on average smaller than females.

The comparison of displays of Pteromalinae with those of other subfamilies of pteromalids yields interesting results. In some groups (Cerocephalinae, Spalanginae) the male's position for courting and for copulating are similar: that is to say, he courts from a caudal position, clasping the female's abdomen with his fore tarsi. At the onset of sexual receptivity, the female's raising of her abdomen seems to provide a direct, tactile stimulus. The male reacts by bending his abdomen to establish genital contact. An antennal receptivity signal is lacking. Antennae also seem to play a minor role in the male's display. Wings, hind- and middle-legs are the appendages most frequently involved in the displays of caudal courtiers. Procedures of this type are found in groups of solitary parasitoids where size differences between the sexes are minor, although males always tend to be somewhat smaller than females. This is probably due to selection for early appearance. As in *Nasonia vitripennis*, male mating potential in other Pteromalinae [e.g. *Lariophagus distinguendus* (Förster), *Anisopteromalus calandrae* (Howard)] is positively correlated to body size, but this feature is unlikely to be under strong selection pressure since even small males seem to have an excess production of sperm at their disposal, considering the most likely number of inseminations they will have to perform (one or very few).

In species of more specialized solitary parasitoids, which adopt hosts which may vary considerably in size (again, *L. distinguendus* and *A. calandrae* are good examples), the egg-laying female is largely responsible for size differences between the sexes. Females appear to discriminate between larger and smaller hosts: a fertilized egg is preferably laid on a large host, an unfertilized one on a small host. This is a very functional choice because the reproductive success of daughters appears to be much more rigidly correlated to body size than that of sons (van den Assem, 1979).

In yet other groups of pteromalids (Miscogastrinae, Asaphinae) intermediate courtship positions occur: males place the fore tarsi near the female's wing bases or on her "shoulders". Again, there is no antennal signal; males seem to project sufficiently far to the rear to perceive abdomen raising by the female directly.

These observations lead to the following hypothetical reconstruction of the evolution of courtship displays in the Pteromalidae. The original position of the courting male is one which coincides with his position for copulating. The male stops courting as soon as the female raises her abdomen (a necessary movement to expose the genital orifice), which provides the male with a direct, tactile stimulus. He stops courting and switches over to copulatory behaviour. The body size of male and female are approximately similar. Apparently this situation was unstable in an evolutionary sense because it led to several more advanced situations (see Fig. 8). The position for courting came to be separated from the position for copulating by a shift to the front, and to be coupled with the use of more frontally located appendages, such as antennae and mouth-parts. Another development was

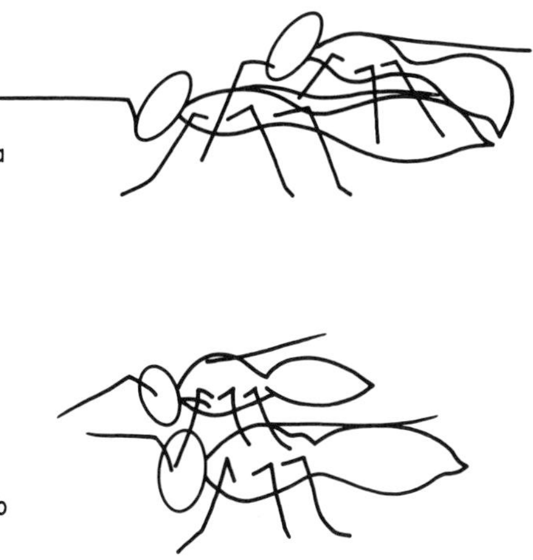

Fig. 8. Schematic drawings of courtship positions in pteromalids. (a) The position of the male *Spalangia endius* Walker which coincides more or less with the position for copulating; (b) The position of a male (subfamily) Pteromalinae [*Nasonia vitripennis* (Walker) belongs to this group]. Such males court from a frontal position and must back up to perform a copulation; (b) is thought to represent a derived (specialized) situation, (a) a more original one (see text).

the female's secondary receptivity signal. A male at the front, who is able to perceive the onset of female readiness by other cues than the primary movement, need not project far to the rear. Reduction of male size is a very general phenomenon in parasitic wasps. Especially in gregarious species, the major advantage of this development is that smaller males eat less food, which means more food is available to female sibs, which, consequently, will develop into larger, more fecund adults. Males frequently copulate with their female sibs: inbreeding is the rule rather than the exception, and thus the increase of female fecundity is to the male's own direct advantage. In the extremely inbred species *Nesolynx albiclavus* (Kerrick), a parasitoid of the pupae of tsetse, Saunders (1961) found an almost perfect correlation between female fecundity and amount of food eaten during larval development.

Separation of the courtship position from the copulation position involves risks, as I have already pointed out. Males need not go through a display procedure to become ready to copulate. The interesting point is that once they are courting they are no longer continuously ready. This state of affairs opens up possibilities for a "sneaker" strategy which is indeed a common feature in species where males display at the front. In some species, specific counter-measures have been developed. Males of the pteromalid *Pteromalus puparum* (L.), for example, court in situations in which there are many rivals; the species parasitizes chrysalids of butterflies from which hundreds of wasps may emerge (Martelli, 1907; Boulétreau, 1977). Such males produce only one courtship cycle at a time, with a single head-nod, and do not wait for a female signal before backing up. If the female is not yet willing to copulate, then the male returns to the front, produces another cycle, rushes back again, etc. This combination of traits is not an original one, rather it seems to be a general adaptation to an unusually high density of rivals. The organization of this display is such that the time required until onset of female receptivity is considerably prolonged, but this is probably evened out by losing less time to "sneakers". A similar display organization has been observed in a completely unrelated species, which does its courting in comparable circumstances.

B. Parallel Developments

A shift of the male's courtship position, development of a frontal receptivity signal and reduction of male size, as outlined for the pteromalids, are paralleled in other large families of Chalcidoidea. Extreme frontal positions occur, for example, in encyrtids. In *Anagyrus pseudococci* (Girault), for instance, the male places his front tarsi on the female's antennae. Moreover, the shift of position is not only found in groups with the male courting on the female, but also in those with the male courting on the substrate, as is the case in several groups of eulophids and encyrtids.

Another general trend is present in all major families of Chalcidoidea. In many groups, males change their posture during a display without shifting the position of the forefeet. The more original situation is one in which the male alternates between two extremes: bringing his head and appendages close to the female's head (called the low-phase posture) and moving it away from the female's head (the high-phase posture). Bending and stretching the front legs is responsible for these postural changes. During a display many such phase-shifts may occur; frequently, the alternations are performed in a very rigid way. The more advanced situation is one in which the male adopts the low-phase posture throughout. Intermediate situations also occur: in the initial stage of a display the male alternates, while in the later stage he permanently adopts the low-phase posture. This has been found in several *Melittobia* species (van den Assem *et al.*, 1982). High and low phases are characterized by the use of different appendages. In general, there is a predominance of wings and legs in the former while in the latter, head, antennae and mouth-parts predominate. Tactile contacts frequently occur during the low phase: a male may embrace the female's antennae with his own antennae, or he may press his mouth-parts to the female's antennae or head (e.g. in many Eulophidae–Tetrastichinae).

C. The Emancipation of Males

Melittobia species provide an example of yet another trend: the emancipation of males. In the majority of chalcidoids, males seem to produce key courtship stimuli periodically. Consequently, females will become receptive somewhere along the way. This is also the normal procedure for *Melittobia clavicornis* (Cameron), the most primitive species of the genus. In all other *Melittobia* spp., displays show qualitative changes as they proceed, and a key-stimulus situation is realized only once, at the finale (van den Assem, 1975). Only then is the male ready to copulate with a receptive female. Rare occurrences of female receptivity at other times do not lead to a copulation because the male will not back up. If a female does not signal at a display's end, then the male may dismount, but he may also start afresh, the next display having the same rigid time course as the previous one. Some species have very long displays: 15 to 30 min are not exceptional, which is in sharp contrast with the matter of seconds which is the rule in most other chalcidoids.

Such time-consuming courtship procedures would seem to put severe constraints on a male's reproductive success, but this is not the case. *Melittobia* spp. males are highly aggressive; competitors are usually eliminated before the bulk of females become available. Frequently they have their heads bitten off after a fatal fight, or at the moment of emerging from their own pupae, when they are most vulnerable and unable to retaliate (Wilhelm, personal communication). Virgin females do not leave a host cocoon: mating takes place inside the host's enclosure, and not outside as is the case in the majority of parasitoid wasps. Moreover, males appear to be remarkably attractive to virgin females, probably due to chemical signals (van den Assem *et al.*, 1982; Gonzalez *et al.*, 1985). This being so, a male may court at leisure, and by doing just that he keeps the rate of inseminations within certain limits, a procedure which apparently prevents sperm depletion. Because of the extreme sex ratio, still further biased by murder, the average *Melittobia* spp. male has to serve a large number of females (100 or more, which appeared adequately inseminated when tested).

The organization of *Melittobia* spp. courtship is such that females never cut short a display (which would result in a rate of copulation influenced by variable female response thresholds) and the maximum output-rate is set by the male alone. A weak point in this reasoning is that males need not court all the time in order to spread out inseminations; they could just as well court at intervals and save energy during periods of non-courtship. There is no satisfactory answer to this. Males appear to be very attractive to females, especially when they are courting. Frequently, females queue up around a courting couple. By courting all the time a male might prevent females from leaving unserved. A striking aspect of the very long displays of species such as *M. australica* Girault is that the male's movements are extremely slow and simple, as if he is "free-wheeling" most of the time; not the kind of movements which are energy-demanding.

D. The Origins of Display Patterns

The final question about courtship behaviour deals with the origins of the usually very ritualized patterns of movements employed. Is it possible to derive these patterns from other, less ritualized ones? I have only a few speculations to offer, which are made, however, on the basis of a comparison of many species. The alternations between low and high phases have already been mentioned above. Both phases are characterized by different sets of postures and movements: head and antennae predominate in the low phase, wings and legs in the high phase. It seems as if periods of close contacts between partners (low) alternate with periods of less contact (high). A plausible inference seems to be that the causative processes underlying these phases may also differ: the shifts between low and high phases would reflect periodic changes in motivational states underlying the display.

Males frequently perform movements with their legs in another context: while fending off rivals which try to displace them from a courtship position. The ritualized leg movements performed during displaying, which are frequent during the high phases with the male in the upright position, could be derived from these defensive responses. Therefore, this part of a display may not belong to the sexual system *sensu stricto*, in contrast to the movements performed during the low phase, at the times of close contacts between partners. The hypothesis of dual motivational sources links up with an earlier conclusion, derived from causal analysis, that periodic changes of a male's readiness to copulate occur during displays.

If this hypothesis has some merit, then a courting male alternates between two major behaviour systems during his performances. Some observations seem to support this theory: for example, the courtship display of the relatively primitive eulophid *Aceratoneuromyia granularis* Domenichini (in den Bosch and van den Assem, 1986): the low-phase behaviour is nothing but a straightforward attempt to copulate (an indication of the predominance of sexual motivation underlying these parts of the display); however, it is not oriented to the female's genital orifice. During the high phases the male is in an upright posture, vibrating his wings and swinging his legs.

The idea of motivational ambivalences underlying courtship behaviour is not a new one. In fact, it is a basic principle in ethology with respect to display behaviour of vertebrates. If it were to apply to insects as well, it would mean that the organization of these displays is, after all, perhaps, not as different from that of vertebrates, contrary to what has been supposed on the basis of their very different central nervous system morphology.

Acknowledgements

My thanks are due to my collaborators F. A. Putters, F. Jachmann and W. Wilhelm for providing a continuous flow of ideas and criticism, and for fruitful discussions. Professor D. J. Kuenen (at that time ecologist at Leiden) introduced me to parasitoid wasps and Dr R. Barrass's 1960, 1961 papers inspired part of my later work on these insects; this intellectual debt is gratefully acknowledged. Comparative work on parasitoid wasps is impossible without the support of taxonomists; I am very grateful to Drs Z. Bouček (London), M. W. R. deV. Graham (Oxford) and M. J. Gijswijt ('s Graveland) for the identification of many specimens. Finally, I want to thank the BBC World Service for English as I have come to know it (only my dullness is to be blamed for residual errors).

REFERENCES

Abraham, R. (1984). Über Insekten aus Vogelnestern. *Vogelkundliche Hefte Edertal* **10,** 7–11.

Alcock, J., Barrows, E. M., Gordh, G., Hubbard, L. J., Kirkendall, L. L., Pyle, D., Ponder, T. L., and Zalom, F. G. (1978). The ecology and evolution of male reproductive behaviour in bees and wasps. *Zoological Journal of the Linnean Society London* **64,** 293–326.

Alexander, R. D. (1964). The evolution of mating behavior in arthropods. *In* "Insect Reproduction" (K. C. Highnam, ed.), pp. 78–94. Royal Entomological Society, London.

Alexander, R. D., and Borgia, G. (1979). On the origin of the male–female phenomenon. *In* "Sexual Selection and Reproductive Competition in Insects" (M. S. Blum and N. A. Blum, eds.), pp. 417–440. Academic Press, New York.

van den Assem, J. (1975). Temporal patterning of courtship behaviour in some parasitic Hymenoptera, with special reference to *Melittobia acasta. Journal of Entomology* **50,** 137–146.

van den Assem, J. (1979). Grote vrouwen, kleine mannetjes, een zaak van belegging en rendement. *In* "Sluipwespen in Relatie tot hun Gastheren" (H. Klomp and J. T. Wiebes, eds.), pp. 64–96. Pudoc, Wageningen.

van den Assem, J., and Feuth-de Bruin, E. (1977). Second matings and their effect on the sex ratio of the offspring in *Nasonia vitripennis* (Hym. Pteromalidae). *Entomologia Experimentalis et Applicata* **21,** 23–28.

van den Assem, J., and Jachmann, F. (1982). The coevolution of receptivity signalling and body size in the Chalcidoidea. *Behaviour* **80,** 96–105.

van den Assem, J., and Putters, F. A. (1980). Patterns of sound produced by courting chalcidoid males and its biological significance. *Entomologia Experimentalis et Applicata* **27,** 293–302.

van den Assem, J., and Vernel, C. (1979). Courtship of *Nasonia vitripennis* (Hym.: Pteromalidae), observations and experiments on male readiness to assume copulatory behaviour. *Behaviour* **68,** 118–135.

van den Assem, J., and Visser, J. (1976). Aspects of sexual receptivity in female *Nasonia vitripennis*. *Biologie du Comportement* **1,** 37–56.

van den Assem, J., in den Bosch, H. A. J., and Prooy, E. (1982). *Melittobia* courtship behaviour, a comparative study of the evolution of a display. *Netherlands Journal of Zoology* **32,** 427–471.

van den Assem, J., Gijswijt, M. J., and Nübel, B. K. (1980a). Observations on courtship and mating strategies in a few species of parasitic wasps (Chalcidoidea). *Netherlands Journal of Zoology* **30,** 208–227.

van den Assem, J., Jachmann, F., and Simbolotti, P. (1980b). Courtship behaviour of *Nasonia vitripennis* (Hym., Pteromalidae): some qualitative, experimental evidence for the role of pheromones. *Behaviour* **75,** 301–307.

van den Assem, J., Putters, F. A., and van der Voort-Vinkestijn, M. J. (1984). Effects of exposure to an extremely low temperature on recovery of courtship behaviour after waning in the parasitic wasp *Nasonia vitripennis*. *Journal of Comparative Physiology* **155,** 233–237.

Barrass, R. (1960a). The courtship behaviour of *Mormoniella vitripennis*. *Behaviour* **15,** 185–209.

Barrass, R. (1960b). The effect of age on the performance of an innate behaviour pattern in *Mormoniella vitripennis*. *Behaviour* **15,** 210–218.

Barrass, R. (1961). A quantitative study of the behaviour of the male *Mormoniella vitripennis* towards two constant stimulus situations. *Behaviour* **18,** 288–312.

Barrass, R. (1976). Inhibitory effects of courtship in the wasp *Nasonia vitripennis* and a new interpretation of the biological significance of courtship in insects. *Physiological Entomology* **1,** 229–234.

Bastock, M. (1967). "Courtship, a Zoological Study." Heinemann Educational, London.

Bergström, G. (1979). Complexity of volatile signals in hymenopteran insects. *In* "Chemical Ecology: Odour Communication in Animals" (F. J. Ritter, ed.), pp. 187–200. Elsevier, Amsterdam.

Blum, M. S. (1974). Pheromonal sociality in the Hymenoptera. *In* "Pheromones" (M. C. Birch, ed.), pp. 223–249. North-Holland Publishing Company, Amsterdam.

Borgia, G. (1979). Sexual selection and the evolution of mating systems. *In* "Sexual Selection and Reproductive Competition in Insects" (M. S. Blum and N. A. Blum, eds.), pp. 19–80. Academic Press, New York.

in den Bosch, H. A. J., and van den Assem, J. (1986). The taxonomic position of *Aceratoneuromyia granularis* (Hym., Eulophidae), as judged by characteristics of its courtship behaviour. *Systematic Entomology* **11,** 19–23.

Boulétreau, M. (1977). "Nutrition Larvaire et Exploitation de l'Hôte chez un Hymenoptère Endoparasite Grégaire *Pteromalus puparum* L. Influence de la Densité de Population Préimaginale et Conséquences sur les Adultes." Thèse, Université Claude Bernard, Lyon.

Bryan, G. (1980). Courtship behaviour, size differences between the sexes and oviposition in some *Achrysocharoides* species (Hym., Eulophidae). *Netherlands Journal of Zoology* **30,** 611–621.

Butler, C. G. (1964). Pheromones in sexual processes in insects. *In* "Insect Reproduction" (K. C. Highnam, ed.), pp. 66–77. Royal Entomological Society, London.

Chapman, R. F. (1983). "The Insects. Structure and Function." 3rd Edn. Hodder and Stoughton, London.

Cousin, G. (1933). Étude biologique d'un Chalcidien, *Mormoniella vitripennis. Bulletin Biologique de France et de Belgique* **67,** 371–400.

Craig, G. B. (1967). Mosquitoes, female monogamy induced by male accessory gland substance. *Science* **156,** 1499–1501.

Dahms, E. C. (1973). The courtship behaviour of *Mellitobia australica. Memoirs of the Queensland Museum* **16,** 411–414.

Dahms, E. C. (1984). A review of the biology of species in the genus *Melittobia* (Hym., Eulophidae) with interpretations and additions using observations on *Melittobia australica. Memoirs of the Queensland Museum* **21,** 337–360.

Dautert-Willemzik, E. (1931). Einige Beobachtungen über das Geschlechtsleben der Männchen der Schlupfwespe *Nasonia brevicornis. Zoologische Anzeiger* **93,** 306–308.

Eller, F. J., Bartelt, R. J., Jones, R. L., and Kulman, H. M. (1984). Ethyl (Z)-9-hexadecenoate, a sex pheromone of *Syndipnus rubiginosus*, a sawfly parasitoid. *Journal of Chemical Ecology* **10,** 291–300.

Gonzalez, J. M., Matthews, R. W., and Matthews, J. R. (1985). A sex pheromone in males of *Melittobia australica* and *Melittobia femorata. Florida Entomologist* **68,** 279–286.

Goodpasture, C. (1975). Comparative courtship and karyology in *Monodontomerus* (Hym., Torymidae). *Annals of the Entomological Society of America* **68,** 391–397.

Gordh, G., and DeBach, P. (1976). Male inseminative potential in *Aphytis lingnanensis* (Hym., Aphelinidae). *Canadian Entomologist* **108,** 583–589.

Gordh, G., and DeBach, P. (1978). Courtship behavior in the *Aphytis lingnanensis* group, its potential usefulness in taxonomy, and a review of sexual behavior in the parasitic Hymenoptera (Chalc., Aphelinidae). *Hilgardia* **46,** 37–75.

Greenfield, M. D. (1981). Moth sex pheromones, an evolutionary perspective. *Florida Entomologist* **64,** 4–17.

Grissell, E. E., and Goodpasture, C. E. (1981). A review of nearctic Podagrionini, with description of sexual behavior of *Podagrion mantis* (Hym., Torymidae). *Annals of the Entomological Society of America* **74,** 226–241.

Hall, J. C. (1979). Control of male reproductive behavior by the central nervous system of *Drosophila.* Dissection of a courtship pathway by genetic mosaics. *Genetics* **92,** 437–457.

Hall, J. C., Siegel, R. W., Tompkins, L., and Kyriacou, C. P. (1980). Neurogenetics of courtship in *Drosophila. University of Missouri, Columbia, Stadler Symposium* **12,** 43–82.

Hamilton, W. D. (1967). Extraordinary sex ratios. *Science* **156,** 477–488.

Hamilton, W. D. (1979). Wingless and fighting males in fig wasps and other insects. *In* "Reproductive Competition and Selection in Insects" (M. S. Blum and N. A. Blum, eds.), pp. 167–220. Academic Press, New York.

Hölldobler, B. (1978). Ethological aspects of chemical communication in ants. *Advances in the Study of Behavior* **8,** 75–115.

Hölldobler, B. (1984). Evolution of insect communication. *In* "Insect Communication" (T. Lewis, ed.), pp. 349–377. Academic Press, London.

Hölldobler, B., and Wilson, E. O. (1983). The evolution of communal nest-weaving in ants. *American Scientist* **71,** 489–499.

Holmes, H. B. (1976). "*Mormoniella* Publications Supplement." Mimeographed edition, circulated by the author.

van Honk, C. G. J., Velthuis, H. H. W., and Röseler, P. F. (1978). A sex pheromone from the mandibular glands in bumblebee queens. *Experientia* **34,** 838–839.

Jachmann, F. (1983). "Over de Balts van *Nasonia vitripennis.*" MSc Thesis, University of Leiden.

Jachmann, F., and Putters, F. A. (in press). A two-factor model for the control of courtship of a parasitic wasp. *Behaviour.*

Jervis, M. A. (1979). Courtship, mating and "swarming" in *Aphelopus melaleucus* (Hym., Dryinidae). *Entomologists' Gazette* **30,** 191–193.

Jones, W. T. (1982). Sex ratio and host size in a parasitoid wasp. *Behavioural Ecology and Sociobiology* **10,** 207–210.

King, P. E. (1960). The passage of sperms to the spermatheca in *Nasonia vitripennis. Entomologists' Monthly Magazine* **96,** 136.

King, P. E., and Radcliffe, N. A. (1969). The structure and possible mode of functioning of the female reproductive system in *Nasonia vitripennis. Journal of Zoology, London,* **157,** 319–344.

King, P. E., Askew, R. R., and Sanger, C. (1969). The detection of parasitized hosts by males of *Nasonia vitripennis* and some possible implications. *Proceedings of the Royal Entomological Society, London* **44,** 85–90.

Kitano, H. (1975). Studies on the courtship behavior of *Apanteles glomeratus.* 2. Role of the male wings during courtship and the release of mounting and copulatory behavior in the males. *Kontyû, Tokyo* **43,** 513–521.

Krebs, J. R., and Davies, N. B. (1978). "Behavioural Ecology, an Evolutionary Approach." Blackwell Scientific Publications, Oxford.

Legner, E. F. (1985). Effects of scheduled high temperature on male production in thelytokous *Muscidifurax uniraptor. Canadian Entomologist* **117,** 383–389.

Leopold, R. A. (1976). The role of male accessory glands in insect reproduction. *Annual Review of Entomology* **21,** 199–221.

Lewis, W. J., Snow, J. W., and Jones, R. L. (1971). A pheromone trap for studying populations of *Cardiochiles nigriceps*, a parasite of *Heliothis virescens. Journal of Economic Entomology* **64,** 1417–1421.

Loher, W., and Huber, F. (1966). Nervous and endocrine control of sexual behaviour in a grasshopper (*Gomphocerus rufus*). *Symposium Society of Experimental Biology* **20,** 381–400.

Manning, A. (1962). A sperm factor affecting the receptivity of *Drosophila melanogaster* females. *Nature* **194,** 252–253.

Martelli, G. (1907). Contribuzioni alla biologia della *Pieris brassicae* e di alcuni suoi parassiti ed iperparassiti. *Bolletino del Laboratorio Zoologia Generale e Agraria di Portici* **3,** 86–149.

Maynard Smith, J. (1978). "The Evolution of Sex." Cambridge University Press, Cambridge.

Merle, J. (1968). Fonctionnement ovarien et réceptivité sexuelle de *Drosophila melanogaster* après implantation de fragments de l'appareil génital mâle. *Journal of Insect Physiology* **14,** 1159–1168.

Miller, M. C., and Tsao, C. H. (1974). Significance of wing vibration in male *Nasonia vitripennis* during courtship. *Annals of the Entomological Society of America* **67,** 772–774.

Nadel, H., and Luck, R. F. (1985). Span of female emergence and male sperm depletion in the female-biased quasi-gregarious parasitoid *Pachycrepoideus vindemiae* (Hym., Pteromalidae). *Annals of the Entomological Society of America* **78,** 410–414.

Obara, M., and Kitano, H. (1974). Studies on the courtship behavior of *Apanteles glomeratus* 1. Experimental studies on releaser of wing vibrating behaviour in the male. *Kontyû, Tokyo* **42,** 208–214.

Orr, D. B., and Borden, J. H. (1983). Courtship and mating behaviour of *Megastigmus pinus* (Hym., Torymidae). *Journal of the Entomological Society of British Columbia* **80,** 20–24.

Parker, G. A. (1974). Courtship persistence and female-guarding as male time-investment strategies. *Behaviour* **48,** 157–184.

Parker, G. A. (1978). Searching for mates. *In* "Behavioural Ecology" (J. R. Krebs and N. B. Davies, eds.), pp. 214–244. Blackwell Scientific Publications, Oxford.

Putters, F. A., Jachmann, F., and van der Poll, R. J. (in press). Time allocation by courting *Nasonia* males. *Behaviour.*

Riede, K. (1983). Influence of courtship song of the acridid grasshopper *Gomphocerus rufus* on the female. *Behavioural Ecology and Sociobiology* **14,** 21–27.

Saunders, D. S. (1961). Laboratory studies of the biology of *Syntomosphyrum albiclavus* (Hym., Eulophidae), a parasite of tsetse flies. *Bulletin of Entomological Research* **52,** 413–429.

Skinner, S. W. (1982). Maternally inherited sex ratio in the parasitoid wasp *Nasonia vitripennis. Science* **215,** 1133–1134.

Southwood, T. R. E. (1957). Observations on swarming in Braconidae (Hymenoptera) and Coniopterygidae (Neuroptera). *Proceedings of the Royal Entomological Society of London* **32,** 80–82.

Tagawa, J., and Hidaka, T. (1982). Mating behaviour of the braconid wasp *Apanteles glomeratus.* Mating sequence and the factor for correct orientation of male to female. *Applied Entomology and Zoology* **17,** 32–39.

Takahashi, S., and Sugai, T. (1982). Mating behavior of the parasitoid wasp *Tetrastichus hagenowii* (Hym., Eulophidae). *Entomologia Generalis* **7,** 287–293.

Thornhill, R., and Alcock, J. (1983). "The Evolution of Insect Mating Systems." Harvard University Press, Cambridge, Massachusetts.

Tinbergen, N. (1963). On the aims and methods of ethology. *Zeitschrift für Tierpsychologie* **20,** 410–433.

Velthuis, H. H. W., Velthuis-Kluppell, F. M., and Bossink, G. A. H. (1965). Some aspects of the biology and population dynamics of *Nasonia vitripennis. Entomologia Experimentalis et Applicata* **8,** 205–227.

Viggiani, G., and Battaglia, D. (1983). Courtship and mating behaviour in a few Aphelinidae. *Bolletino del Laboratorio Agraria Filippo Silvestri di Portici* **40,** 89–96.

Werren, J. H. (1980). Sex ratio adaptations to local mate competition in a parasitic wasp. *Science* **208,** 1157–1159.

Werren, J. H. (1983). Sex ratio evolution under local mate competition in a parasitic wasp. *Evolution* **37,** 116–124.

Werren, J. H., and van den Assem, J. (unpublished). Experimental analysis of a paternally inherited extrachromosomal factor in a haplodiploid insect.

Werren, J. H., Skinner, S. W., and Charnov, E. L. (1981). Paternal inheritance of a daughterless sex ratio factor. *Nature* **293,** 467–468.

Whiting, A. R. (1967). The biology of the parasitic wasp *Mormoniella vitripennis* (= *Nasonia brevicornis*). *Quarterly Review of Biology* **2,** 333–406.

Wilkes, A. (1965). Sperm transfer and utilization by the arrhenotokous wasp *Dahlbominus fuscipennis. Canadian Entomologist* **97,** 647–657.

Yoshida, S., and Hidaka, T. (1979). Determination of the position of courtship display of the young unmated male *Anisopteromalus calandrae. Entomologia Experimentalis et Applicata* **26,** 115–120.

6

The Genetic and Coevolutionary Interactions between Parasitoids and their Hosts

M. BOULETREAU

I. INTRODUCTION

In recent years, increasing attention has been directed towards genetic and evolutionary aspects of the relationship between hosts and parasites (Day, 1974; Taylor and Muller, 1976; Slatkin and Maynard Smith, 1979; Price, 1980; Thompson, 1982; Futuyma and Slatkin, 1983; Barrett, 1984; Pimentel,

1984) which is for several reasons an important subject in the field of evolutionary biology. Firstly, all living species, both procaryotes and eucaryotes, are associated with parasitism either as parasites (more than half of all living species; Price, 1980) or as hosts. Secondly, parasitism plays an important role in the limitation of natural host populations as well as in the balance of ecosystems. Thirdly, parasitism is the most specialized of interspecific associations and involves a certain co-adaptation between two partners. What is the role of genetic interactions in the acquisition, maintenance and development of this co-adaptation?

This chapter examines genetic interactions in a particular kind of parasitic relationship, that which occurs between insects and their parasitoids. After reviewing some basic ideas and concepts, I will analyse the biological data from the literature and from our experiments on the *Drosophila*-parasitoid system.

A. Some Basic Ideas: Coevolutionary Interactions between Parasites and Hosts

Parasitism represents an intimate interspecific association with a negative impact on one of the partners (Price, 1980). The parasite reduces the fitness of individual hosts and thus may act as a selective factor to instigate evolutionary changes in the host population. Selection can operate in both directions and the maintenance of the parasitic association is subject to the balance between these reciprocal constraints.

We often forget that parasitism is asymmetric. In the case of obligate parasites (Day, 1974), parasites cannot avoid the constraints exerted by the host. In contrast, for the hosts, parasitism and its associated constraints are contingent upon infection. This asymmetry has the following consequences:

1. The host exerts continuous effects on all parasitic individuals, while reciprocal effects are discontinuous and concern only a part of the host population
2. The effect of a parasite on its host can be of more variable intensity, and can even be negligible. The number of parasites available to a given host is generally greater than the number of hosts available to a given parasite
3. Finally, at the methodological level, it is in principle easy to study the effects of parasites on their hosts, as one can find or create situations where these effects are modified or eliminated. However, effects of hosts on their parasites are more difficult to evaluate because we can tamper only slightly with their nature and intensity.

Direct and reciprocal genetic effects are most frequently evoked when we speak of interspecific genetic interactions. They are illustrated by the "gene for gene" relation established by Flor (1956) in the flax-rust association, modelled by Person (1966), and found in other plant-fungus associations (Day, 1974; Barrett, 1983, 1984) as well as in plant-insect associations (Gallun, 1977): a series of genes or alleles controlling the resistance in the host corresponds to a series of genes or alleles controlling the virulence in the parasite. This mechanism brings to mind the antigenic variability in trypanosomes (Turner, 1982) of which the molecular base is not well known (Kolata, 1984). These selective, reciprocal effects influence the characteristics directly involved in the parasitism, bound to the fitness of the individual, and of which the variability is genetically determined. When such links exist between the variations of the two partners, each variation in partner A involves a variation in partner B, which then reacts on partner A. This is the coevolutionary system, in the restrictive sense of Labeyrie (1978) and Janzen (1980), which implies a retroactive loop.

As well as these direct interactions, other less evident interactions can exist in which reciprocity rests on pleiotropic effects or on the correlation between characters (Clarke, 1976). Group selection may also operate, with consequences different from those predicted by individual selection. Price (1980) suggests that numerous species, and in particular parasites with their highly subdivided populations, could be subject to this type of selection. It is generally by mechanisms of this type (Levin and Pimentel, 1981) that we explain the evolution towards the intermediate form of virulence as found in the myxomatosis virus of rabbits (Fenner and Ratcliffe, 1965).

The theoretical and mathematical basis of coevolutionary interactions have been analysed in numerous studies (Slatkin and Maynard Smith, 1979; Anderson and May, 1982; Levin, 1983; May and Anderson, 1983). Deviating from the popular idea that evolution towards an attenuated form of virulence is common in parasitic associations, May and Anderson (1983) and Barrett (1984) suggested that this evolution depends on the interaction between the parasite's virulence and its transmissibility, and the cost of resistance acquisition in the host. According to the particular circumstances and the biology of the partners, we can end up with polymorphisms which are stable, cyclic or chaotic, with various degrees of resistance or virulence.

As with all selective processes (Lewontin, 1974), coevolutionary interactions can only take place if at least two conditions are met: firstly, the existence, in the populations concerned, of variability in characters associated with parasitism and secondly, the existence of genetic determinants to this variability. The importance of genetic variability in the interactions between hosts and parasites was emphasized by Clarke (1976) and Barrett (1984). In practice, this variability can be studied using several methodological approaches, each of which has its limitations:

1. Analysis of genetic variability in natural populations
2. Study of variation in the host and parasite populations and its consequence for different combinations of variants
3. Research on geographical variation in the two partners, and parallels between them
4. Laboratory study of evolution over several generations of hosts and parasites in experimental systems.

II. THE UNIQUE STATUS OF INSECT PARASITOIDS

The distinction between parasites and parasitoids is based largely on the nutritional aspects of the relationship with the host. From the genetic and evolutionary point of view, insect parasitoids distinguish themselves by a number of characteristics.

A. The Similarity of the Two Partners

In the most frequently studied coevolutionary associations (e.g. plant-fungus or mammal-virus), the two partners originate from taxa that are very different (Mayr, 1963). However, insect parasitoids usually come from the same taxonomic class as their hosts. As an initial consequence, the genetic complexity of both partners should be the same. Secondly, both partners have similar biologies: their generation time, reproductive rate and powers of dispersion are of the same order of magnitude. Thus, their evolutionary capacities are the same. The situation in other parasitic associations is generally quite different: the speed of evolution of many helminths is for example, slower than that of their hosts (Holmes, 1983). The inverse frequently occurs (Price, 1977, 1980) as in the rabbit-virus association, where the parasite evolves much quicker than its host (Anderson and May, 1982). On the coevolutionary level, the large biological and genetic similarity between insect parasitoids and their hosts allows us to consider both partners as being “equally armed”.

B. The Responses of Insect Parasitoids to their Hosts

The key conditions of successful parasitism are reviewed by Smith Trail (1980), and are easily applied to parasitoids. For a parasitoid-host interaction to persist, parasitoids must ensure the disperson of propagules to new hosts, the acquisition of sufficient energy from the host to assure growth, and the survival of the host until reproductive maturity of the parasitoid. Hosts, in turn, must be able to limit the effects of reduced fitness arising from parasitism. This may involve avoiding infestation or, for the infested host, killing the parasitoid. Protection of related host individuals is also conceivable.

The perpetuity of the association is only assured if compatible solutions to these questions are achieved by both partners. As a result, there are a number of constraints on parasitism: habitat and microhabitat overlap of parasitoid and host (ecological level); a dispersion system that puts the two partners in contact (behavioural level); immunological compatibility in the broad sense (immunological level); and physiological compatibility (e.g. biochemical, endocrine, etc.—the physiological level). The responses of insect parasitoids to their hosts on these different levels have been extensively reviewed (Salt, 1935; Flanders, 1953; Doutt, 1959; Vinson, 1976; Vinson and Iwantsch, 1980a, 1980b; see also Chapters 2 and 4). I will now examine possible genetic variability in parasitoids and hosts at the ecological, behavioural, immunological and physiological levels and their consequences for the parasitoid-host interaction.

III. ECOLOGICAL AND BEHAVIOURAL ASPECTS

A. Reactions of Parasitoids to the Habitat of their Hosts

The presence of parasitoid and host in the same habitat is the result of the active behaviour of the parasitoids. Stimuli emanating from the host or from its food substrate play a decisive role in host finding (see reviews in Vinson, 1975, 1976, 1981, 1984b; Price, 1981). The diverse plant hosts used by the same phytophagous insect can have very different attractive effects on the parasitoids. For example, Arthur (1962) has demonstrated that *Itoplectis conquisitor* Say is more attracted by the odour of Scots pine than by that of red pine. Correspondingly, it attacks its host *Rhyacionia buoliana* (Denis and Schiffermüller) more on the first tree species than on the second. In the same fashion, the relative attractiveness of collard and beets to the parasitoid *Diaeretiella rapae* (McIntosh) explains different levels of parasitism of the host, *Myzus persicae* (Sulzer), reared on both plant species (Read *et al.*, 1970). Therefore it seems likely that for a generalist phytophagous insect, parasitism rate varies in relation to the plant species consumed. If polymorphism for host plant or habitat preference exists in generalist species (niche variation hypothesis) as suggested by van Valen (1965), then this polymorphism is not neutral with respect to parasitism. A reciprocal selective process can develop, affecting the behavioural response of host and parasitoid towards the stimuli emanating from the host plant.

This scenario neglects the role of possible competitors (hosts or parasitoids) and supposes the existence of a genetic determinant in the variability of the host behaviour. We are aware of some examples of genetic variations in the plant-herbivore association between populations of Colorado beetle, *Leptinotarsa decemlineata* (Say), (Hsiao, 1982) and *Drosophila tripunctata* Loew (Jaenike and Grimaldi, 1983), and within the populations of *D. tripunctata* (Jaenike and Grimaldi, 1983), *D. melanogaster* Meigen (Hoffman *et al.*, 1984), and *Liriomyza brassicae* (Riley) (Tavormina, 1982). A complete analysis is given by Futuyma and Peterson (1985). It is astonishing that none of these authors have suggested the possible role of parasitoids in the maintenance of this variability.

The interactions between the host's habitat and the parasitoid are not limited to behavioural aspects, but also include physiological aspects of which the parasitoids of *Drosophila* provide a good example. *Drosophila* species which exploit fermenting substrates, notably *D. melanogaster* and *D. lebanonensis* Wheeler, demonstrate a strong resistance to elevated concentrations of ethanol (up to 15%), which is considered a result of the process of natural selection (David and van Herrewege, 1983). Carton (1976) has shown that the cynipid *Leptopilina boulardi* (as *Cothonaspis* sp.) Barbotin, Carton & Kelner-Pillault, a parasitoid of *D. melanogaster* larvae, was attracted by ethanol and demonstrated, like its host, a certain tolerance for this product. A more complete study (Boulétreau and David, 1981) compared the ethanol tolerance of various species of parasitoids which are dependent on *Drosophila* (Cynipidae, Braconidae, Pteromalidae) in diverse and geographically distinct populations. In general, larval parasitoids were found to be more resistant to alcohol than pupal parasites. Species specific to *D. melanogaster* are more tolerant than polyphagous species, and a net parallelism exists for some species between the variations in ethanol tolerance of the parasitoid and those that are known in its particular host (David and Bocquet, 1975). There is also a correlation between the resistance of parasitoids and the intensity of their contacts with ethanol (David and van Herrewege, 1983). The relative protection from which *Drosophila* larvae benefit in these toxic environments is overcome by the specific parasitoids which suffer the same selective environmental pressures as their hosts and have themselves evolved a substantial resistance to ethanol.

B. Direct Behavioural Interactions between Parasitoids and Hosts

The mechanisms involved in the discovery of a host, as well as its acceptance by the parasitoid, are extraordinarily complex and sometimes result in a narrow specificity (Vinson, 1975; Arthur, 1981; Weseloh, 1981). Jones *et al.* (1971) identified the substances emitted by *Heliothis zea* (Boddie) which elicit an antennal response by the parasitoid *Microplitis croceipes* (Cresson). A small molecular change (switching a methyl group) is sufficient to reduce the effectiveness of the product. Some analogous results were reported by Vinson *et al.* (1975) in the association between *Cardiochiles nigriceps* Viereck and *Heliothis virescens* (F.). Dosage effects can complicate the situation (Read *et al.*, 1970; Corbet, 1973).

In view of such specific associations, it seems inevitable that genetic variations exist in the host (e.g. in the nature and doses of emitted substances) as well as in the parasitoid (e.g. in the sensitivity of the receptors). For example, Lanier *et al.* (1972) demonstrated differences in the reactions of the pteromalid *Tomicobia tibialis* Ashmead to kairomones of hosts (*Ips pini* Say) from different geographical origins. In natural populations, polymorphisms caused by such critical systems could lead to a variability in the infestation rate and be of considerable evolutionary interest. Unfortunately, as Price (1981) mentioned, we still lack data in this area of research.

Genetic analysis by the isofemale strains method (Parsons, 1980) has demonstrated significant heterogeneity, under experimental conditions, in the rate of parasitism by the cynipids *Leptopilina boulardi* and *L. heterotoma* (Thomson) (Wajnberg, 1986; Boulétreau and Wajnberg, in press) of a *D. melanogaster* population of Tunisian origin. One explanation of this variation may be the behavioural variability of *Drosophila* larvae. Bauer and Sokolowski (1984) established the existence of genetically based, intrapopulation variability in larval behaviour, while Carton and David (1985) found a correlation between the burrowing depth of larvae and parasitism rates by *L. boulardi*.

Chabora (1972) and Olson and Pimentel (1974) tried to vary, using directional selection, the rate of parasitism of *Musca domestica* L. pupae by the pteromalid *Nasonia vitripennis* (Walker). These experiments incorporated both physiological and behavioural aspects and will be analysed later. Note, however, that Olson and Pimentel (1974) succeeded in lowering very appreciably the parasitism rate of hosts over 40 generations. In the same fashion, experimental populations of parasitoids and hosts, followed for several generations, showed a decline in effectiveness of the parasitoids for *Musca-Nasonia* associations (Pimentel *et al.*, 1963; Pimentel and al-Hafidh, 1965; Pimentel and Stone, 1968; Zareh *et al.*, 1980) as well as for *Drosophila-*

Leptopilina associations (see later discussion). This reduction of attack rates seems due, at least in part, to the effect of host evolution on the behavioural response of the parasitoid.

Considered as a whole, the process of host selection is very important in parasitoid-host associations and its genetic study is difficult. Numerous non-genetic influences are superimposed on the genetic determinants, and individual behaviour is susceptible to an equally large flexibility due to conditioning and associative learning (see Vinson, 1984a; and Chapter 2). Research on genetic determinants of behaviour aims to confirm the existence of strong selective pressures at this level and the possibility of adaptive responses in both partners.

IV. IMMUNOLOGICAL ASPECTS

Only endoparasitic parasitoid species are concerned with the immune defences of the hosts. Compared to those of vertebrates, the immune mechanisms of insect hosts are characterized by their weak specificity and their non-inducibility (Lackie, 1980; Holmes, 1983). Thus, the probability of a parasitoid eliciting an immune response from a host is independent of previous challenges. It seems logical to assume that variation in these mechanisms is genetically based.

Several defensive mechanisms are known in insects but the most frequently evolved is the formation of a haemocyte capsule around the parasitoid egg or larva (Salt, 1970). This capsule, normally pigmented, kills the parasitoid by asphyxia or by starvation (Fisher, 1971). The parasitoid can escape these defences in two ways: it can be placed in tissues where an immunological response is not elicited (an hypothesis suggested by Salt, 1963, 1968) or it can exert an immunodepressive effect, activated perhaps by intermediate viral particles (Stoltz and Vinson, 1979; Edson *et al.*, 1981) or by specific substances injected at the time of oviposition (Rizki and Rizki, 1984), or secreted by the eggs or larvae (Boulétreau and Quiot, 1972; Kitano and Nakatsuji, 1978; see also Chapter 4).

The efficiency of the host's defensive system may be measured by the proportion of rejected parasitoids and depends on both the host's ability to encapsulate the parasitoid and the parasitoid's ability to avoid this reaction. Much research has shown that the efficiency of a defensive system for a given parasitoid-host association can vary, without appearing to follow a general pattern, according to the conditions of the infestation (e.g. age and health of the hosts, superparasitism) and environmental conditions (e.g. temperature) (Muldrew, 1953; van den Bosch, 1964; Salt and van den Bosch, 1967; Morris, 1976; Blumberg and DeBach, 1981; Nappi and Silvers, 1984).

A. Variations in Field Situations

The intensity of the host's defensive reaction can vary over time. The evolution of the immune response of the tenthredinid *Pristiphora erichsonii* (Hartig) against its parasitoid *Mesoleius tenthredinis* Morley, introduced into Canada in 1910, is a classic example of a parasitoid losing effectiveness because of the development of an immunological resistance by the host (Muldrew, 1953; Turnbull and Chant, 1961). However, this rapid evolution was not uniform, as the parasitoid has remained effective in certain regions of Canada. Furthermore, other strains of the same parasitoid are able to develop in hosts which are resistant to the first strain (Messenger and van den Bosch, 1971).

Bathyplectes curculionis (Thomson), a parasitoid of *Hypera postica* (Gyllenhal) and *H. brunneipennis* (Boheman), provides a reciprocal example of the evolution of immunological interactions. Analyses by van den Bosch (1964) and Messenger and van den Bosch (1971) demonstrated that the encapsulation rate of *B. curculionis* by a new host, *H. brunneipennis*, decreased over a period of 15 years from 35–40% to 5%. This evolution was accompanied in the parasitoid by a partial loss of resistance to the immune defences of the original host, *H. postica*.

The evolution of immune responses in such newly created parasitic associations may be explained by the selective advantage conferred on hosts capable of rejecting the parasitoid, and the advantage to parasitoids of avoiding rejection. This hypothesis was prudently suggested by van den Bosch (1964) and then reiterated by Salt and van den Bosch (1967) and by Messenger and van den Bosch (1971). It is currently accepted despite the lack of experimental evidence and despite its inability to explain certain findings.

In the case of one natural parasitic association, Morris (1976) established that the percentage encapsulation of *Hyphantria cunea* (Drury) parasitoids varies in different regions of Canada, and that this level is not consistent from one year to the next. These variations were found to correlate with short-term changes in the genetic structure of the hosts, caused by annual climatic fluctuations, and therefore are not directly related to the effect of parasitoids. In addition, Bartlett and Ball (1966) have shown that encapsulation rates of *Metaphycus luteolus* (Timberlake) vary according to the geographical origin of its host, *Coccus hesperidum* L. These variations do not appear to be related to the history of the interaction.

These examples demonstrate that in newly created parasitic associations, the immune phenomena are not definitely fixed and evolution is possible. This evolution can follow various paths, depending on the region considered, so that a differentiation between local populations can occur.

B. Studies on Parasitoids of *Drosophila* spp.

Genetic analysis of the determinants of encapsulation rate has been undertaken for the parasitoids of *Drosophila* spp. Walker (1959) and Hadorn and Walker (1960) found that Swiss and Egyptian populations of *D. melanogaster* encapsulated the cynipid *Leptopilina heterotoma* (as *Pseudeucoila bochei* Weld) at very different rates (5–95%). In addition, the hybrids between these two strains have characteristics which are intermediate between the parents (Hadorn and Walker, 1960). Artificial selection of these hybrids over 12 generations resulted in a considerable increase in the encapsulation rate of the parasitoid, from 55 to 85%. Walker (1962) showed that in crosses between strains of parasitoids which differed in encapsulation rate, first-generation parasitoids were identical to their mothers. Male descendants of unmated females from the population of F1 hybrids were identical and intermediate between their grandparents. Thus, the parasitoid must contain a semi-dominant gene that inhibits encapsulation by the host. Physiological and biochemical support for this hypothesis was found by Rizki and Rizki (1984), who demonstrated that parasitoid eggs are protected from host immune actions by a specific immunodepressive substance (lamellolysin), which is a product of maternal genome, injected by the female at the moment of oviposition.

In the association between *D. melanogaster* and *L. boulardi*, we find clear local differences with regard to the encapsulation rate of the parasitoid. This rate is zero in Tunisia (Wajnberg *et al.*, 1985), weak in France (Boulétreau and Fouillet, 1982), and more elevated in tropical Africa (Carton and Boulétreau, 1985). Furthermore, genetic analysis of isofemale lines has shown that wherever the encapsulation reaction exists, there is a significant variability in *D. melanogaster* populations (Fig. 1).

Taken together, the results presented above strongly suggest that the ability of *D. melanogaster* to encapsulate parasitoids is genetically determined, and that natural populations are not homogeneous with regard to this character. Furthermore, the presence of a haemocyte capsule in the body of an adult fly reduces its reproductive potential (Carton and David, 1983). Thus the conditions for a coevolutionary interaction, in the restrictive sense of Janzen (1980), are satisfied; the characters are genetically determined, they are variable, and they are associated with the individual fitness of both host and parasitoid. Does coevolution occur under natural conditions?

Table I gives the test results of cross infestation between strains of *D. melanogaster* and *L. boulardi* from different geographical origins. Clear differences exist in the intensity of reaction depending on the origin of the host. Populations can be classified by order of increasing host reactivity:

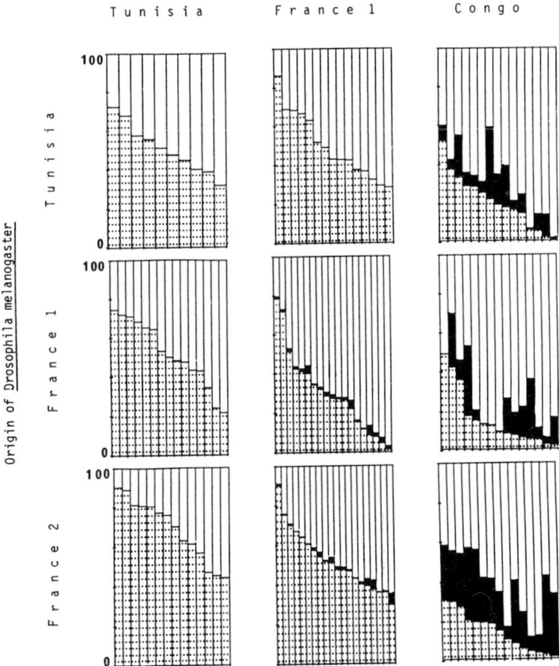

Fig. 1. Variability of success of *Leptopilina boulardi* development in larvae of *Drosophila melanogaster* (different strains). Each column represents one isofemale strain of *Drosophila* and is based on three replicates of 100 hosts. Lines are ordered according to percentage of wasp emergence. Stippled: emerged wasps; black: emerged flies with a capsule; white: failure of host and parasitoid. Results are expressed as percentages of infested hosts.

Table I. Rate of encapsulation of *Leptopilina boulardi* by *Drosophila melanogaster* larvae with respect to the origin of parasitoids and hosts

	Origin of parasitoids		
Origin of hosts	Tunisia	France 1	Congo
Tunisia	0	0	10.8%
France 1	0	1.5%	11.7%
France 2	0	1.7%	24.0%
Congo	0	—	51.8%

Tunisia < France 1 < France < Congo

In the same way, parasitoid populations can be classed by increasing order of protection against the host's reactions:

Congo < France < Tunisia

The rate of capsule formation in the sympatric interactions shows the same order as that of the reactivity of the hosts:

Tunisia/Tunisia < France/France < Congo/Congo

Consequently, in Tunisia *D. melanogaster* with weak defensive reactions are associated with strongly protected parasitoids. Conversely, in the Congo *D. melanogaster* with strong defensive reactions are associated with weakly protected parasitoids. This paradoxical situation is contrary to the expected coevolutionary mechanisms which allow prediction of a certain harmony between reactivity in the hosts and protection in the parasitoids.

How can the observed differences between populations be explained? One possible hypothesis is that these geographical variations are simply the pleiotropic effects of local geographical differentiations in the host (Oakeshott *et al.*, 1982) and in the parasitoid (Nordlander, 1980). They would thus correspond with mechanisms similar to those which Morris (1976) found in *Hyphantria cunea* (Drury), and would not be associated with any coevolutionary interaction.

A second hypothesis considers the *Drosophila*-parasitoid association on a broader perspective. In the case of the Tunisian population, we know that *L. boulardi* exploits not only *D. melanogaster* larvae but also those of *D. simulans* Sturtevant, which have an elevated encapsulation rate (9.6%). Perhaps the strong reactivity of *D. simulans* larvae is responsible for the acquisition by the *L. boulardi* larvae of an increased level of protection. This protection, while insufficient for totally escaping encapsulation by *D. simulans*, would be efficient against encapsulation by *D. melanogaster* larvae, which are less reactive. Similarly, the strong reactivity of Congo strains may be attributable to the local selective effects of a parasitoid other than *L. boulardi*. This hypothesis would assume that host immunological responses and parasitoid resistance are weakly specific in a taxonomic sense, which is consistent with the findings of Streams and Greenberg (1969).

In summary, the existence of intraspecific variations in the encapsulation rate of parasitoids by their hosts is well-established. The efficiency of encapsulation observed in nature is a local characteristic, susceptible to evolution. The causes of this evolution are still poorly understood and there is probably no general rule. The role of coevolutionary interactions in the host-parasitoid systems so far considered remains to be demonstrated. Multi-species interactions may be important, as may processes independent of parasitoid attack, as suggested by Bartlett and Ball (1966). Thus, at present it is not possible to predict how a newly created parasitic association will evolve, and it may be fruitless to look for a correlation between the observed encapsulation rate and the age of the parasitic association.

V. PHYSIOLOGICAL ASPECTS

The suitability of a host to a parasitoid larva which has escaped encapsulation depends on numerous biochemical, physiological, nutritional and endocrinal factors which have recently been reviewed (Vinson and Iwantsch, 1980a, 1980b; Vinson, 1984a; Beckage, 1985). Of the many known causes of variation in host suitability we are concerned only with those which have a genetic basis.

The work of Utida (1957) on a *Callosobruchus-Heterospilus* system, and that of Takahashi (1963) on an *Ephestia-Nemeritis* system suggest a selective effect of the parasitoid on the biotic potential of the host. However, the most detailed analyses to date are studies by Pimentel's group on the *Musca domestica* L. (or *Lucilia sericata* (Meigen))—*Nasonia vitripennis* (Walker) system, and our work on the *Drosophila melanogaster*—*Leptopilina boulardi* system. Similar experimental approaches have been adopted in both projects, including research on variability within and between populations, attempts at directional selection and maintenance of mixed experimental populations over several generations. This last approach provides complex results, which demand delicate interpretation and are subject to controversy. These experimental approaches do not always permit the clear distinction between the physiological and behavioural components of the parasitoid's success.

A. Studies on *Nasonia vitripennis*

Under the impetus of Pimentel, a series of studies was undertaken on the association between the chalcid *Nasonia vitripennis* (a gregarious ectoparasitoid of Diptera pupae), and *Musca domestica* (Muscidae) or *Lucilia sericata* (Calliphoridae). The experiments were designed to test Pimentel's hypothesis (1961) of the genetic feedback mechanism, which may be summarized as follows: the population of the consumed species (plant, herbivore, etc.) is put under selective pressure by the consumer population, which leads to the evolution of a form of resistance. This evolution in the consumed species results in a reduced effectiveness for the consumer and thus diminishes its numbers and the corresponding selective pressure. With lower selective pressure by the consumer, the consumed species may lose some of its resistance, and the consumer may increase in number again. The resulting feedback loop would be capable of regulating, or at least, in association with density dependence, of contributing to, the regulation of population size (Huffaker *et al.*, 1984).

In these experiments, the parasitoid was used as the principal selective factor and its activity was maintained at a constant, elevated level. Some of these systems were spatially partitioned (Pimentel *et al.*, 1963; Pimentel and Stone, 1968; Chabora and Pimentel, 1970) which allowed both species to fluctuate freely in numbers. Other systems were not partitioned and fluctuation was possible either in the parasitoid population only (Pimentel and al-Hafidh, 1965) or in both species (Zareh *et al.*, 1980). These different systems have been maintained over a number of generations and have yielded the same results: rapid reduction in the parasitoid's impact and the attenuation of fluctuations in host numbers (Pimentel *et al.*, 1963; Pimentel and al-Hafidh, 1965; Pimentel and Stone, 1968; Zareh *et al.*, 1980). Parallel tests show that the decline in the parasitoid population can be attributed to a qualitative evolution in the hosts. These hosts are less suitable to parasitoids for nourishment of the adults (i.e. in the nutritive value of body fluids imbibed by the females and used for egg production), for oviposition (pupal cuticle characteristics), and for the development of pre-imaginal stages. This evolution of the host is accompanied by a reduction in the host's biotic potential (Zareh *et al.*, 1980); it is interpreted as a form of resistance to the parasitoid, as in the sense of Pimentel's hypothesis (1961).

To confirm the role of the host's genotype in the general reproductive capacity of the parasite, Chabora (1970a, 1970b, 1970c) has compared the performances of *N. vitripennis* on strains of *M. domestica* and *L. sericata* with different geographical origins. The production of descendants on *M. domestica* was eight times higher on a host strain from Florida than on one from New York. Further, Chabora and Chabora (1971) compared the success of the two parasitoids, *N. vitripennis* and *Muscidifurax raptor* (Girault & Saunders), on both strains of *Musca domestica* and their F1 hybrids. The two parasitoid species showed a significantly higher reproductive success on the F1 hybrids than on any of the parental strains. Thus, *M. domestica* and *L. sericata* demonstrate geographical variation in their response to parasitism and these variations appear to be the result of differences in nutritive value, related to their genetic make-up. These variations could not be associated with local selective effects of the parasitoid: the natural infestation rate on *L. sericata* is not known (Chabora, 1970a), and the parasitoid does not attack *M. domestica* in nature (Chabora, 1970b). It is more probable that the variations are due to the pleiotropic effects of local genetic differentiations (Chabora, 1970a, 1970b).

Finally, Chabora (1972) attempted to vary the host's characteristics by directional selection on the Florida and New York strains of *M. domestica.* Over 10 generations, hosts were selected for their increased ability to escape parasitism. In each generation, the experimental design was arranged so that the rate of parasitism reached 50%. Evolution occurred in the direction expected but was not the same in the two different strains. In addition, the changes occurred in non-selected characters associated with the nutritive value of host pupae: female parasitoid longevity and mortality during pre-imaginal stages. Lastly, in the Florida strain the non-selected reference line demonstrated the same variation as the selected line, which may be attributable to the ability of *M. domestica* to respond adaptively to laboratory conditions.

The same experiments have been repeated by Olson and Pimentel (1974) on the New York strain of *M. domestica* using a strong selection pressure (90% of host pupae infested in each generation). Over 10 generations the rate of parasitism among exposed pupae dropped considerably. The mean number of descendants per female parasitoid was reduced from 269 to 89 and the longevity of females decreased from 11 to six days. This evolution continued until the 20th generation and then stabilized. By the 40th generation 75% reduction in the parasitoid's reproductive potential was achieved.

Taken together, these studies establish that the reproductive success of *Nasonia vitripennis*, with respect to the nutritional needs of the parasitoid adults and larvae, is strongly influenced by the genotype of the host. Host variation can be produced in response to selective pressure exerted by the parasitoid, which conforms to the hypothesis of a genetic feedback mechanism. In addition, the experiments demonstrate that the host can also evolve without the intervention of parasitism, as per Chabora's hypothesis of pleiotropic effects (1972).

Pimentel's hypothesis (1961) of a genetic feedback mechanism and the research on *N. vitripennis* suggest that exotic parasitoids, involved with a new parasitic association, will be better control agents than indigenous parasitoids. This hypothesis has produced a controversy among biological control specialists (Huffaker *et al.*, 1971, 1976) that does not appear to be settled (Hokkanen and Pimentel, 1984; see Chapter 10).

B. Studies on Parasitoids of *Drosophila* spp.

The *Drosophila-Leptopilina* system has the advantage that it involves a solitary larval endoparasitoid, for which we expect a closer adjustment between the physiology of parasitoid and host and a better opportunity for discrimination of the behavioural and physiological aspects.

The degree of physiological adaptation between the host larva and parasitoid larva can be quantified in terms of the success of parasitoid development. In the absence of an immune reaction, the parasitized larva can either allow the emergence of the adult parasitoid (developmental success) or die, thus killing the parasitoid. The ratio of parasitoids obtained to parasitized larvae is the key important quantitative parameter. It is, however, very sensitive to environmental conditions and depends strongly on the nutritional state of the hosts. Undernourishment of the host larvae results in a twofold augmentation in developmental success for *Leptopilina boulardi* (Prévost, 1985; Wajnberg *et al.*, 1985; Boulétreau and Wajnberg, in press) and a clear increase for *L. heterotoma* (Wajnberg, 1986; Boulétreau and Wajnberg, in press). Also, despite experimental precautions to standardize the individuals tested and their parents (physiological state, age, etc.), a variability appears between successive experiments. Genetic study of such a parameter is only possible at the price of numerous replicates, and only simultaneous tests offer reliable conclusions.

1. Intrapopulation Variability

The method of isofemale lines has revealed a strong interfamilial variability in *Drosophila* populations (Boulétreau and Fouillet, 1982; Wajnberg *et al.*, 1985; Wajnberg, 1986; Boulétreau and Wajnberg, in press). This variability also expresses itself in *L. heterotoma* and appears in all populations tested, whether the parasitoid is of sympatric origin or not (Fig. 1). In the same population, individuals from different families are not equally suitable to parasitoid development and in certain cases, the interfamilial distances are great (between 5 and 80% success in the France 1 population) and remain so over at least two generations. Similar variability has been found in the Tunisian population of the parasitoid *L. boulardi*.

2. Comparisons Between Populations

Table II compares the results of tests involving three *D. melanogaster* populations from Tunisia, the south of France (France 1) and central France (France 2, a population not infested by *L. boulardi* in nature) and three populations of *L. boulardi*: Tunisia, south of France (France 1) and the

Table II. Success of development of *Leptopilina boulardi* in *Drosophila melanogaster* larvae with respect to the origin of parasitoids and hosts. Means are given as percentages with their standard errors and (in brackets) the number of batches of 100 hosts. Only parasitoids which have evaded encapsulation are considered

	Origin of parasitoids		
Origin of hosts	Tunisia	France 1	Congo
Tunisia	50.7 ± 4.1 (30)	47.2 ± 4.4 (45)	27.1 ± 4.1 (45)
France 1	52.0 ± 4.4 (45)	30.1 ± 4.5 (57)	18.6 ± 4.5 (45)
France 2	69.0 ± 4.3 (43)	52.9 ± 3.4 (60)	20.4 ± 3.7 (45)

Congo. Each test involved 5000 to 10,000 host larvae, and it was not possible to undertake all tests simultaneously. We must therefore be content with a general analysis of the results. The developmental success of a parasitoid depended on the origin of both host and parasite. The hosts can be classified by order of increasing suitability (France 1 < Tunisia < France 2) and the parasitoids in order of increasing exploiting efficiency (Congo < France 1 < Tunisia).

A striking result of these experiments is that parasitoid developmental success in sympatric infestations (parasitoids and hosts of the same origin) showed considerable variation between the populations and was not better than in non-sympatric situations. The *Drosophila* population France 2, which is not naturally infested with *L. boulardi*, showed the best parasitoid development. This result contradicts Carton's hypothesis (1984) of genetic co-adaptation in the two partners of sympatric populations. Furthermore, it seems to indicate that, as for immune phenomena, there is no simple relation between parasitic development success and the age of the association between the two species.

3. *Selection Studies*

A Tunisian population of *D. melanogaster* was selected for extremes of developmental success with respect to parasitism by its sympatric parasitoid *L. boulardi* (Fig. 2). Because of the duration of parasitoid development (18 to 20 days), selection occurs only in each second generation of *D. melanogaster*. After the fourth and fifth generations selected (eighth and 10th real generations), the two groups had been totally separated. Continuing the same

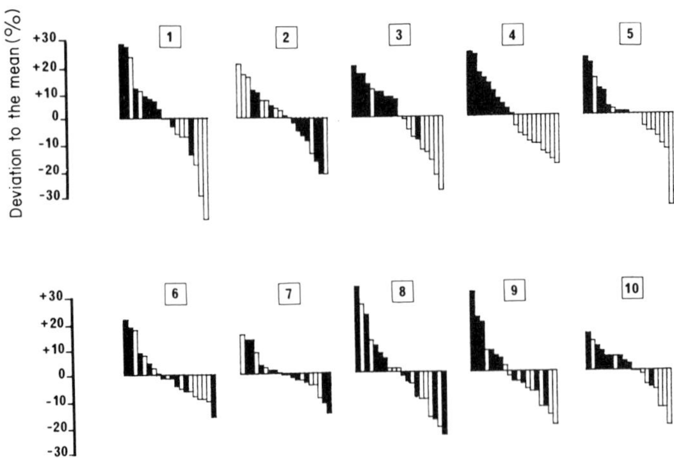

Fig. 2. Selection of *Drosophila melanogaster* for suitability to the development of *Leptopilina boulardi*. At each generation (1–10) families are tested in the higher line (black columns) and the lower line (white columns) for their suitability to the parasitoid. Within the higher and the lower lines, respectively, the two best families and the two worst ones are used as founders of the next generation.

selective pressure did not maintain this distinction during the next three generations, but it reappeared in the ninth and 10th selected generations. The reciprocal F1 descendants of this 10th generation demonstrated a significant maternal effect.

Predictably, this experiment does not reveal simple genetic determinants in the host for adaptation to the parasitoid, but does suggest that at least some of the characters involved have genetic determinants.

4. The Evolution of Experimental Populations

Population cages for *Drosophila* spp. permit the indefinite maintenance of mixed populations of *D. melanogaster* and *L. boulardi* in an unpartitioned space, where the numbers of both species fluctuate freely with overlapping generations, within the limits of the cage.

Two experimental populations (A and B) were set up with Tunisian hosts and parasitoids and maintained for 87 weeks (about 50 host and 25 parasitoid generations). Figure 3 presents the host and parasitoid production in each cage (estimated by weekly census of dead adults). The parasitoid numbers were much higher than those of the hosts and showed strong fluctuations with a high peak around weeks 17–20. The numbers of *D. melanogaster* were less variable, with an unremarkable maximum at week 35.

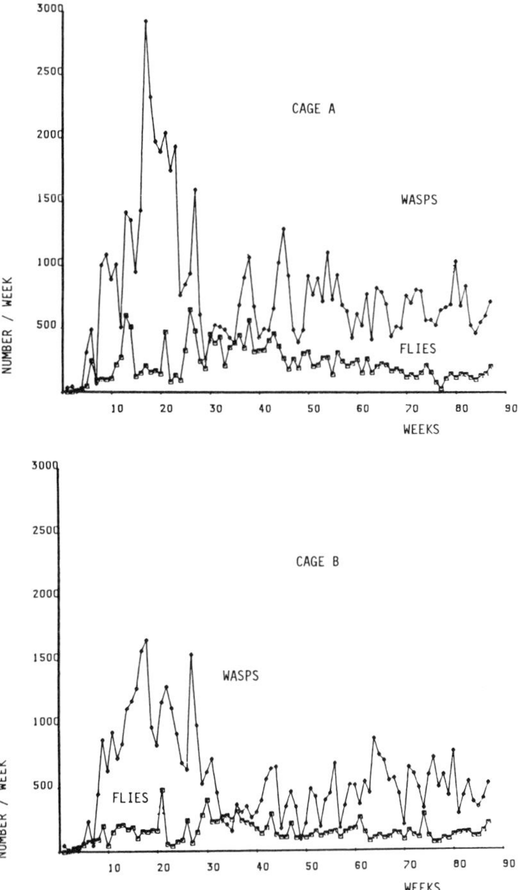

Fig. 3. Weekly production of *Drosophila melanogaster* and *Leptopilina boulardi* adults in experimental populations.

There was no correlation between the numbers of hosts and parasitoids. In particular, the decrease in parasitoid numbers after the 20th week was not accompanied by an increase in the number of hosts.

In week 60, strains of flies and of wasps were extracted from populations A and B, and from control cultures (CTRL) kept free of any prolonged contact between host and parasitoid populations. One generation was reared outside of the cages in order to eliminate possible developmental effects. Self- and cross-infestation tests were carried out simultaneously; each of them involved 10 batches of 100 hosts, each infested by one parasitoid female over a 24-h period. The experiment was repeated later with insects extracted from cages in week 70. The whole experiment involved 21,000 hosts. The degree of infestation (percentage of parasitized hosts) and rate of parasitoid development success (ratio of emerged wasps to parasitized hosts) were measured in each replicate. Results were similar in weeks 60 and 70 and are gathered for analysis.

In self-infestations (Fig. 4), the degrees of infestation were lower in the experimental populations than in the control. Parasitoid development success was slightly increased but the differences were not significant. The experimental procedure permits the determination of the relative contributions of hosts and parasites to the overall variation. With respect to the origin of hosts (Fig. 5a), hosts from experimental populations were far less infested by parasitoids of any origin (A, B or CTRL) than control hosts and they were significantly less suitable for parasitoid development. With respect to the origin of parasitoids (Fig. 5b), parasitoids from the experimental populations

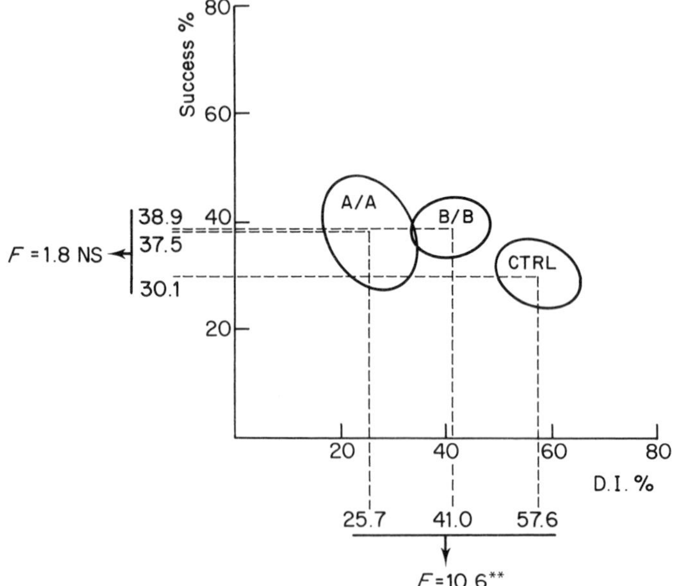

Fig. 4. Self-infestation experiments [hosts A × parasitoids A, hosts B × parasitoids B, hosts control (CTRL) × parasitoids CTRL] comparing the percentage of infested (parasitized) larvae (D.I.) and the percentage of infested larvae producing adult wasps (Success). Data from experimental populations (A and B) obtained from samples made in weeks 60 and 70 of the parasitoid-host interaction. Ellipses give the 95% confidence level around the means, which are given on the axes. F values are given for an analysis of variance between the means of each character (*significant at 5% level; **1% level). See text for further explanation.

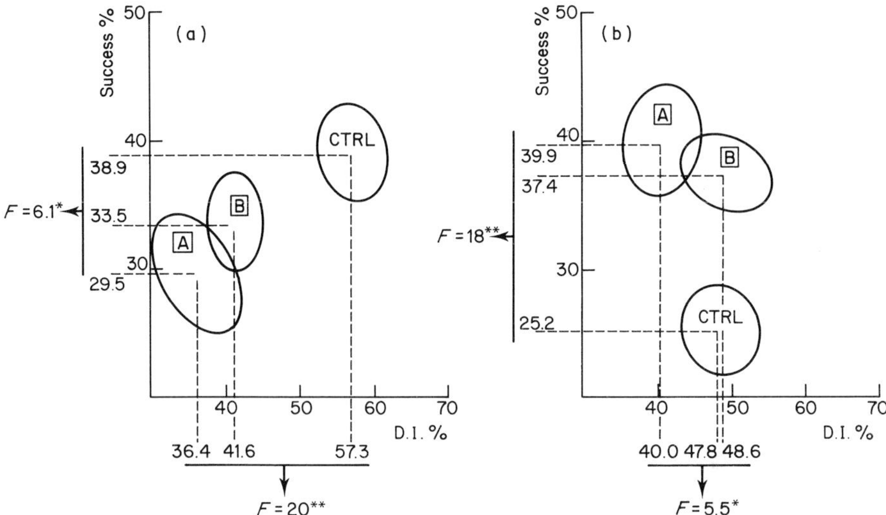

Fig. 5. Percentage infestation (D.I.) and successful development (Success) as a function of the origin (a) of hosts [A, B, or control (CTRL)] and (b) of parasitoids. Thus, for (a), means for A, B and CTRL hosts represent pooled data for exposure of the particular host population to all three parasitoid populations (A, B and CTRL). In (b), each parasitoid population is compared with respect to its overall performance on all three host populations, again pooled for the analysis. See Fig. 4 and text for more details.

infested hosts at a slightly lower rate (population A only) than the controls, but compared to the controls parasitoid larvae from both experimental populations thrived much better in hosts of any origin.

Thus, in the *D. melanogaster—L. boulardi* association, the characters of hosts and parasitoids may change over a short period of time. The system evolves towards a lower rate of attack by parasitoids, which seems to be caused by changes in the behavioural response of the parasitoids to features of the host, and which needs further analysis. This result is consistent with the decline of parasitoid numbers in cages, and is in line with observations on systems mentioned previously (Utida, 1957; Takahashi, 1963; Pimentel *et al.*, 1963; Pimentel and al-Hafidh, 1965; Pimentel and Stone, 1968; Zareh *et al.*, 1980).

The lack of an increase in the host's number in response to a reduced parasitoid impact indicates that the competitive ability of parasitized larvae is much higher than that of healthy larvae (Prévost, 1985). The decrease in the infestation rate of host larvae was accompanied by a drop in the production of adult flies from the cage in question.

Despite considerable change in parasitoids and hosts in the experimental populations (Fig. 5), the developmental success of parasitoids ended up the same as that in the controls. This is the result of complementary evolution in both hosts and parasitoids. The decrease of the host's suitability was approximately compensated by the increased exploiting efficiency of the parasitoid. The result is that parasitoid developmental success does not seem to change with the age of the interaction, moving from new (control) to "old" (experimental) systems.

Our results do not permit us to affirm that the observed changes in each partner are exclusively the result of selective pressure exerted by the other partner. We cannot ignore the selective effects caused by laboratory conditions (Mackauer, 1976). However, these results demonstrate that, beginning with a host and a specific parasitoid with a long natural association which has resulted in a certain balance, we *can* facilitate rapid changes in the two partners.

5. *Conclusions*

1. At least some of the characters involved in host suitability for parasitism are under genetic control in both host and parasitoid
2. Characters contributing to host suitability are not evenly distributed in host and parasitoid population
3. Between distinct populations, there are geographical variations in the host as well as in the parasitoid which are possibly simple pleiotropic effects of local genetic differentiation
4. In a given parasitoid-host association, the characters of the two partners can evolve over time in response to selective pressures, whether associated with parasitism or not
5. A clear relation does not appear to exist between the developmental success of a parasitoid and the age of its association with the host. As in immunological reactions, sympatric associations do not seem to be associated with any unique pattern of adaptations.

VI. NON-RECIPROCAL GENETIC EFFECTS BETWEEN PARASITOIDS AND THEIR HOSTS

Up to now, we have only discussed the characters directly involved in the functioning of the parasitoid-host interaction. As suggested by Price (1975), we can expect there to be more extensive genetic effects of parasitoids on the host population. From a theoretical viewpoint, Haldane (1949) and Clarke (1979) outlined the possible role of parasites in the maintenance of polymorphism in natural populations. We know of examples where the presence of a parasitoid modifies the population genetic structure at a locus not directly related to parasitism. Takahashi and Pimentel (1967) have shown that the female *Nasonia vitripennis* had a weak preference for the pupae of *Musca domestica* homozygous for a recessive allele responsible for the black colour of pupa. The gaps between selective values of diverse genotypes were modified in the parasitoid's presence, and the study of experimental populations have shown a faster decline of the recessive allele in the parasitoid's presence. This discriminatory destruction by the parasite has, therefore, effects which are related to those of apostatic selection (Clarke, 1969).

Mixed populations of *D. melanogaster* and *L. boulardi* offer a second example. Fly populations, polymorphic for wild type and a recessive, dark-eyed "sepia" mutation, have been infested with *L. boulardi* with the introduction of 200 females each week. The experiment was designed to permit the emergence of *D. melanogaster* but not of parasitoids in the cage. Therefore the infestation was at a constant level, both qualitatively and quantitatively. Figure 6 illustrates that whatever the initial frequency of the mutant allele (0.9 or 0.1), the infested populations converged towards a higher balanced frequency (0.35) than the non-infested control populations (0.20) (Boulétreau *et al.*, 1984). Wajnberg (1986) tested other genetic markers and found that the same result was obtained with some but not all mutations.

Choice tests for the parasitoid between hosts of genotypes se/se, se+/se,

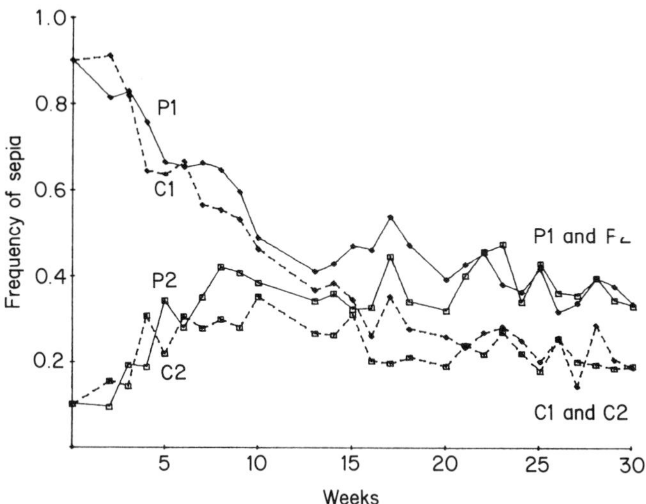

Fig. 6. Allelic frequencies at the sepia locus in infested (P1 and P2) and control (C1 and C2) experimental populations of *Drosophila melanogaster*.

se+/se+ did not reveal any significant difference in preference. The parasitoid's effect on the evolution of polymorphism does not appear to be caused by discriminatory destruction. The most likely explanation of these changes emerges from a comparison of the numbers of adult insects in both infested and non-infested populations, and by measuring adult flies. Table III gives an indication of host and adult parasitoid numbers in each of four cages, as measured by the daily production of adults (measured by census of dead adults). The total number of insects is four times higher in the infested cages. This indicates that the presence of parasitoids multiplies by at least four the number of *Drosophila* larvae which can develop (at least to the pupal stage) in cages that are otherwise identical.

Table IV shows that the size of living adult flies which have escaped

Table III. Numbers of flies and wasps in infested (P1 and P2) and control (C1 and C2) cages for the experiments on wild type and sepia *Drosophila* mutants. Numbers are based on weekly counts of dead adults, averaged over six weeks. Mean values are given with their standard errors

	P1	P2	C1	C2
Flies	126 ± 17	120 ± 16	345 ± 44	296 ± 23
Wasps	1297 ± 125	1162 ± 93	—	—
Total	1423	1282	345	296

Table IV. Thorax length (1/100 mm) and fresh weight (mg) of adult males of *Drosophila melanogaster* in cages infested or not (control) by *Leptopilina boulardi*. Wild=wild phenotype; sepia=sepia phenotype. Means are given with their standard errors and (in brackets) the number of replicates. Differences between infested and control populations tested by *t*-test

	Infested	Control	≠
Wild			
Thorax length	84.13 ± 0.87	75.57 ± 1.19	**
	(30)	(30)	
Fresh weight	0.63 ± 0.02	0.51 ± 0.02	**
Sepia			
Thorax length	80.73 ± 0.91	74.29 ± 1.72	**
	(26)	(16)	
Fresh weight	0.64 ± 0.02	0.52 ± 0.04	**

**Significant at 1% level.

parasitism is quite different. In infested populations, the flies are significantly larger than those in control populations, even though the number of larvae developing in the former was much greater. This paradoxical result appeared in all populations tested (Wajnberg, 1986). It is consistent with the finding that parasitized larvae are superior competitors (Prévost, 1985). The presence of parasitized individuals among the host larvae introduces a heterogeneity that modifies the interactions between the larvae and reduces the effects of competition between non-parasitized larvae, thus explaining both the greater size of adults and the greater frequency of counter-selected alleles. The sepia mutation is well known for reducing the pre-imaginal viability (Anxolabéhère, 1980).

These laboratory results may be merely caricatures of the real world. However, they demonstrate the existence of a new category of genetic effects which parasitoids may inflict on a host population, whereby the introduction of a heterogeneity upsets the host's functioning and thus modifies its adaptive responses to the environment. Like predation (Harvey and Greenwood, 1978), parasitism interacts strongly with other selective factors of the environment.

VII. CONCLUDING REMARKS

As suggested by Barrett (1984), the only parasitic associations we can observe are those that exist now, thus those in which a certain balance exists. It is a biased sample of the associations which could be tested in the course of evolution and which were not always persistent. As in the predator-prey associations (Schaffer and Rosenzweig, 1978), the condition of the persistence is the non-elimination of both host and parasite. Only slight variations can be tolerated between the two partners and these will only be expressed at moments of weak perturbation when they are subject to natural selection. Clarke (1976) and Barrett (1984) enumerated the criteria for this process: the existence of population variations in both partners, genetic determination of these variations, and a relation between these variations and the production of descendants.

The biological data presently available concern only a small number of insect species, but in general these conditions are fulfilled and the experimental intensification of selective effects can rapidly modify the characteristics of the association. Thus, it seems that the reciprocally selective phenomena could take place in natural parasitoid-host interactions, and lead to a better local adjustment and a better co-adaptation between the two species. How can it be explained, then, that sympatric associations do not demonstrate, in a systematic fashion, a better compatibility between host and parasitoid?

One hypothesis is that the selective pressures between the two partners are weak and/or discontinuous, such that the observed inter- or intrapopulation variation in parasitoids and hosts is caused by pleiotropic effects resulting from other selective processes. Given that these variations do not exceed proscribed limits, the parasitoid-host association persists. This hypothesis was proposed by van den Bosch (1964) to explain certain puzzling questions and was reconsidered by Chabora (1972) and Morris (1976).

A second hypothesis places the parasitic association in a larger selective and ecological context, and derives from the suggestion of Bartlett and Ball (1966) that an infallible or reliable index does not exist for the overall compatibility of hosts and parasitoids. Three levels of selection have been distinguished in the functioning of parasitic associations: the ecological or behavioural level, the immunological level and the physiological level. All can serve as targets for selection, but their relative weights are unknown and can vary locally or with time depending on the overall selective context. Furthermore, selection on any level will be limited by bioclimatic and nutritional constraints, constraints arising from other host and parasitoid species integrated in the same parasitic complex, and constraints associated with demographic variation and population structure. The resulting parasitoid-host association reflects the interplay of a group of selective pressures bearing on diverse characters, the results of which can vary locally in intensity and direction. A specific measure, based on a restricted number of characters, could lead to erroneous interpretations of the evolutionary state of the host-parasitoid association in question.

In other respects, it is clear that parasitism interacts strongly with the other selective factors of the environment. We have examples that demonstrate that parasitoids can modify selective values in a polymorphic host population, and thus indirectly modify the adaptive response of that population to its environment. An understanding of coevolutionary interactions between hosts and parasitoids demands therefore not only an integration of the demographic and genetic aspects of their relationship, as proposed by May and Anderson (1983) and Barrett (1984), but also a broader view of the genetic phenomena which control the populations of both partners. This synthetic approach should be applied to biological data integrating ecological, demographic and genetic aspects. For parasitoids and their hosts, as for many organisms, these data are currently lacking and are difficult to acquire.

Acknowledgements

I am indebted to P. Fouillet, G. Prévost and E. Wajnberg for allowing me to use unpublished data, and to R. Allemand and C. Bernstein for their helpful comments on an earlier draft. I express my gratitude to J. Waage, who corrected the manuscript with extreme care, and who greatly improved it. Experimental work was supported by C.N.R.S. (UA 243 and GRECO 44).

REFERENCES

Anderson, R. M., and May, R. M. (1982). Coevolution of hosts and parasites. *Parasitology* **85,** 111–126.

Anxolabéhère, D. (1980). The influence of sexual and larval selection on the maintenance of polymorphism at the sepia locus in *Drosophila melanogaster*. *Genetics* **95,** 743–755.

Arthur, A. P. (1962). Influence of host tree on abundance of *Itoplectis conquisitor* (Say) (Hymenoptera; Ichneumonidae), a polyphagous parasite of the European pine shoot moth *Rhyacionia buoliana* (Schiff) (Lepidoptera; Olethreutidae). *Canadian Entomologist* **94,** 337–347.

Arthur, A. P. (1981). Host acceptance by parasitoids. *In* "Semiochemicals: Their Role in Pest Control" (D. A. Nordlund, R. L. Jones and W. J. Lewis, eds.), pp. 97–120. John Wiley, New York.

Barrett, J. A. (1983). Plant-fungus symbioses. *In* "Coevolution" (D. J. Futuyma and M. Slatkin, eds.), pp. 137–160. Sinauer, Sunderland.

Barrett, J. A. (1984). The genetics of host-parasite interaction. *In* "Evolutionary Ecology" (B. Shorrocks, ed.), pp. 275–294. Blackwell, Oxford.

Bartlett, B. R., and Ball, J. C. (1966). The evaluation of host suitability in a polyphagous parasite with special reference to the role of parasite egg encapsulation. *Annals of the Entomological Society of America* **59,** 42–45.

Bauer, S. J., and Sokolowski, M. B. (1984). Larval foraging behavior in isofemale lines of *Drosophila melanogaster* and *D. pseudoobscura*. *Journal of Heredity* **75,** 131–134.

Beckage, N. (1985). Endocrine interactions between endoparasitic insects and their hosts. *Annual Review of Entomology* **30,** 371–413.

Blumberg, D., and DeBach, P. (1981). Effects of temperature and host age upon the encapsulation of *Metaphycus stanleyi* and *Metaphycus helvolus* eggs by brown soft scale *Coccus hesperidum*. *Journal of Invertebrate Pathology* **37,** 73–79.

van den Bosch, R. (1964). Encapsulation of the eggs of *Bathyplectes curculionis* (Thomson) (Hymenoptera: Ichneumonidae) in larvae of *Hypera brunneipennis* (Boheman) and *Hypera postica* (Gyllenhal) (Coleoptera: Curculionidae). *Journal of Invertebrate Pathology* **6,** 343–367.

Boulétreau, M., and David, J. R. (1981). Sexually dimorphic response to host habitat toxicity in *Drosophila* parasitic wasps. *Evolution* **35,** 395–399.

Boulétreau, M., and Fouillet, P. (1982). Variabilité génétique intrapopulation de l'aptitude de *Drosophila melanogaster* à permettre le développement d'un Hyménoptère parasite. *Comptes-Rendus de l'Académie des Sciences Paris* **295,** 775–778.

Boulétreau, M., and Quiot, J. M. (1972). Effet toxique des larves d'un hyménoptère parasite *Pteromalus puparum* L. (Chalc.) sur les cultures cellulaires de lépidoptères. *Comptes-Rendus de l'Académie des Sciences* **275,** 233–234.

Boulétreau, M., and Wajnberg, E. (in press). Comparative responses of two sympatric parasitoid cynipids to the genetic and epigenetic variations of their host, *Drosophila melanogaster* larvae. *Entomologia Experimentalis et Applicata*.

Boulétreau, M., Fouillet, P., Wajnberg, E., and Prévost, G. (1984). A parasitic wasp changes genetic equilibrium in *Drosophila melanogaster* experimental populations. *Drosophila Information Service* **60,** 68–69.

Carton, Y. (1976). Attraction de *Cothonaspis* sp. par le milieu trophique de son hôte *Drosophila melanogaster*. *Colloques internationaux du C.N.R.S.* **265,** 285–303.

Carton, Y. (1984). Analyse expérimentale de trois niveaux d'interaction entre *Drosophila melanogaster* et le parasite *Leptopilina boulardi*. *Génétique Sélection Evolution* **16,** 417–430.

Carton, Y., and Boulétreau, M. (1985). Encapsulation ability of *Drosophila melanogaster*: a genetic analysis. *Developmental and Comparative Immunology* **9,** 211–219.

Carton, Y., and David, J. R. (1983). Reduction of fitness in *Drosophila* adults surviving parasitization by a cynipid wasp. *Experientia* **39,** 231–233.

Carton, Y., and David, J. R. (1985). Relation between the genetic variability of digging behavior of *Drosophila* larvae and their susceptibility to a parasitic wasp. *Behavior Genetics* **15,** 143–154.

Chabora, P. C. (1970a). Studies of parasite-host interaction using geographical strains of the blow fly *Phaenicia sericata* and its parasite *Nasonia vitripennis*. *Annals of the Entomological Society of America* **63,** 495–501.

Chabora, P. C. (1970b). Studies of parasite-host interaction. II. Reproductive and development response of the parasite *Nasonia vitripennis* (Hymenoptera: Pteromalidae) to strains of the house fly host *Musca domestica*. *Annals of the Entomological Society of America* **63,** 1632–1636.

Chabora, P. C. (1970c). Studies in parasite-host interaction. III. Host race effect on the life table and population growth statistics of the parasite *Nasonia vitripennis*. *Annals of the Entomological Society of America* **63,** 1637–1642.

Chabora, P. C. (1972). Studies in parasite-host interaction. IV. Modifications of parasite *Nasonia vitripennis* responses to control and selected host *Musca domestica* populations. *Annals of the Entomological Society of America* **65,** 323–328.

Chabora, P. C., and Chabora, A. J. (1971). Effects of an intrapopulation hybrid host on parasite population dynamics. *Annals of the Entomological Society of America* **64,** 558–562.

Chabora, P. C., and Pimentel, D. (1970). Patterns of evolution in parasite-host systems. *Annals of the Entomological Society of America* **63,** 479–486.

Clarke, B. (1969). The evidence for apostatic selection. *Heredity* **24,** 347–352.

Clarke, B. (1976). The ecological genetics of host-parasite relationships. *In* "Genetic Aspects of Host-Parasite Relationships" (A. E. R. Taylor and M. Muller, eds.), pp. 87–103. Blackwell, Oxford.

Clarke, B. (1979). The evolution of genetic diversity. *Proceedings of the Royal Society* **205,** 453–474.

Corbet, S. A. (1973). Concentration effects and the response of *Nemeritis canescens* to a secretion of its host. *Journal of Insect Physiology* **19,** 2119–2128.

David, J., and Bocquet, C. (1975). Similarities and differences in the latitudinal adaptation of two *Drosophila* sibling species. *Nature* **257,** 588–590.

David, J., and van Herrewege, J. (1983). Adaptation to alcoholic fermentation in *Drosophila* species: relationships between alcohol tolerances and larval habitats. *Comparative Biochemical Physiology* **74,** 283–288.

Day, P. R. (1974). "Genetics of Host-Parasite Interaction." W. H. Freeman, San Francisco.

Doutt, R. L. (1959). The biology of parasitic Hymenoptera. *Annual Review of Entomology* **4,** 161–182.

Edson, K. M., Stoltz, D. B., and Summers, M. D. (1981). Virus in a parasitoid wasp: suppression of the cellular immune response in the parasitoid's host. *Science* **211,** 582–583.

Fenner, F., and Ratcliffe, F. N. (1965). "Myxomatosis." Cambridge University Press, London.

Fisher, R. C. (1971). Aspects of the physiology of endoparasitic Hymenoptera. *Biological Review* **46,** 243–278.

Flanders, S. E. (1953). Variations in susceptibility of citrus-infesting coccids to parasitization. *Journal of Economic Entomology* **46,** 266–269.

Flor, H. H. (1956). The complementary genic systems in flax and flax rust. *Advances in Genetics* **8,** 29–54.

Futuyma, D. J., and Peterson, S. C. (1985). Genetic variation in the use of resources by insects. *Annual Review of Entomology* **30,** 217–238.

Futuyma, D. J., and Slatkin, M. (1983). "Coevolution." Sinauer, Sunderland.

Gallun, R. L. (1977). Genetic basis of Hessian fly epidemics. *Annals of the New York Academy of Sciences* **287,** 223–229.

Hadorn, E., and Walker, I. (1960). *Drosophila* und *Pseudeucoila.* I. Selektionsversuche zur Steigerung der Abwehrreaktion des Wirtes gegen den Parasiten. *Revue Suisse de Zoologie* **67,** 216–225.

Haldane, J. B. S. (1949). Disease and evolution. *Ricerca Scientifica* **19** (suppl), 68–76.

Harvey, P. H., and Greenwood, P. J. (1978). Anti-predator defence strategies: some evolutionary problems. *In* "Behavioural Ecology, an Evolutionary Approach" (J. R. Krebs and N. B. Davies, eds.), pp. 129–151. Blackwell, Oxford.

Hoffman, A. A., Parsons, P. A. and Nielsen, K. M. (1984). Habitat selection olfactory response of *Drosophila melanogaster* depends on resources. *Heredity* **53,** 139–143.

Hokkanen, H., and Pimentel, D. (1984). New approach for selecting biological control agents. *Canadian Entomologist* **116,** 1109–1121.

Holmes, J. C. (1983). Evolutionary relationships between parasitic Helminths and their hosts. *In* "Coevolution" (D. J. Futuyma and M. Slatkin, eds.), pp. 161–185. Sinauer, Sunderland.

Hsiao, T. H. (1982). Geographic variation and host plant adaptation of the Colorado potato beetle. *Proceedings of the 5th International Symposium on Insect-Plant Relationships* 315–324, Wageningen.

Huffaker, C. B., Messenger, P. S., and DeBach, P. (1971). The natural enemy component in natural control and the theory of biological control. *In* "Biological Control" (C. B. Huffaker, ed.), pp. 16–67. Plenum, New York.

Huffaker, C. B., Simmonds, F. J., and Laing, J. E. (1976). The theoretical and empirical basis of biological control. *In* "Theory and Practice of Biological Control" (C. B. Huffaker and P. S. Messenger, eds.), pp. 41–78. Academic Press, New York.

Huffaker, C. B., Berryman, A. A., and Laing, J. E. (1984). Natural control of insect populations. *In* "Ecological Entomology" (C. B. Huffaker and R. L. Rabb, eds.), pp. 359–398. John Wiley, New York.

Jaenike, J., and Grimaldi, D. (1983). Genetic variation for host preference within and among populations of *Drosophila tripunctata. Evolution* **37,** 1023–1033.

Janzen, D. H. (1980). When is it coevolution? *Evolution* **34,** 611–612.

Jones, R. L., Lewis, W. J., Bowman, M. C., Beroza, M., and Bierl, B. A. (1971). Host-seeking stimulant for parasite of corn earworm: isolation, identification and synthesis. *Science* **173,** 842–843.

Kitano, H., and Nakatsuji, N. (1978). Resistance of *Apanteles* eggs to the haemocytic encapsulation by their habitual host *Pieris. Journal of Insect Physiology* **24,** 261–271.

Kolata, G. (1984). Scrutinizing sleeping sickness. *Science* **226,** 956–959.

Labeyrie, V. (1978). Comportement et écologie. *Annales de Zoologie Ecologie Animale* **10,** 381–386.

Lackie, A. M. (1980). Invertebrate immunity. *Parasitology* **80,** 393–412.

Lanier, G. N., Birch, M. C., Schmitz, R. F., and Furniss, M. M. (1972). Pheromones of *Ips pini* (Coleoptera; Scolytidae): variations in response among three populations. *Canadian Entomologist* **104,** 1917–1923.

Levin, S. A. (1983). Some approaches to the modelling of coevolutionary interactions. *In* "Coevolution" (M. H. Nitecki, ed.), pp. 21–65. University of Chicago Press, Chicago.

Levin, S., and Pimentel, D. (1981). Selection of intermediate rates of increase in parasite-host systems. *American Naturalist* **117,** 308–315.

Lewontin, R. C. (1974). "The Genetic Basis of Evolutionary Change". Columbia University Press, New York.

Mackauer, M. (1976). Genetic problems in the production of biological control agents. *Annual Review of Entomology* **21,** 369–385.

May, R. M., and Anderson, R. M. (1983). Epidemiology and genetics in the coevolution of parasites and hosts. *Proceedings of the Royal Society of London* **219,** 281–313.

Mayr, E. (1963). "Animal Species and Evolution." Harvard University Press, Cambridge.

Messenger, P. S., and van den Bosch, R. (1971). The adaptability of introduced biological control agents. *In* "Biological Control" (C. B. Huffaker, ed.), pp. 68–92. Plenum Press, New York.

Morris, R. F. (1976). Influence of genetic changes and other variables on the encapsulation of parasites by *Hyphantria cunea. Canadian Entomologist* **108,** 673–684.

Muldrew, J. A. (1953). The natural immunity of the larch sawfly (*Pristiphora erichsonii* Htg.) to the introduced parasite *Mesoleius tenthredinis* Morley in Manitoba and Saskatchewan. *Canadian Journal of Zoology* **31,** 313–332.

Nappi, A. J., and Silvers, M. (1984). Cell surface change associated with cellular immune secretions in *Drosophila. Science* **22,** 1166–1168.

Nordlander, G. (1980). Revision of the genus *Leptopilina* Forster; 1969, with notes on the status of some other genera (Hymenoptera; Cynipoidae: Eucoilidae). *Entomologica Scandinavica* **11,** 428–453.

Oakeshott, J. G., Gibson, J. B., Anderson, P. R., Knibb, W. R., Anderson, D. G., and Chambers, G. K. (1982). Alcohol dehydrogenase and glycerol-3-phosphate dehydrogenase clines in *Drosophila melanogaster* on different continents. *Evolution* **36,** 86–96.

Olson, D., and Pimentel, D. (1974). Evolution of resistance in a host population to attacking parasite. *Environmental Entomology* **3,** 621–624.

Parsons, P. A. (1980). Isofemale strains and evolutionary strategies in natural populations. *In* "Evolutionary Biology" (M. Hecht, W. Steere and B. Wallace, eds.), pp. 175–217. Plenum, New York.

Person, C. (1966). Genetic polymorphism in parasitic systems. *Nature* **212,** 266–267.

Pimentel, D. (1961). Animal population regulation by the genetic feed-back mechanism. *American Naturalist* **95,** 65–79.

Pimentel, D. (1984). Genetic diversity and stability in parasite-host systems. *In* "Evolutionary Ecology" (B. Shorrocks, ed.), pp. 295–311. Blackwell, Oxford.

Pimentel, D., and al-Hafidh, R. (1965). Ecological control of a parasite population by genetic evolution in the parasite-host system. *Annals of the Entomological Society of America* **58,** 1–6.

Pimentel, D., and Stone, F. A. (1968). Evolution and population ecology of parasite-host systems. *Canadian Entomologist* **100,** 655–662.

Pimentel, D. W., Nagel, P., and Madden, J. L. (1963). Space-time structure of the environment and the survival of parasite-host systems. *American Naturalist* **97,** 141–167.

Prévost, G. (1985). "Etude expérimentale des interactions entre parasitisme et compétition larvaire chez *Drosophila melanogaster* Meigen." Thesis No 1628, Lyon.

Price, P. W. (1975). The parasitic way of life and its consequences. *In* "Evolutionary Strategies of Parasitic Insects and Mites" (P. W. Price, ed.), pp. 1–13. Plenum, New York.

Price, P. W. (1977). General concepts in the evolutionary biology of parasites. *Evolution* **31,** 405–420.

Price, P. W. (1980). "Evolutionary Biology of Parasites." Princeton University Press, Princeton.

Price, P. W. (1981). Semiochemicals in evolutionary time. *In* "Semiochemicals: their Role in Pest Control" (D. J. Nordlund, R. L. Jones and W. J. Lewis, eds.), pp. 251–279. John Wiley, New York.

Read, D. P., Feeny, P. P., and Root, R. B. (1970). Habitat selection by the aphid parasite *Diaeretiella rapae* (Hymenoptera: Braconidae) and hyperparasite *Charips brassicae* (Hymenoptera; Cynipidae). *Canadian Entomologist* **102,** 1567–1578.

Rizki, R. M., and Rizki, T. M. (1984). Selective destruction of a host blood cell type by a parasitoid wasp. *Proceedings of the National Academy of Sciences* **81,** 6154–6158.

Salt, G. (1935). Experimental studies in insect parasitism. III. Host selection. *Proceedings of the Royal Society* **1175,** 413–435.

Salt, G. (1963). The defense reactions of insects to metazoan parasites. *Parasitology* **53,** 527–642.

Salt, G. (1968). The resistance of insect parasitoids to the defence reactions of their hosts. *Biological Review* **43,** 200–232.

Salt, G. (1970). "The Cellular Defence Reactions of Insects". Cambridge University Press, London.

Salt, G., and van den Bosch, R. (1967). The defence reactions of three species of *Hypera* (Coleoptera: Curculionidae) to an ichneumon wasp. *Journal of Invertebrate Pathology* **9,** 164–177.

Schaffer, W. M., and Rosenzweig, M. L. (1978). Homage to the Red Queen. I. Coevolution of predators and their victims. *Theoretical Population Biology* **14,** 135–157.

Slatkin, M., and Maynard Smith, J. (1979). Models of coevolution. *Quarterly Review of Biology* **54,** 233–263.

Smith Trail, D. R. (1980). Behavioral interactions between parasites and hosts: host suicide and the evolution of complex life cycles. *American Naturalist* **116,** 77–91.

Stoltz, D. B., and Vinson, S. B. (1979). Viruses and parasitism in insects. *Advances in Virus Research* **24,** 125–171.

Streams, F. A., and Greenberg, L. (1969). Inhibition of the defence reaction of *Drosophila melanogaster* parasitized simultaneously by the wasps *Pseudeucoila bochei* and *Pseudeucoila mellipes. Journal of Invertebrate Pathology* **13,** 371–377.

Takahashi, F. (1963). Changes in some ecological characters of the almond moth caused by the selective action of an ichneumon wasp in their interacting system. *Researches on Population Ecology* **5,** 117–129.

Takahashi, F., and Pimentel, D. (1967). Wasp preference for black brown and hybrid-type pupae of the house fly. *Annals of the Entomological Society of America* **60,** 623–625.

Tavormina, S. J. (1982). Sympatric genetic divergence in the leaf-mining insect *Liriomyza brassicae* (Diptera: Agromyzidae). *Evolution* **36,** 523–534.

Taylor, A. E. R., and Muller, R. (1976). "Genetic Aspects of Host-Parasite Relationships." Blackwell, Oxford.

Thompson, J. N. (1982). "Interaction and Coevolution." John Wiley, New York.

Turnbull, A. L., and Chant, D. A. (1961). The practice and theory of biological control of insects in Canada. *Canadian Journal of Zoology* **39,** 697–753.

Turner, M. (1982). Antigenic variation in the trypanosome. *Nature* **298,** 606–607.

Utida, S. (1957). Population fluctuation: an experimental and theoretical approach. *Cold Spring Harbour Symposium on Quantitative Biology* **22,** 139–151.

van Valen, L. (1965). Morphological variation and width of ecological niche. *American Naturalist* **99,** 377–390.

Vinson, S. B. (1975). Biochemical coevolution between parasitoids and their hosts. *In* "Evolutionary Strategies of Parasitic Insects and Mites" (P. W. Price, ed.), pp. 14–48. Plenum, New York.

Vinson, S. B. (1976). Host selection by insect parasitoids. *Annual Review of Entomology* **21,** 109–133.

Vinson, S. B. (1981). Habitat location. *In* "Semiochemicals: their Role in Pest Control" (D. A. Nordlund, R. L. Jones, and W. J. Lewis, eds.), pp. 51–77. John Wiley, New York.

Vinson, S. B. (1984a). Parasite-host relationship. *In* "Chemical Ecology of Insects" (W. J. Bell and R. T. Cardé, eds.), pp. 205–232. Chapman and Hall, London.

Vinson, S. B. (1984b). How parasitoids locate their hosts: a case of insect espionage. *In* "Insect Communication" (T. Lewis, ed.), pp. 325–348. Academic Press, London.

Vinson, S. B., and Iwantsch, G. F. (1980a). Host suitability for insect parasitoids. *Annual Review of Entomology* **25,** 397–419.

Vinson, S. B., and Iwantsch, G. F. (1980b). Host regulation by insect parasitoids. *Quarterly Review of Biology* **55,** 143–165.

Vinson, S. B., Jones, R. L., Sonnet, P. E. Bierl, B. A., and Beroza, M. (1975). Isolation, identification and synthesis of host-seeking stimulant for *Cardiochiles nigriceps*, a parasitoid of tobacco budworm. *Entomologia Experimentalis et Applicata* **18,** 443–450.

Wajnberg, E. (1986). "Interactions Démographiques et Génétiques entre la Drosophile (*Drosophila melanogaster* Meigen) et ses Parasites Larvaires Hymenoptères." Thesis, Lyon.

Wajnberg, E., Prévost, G., and Boulétreau, M. (1985). Genetic and epigenetic variation in *Drosophila* larvae suitability to a hymenopterous endoparasitoid. *Entomophaga* **30,** 187–190.

Walker, I. (1959). Die Abwehrreaktion des Wirtes *Drosophila melanogaster* gegen die zoophage Cynipidae *Pseudeucoila bochei* Weld. *Revue Suisse de Zoologie* **66,** 569–632.

Walker, I. (1962). *Drosophila* und *Pseudeucoila.* III Selektionsversuche zur Steigerung der Resistenz des Parasiten gegen die Abwehrreaktion des Wirtes. *Revue Suisse de Zoologie* **69,** 209–227.

Weseloh, R. M. (1981). Host location by parasitoids. *In* "Semiochemicals: their Role in Pest Control" (D. A. Nordlund, R. L. Jones and W. J. Lewis, eds.), pp. 79–95. John Wiley, New York.

Zareh, N., Westoby, M., and Pimentel, D. (1980). Evolution in a laboratory host-parasitoid system and its effects on population kinetics. *Canadian Entomologist* **112,** 1049–1106.

7

Parasitoids and Population Regulation

M. P. HASSELL

I. INTRODUCTION

The importance of parasitoids in the regulation of their insect hosts has recently been questioned on several fronts. On the one hand, there has been the broad debate on the relative importance of competition and predation in structuring animal communities, with views sometimes becoming quite polarized into "competition" and "anti-competition" camps (Lewin, 1983a, 1983b). At least for insects, the middle ground seems most reasonable where competition is of paramount importance in some groups (e.g. ants, bees, dragonflies) and far less so in others, such as insect herbivores, where natural enemies and other environmental factors appear to play a much greater role in keeping populations at relatively low levels (Lawton and Strong, 1981; Lawton and Hassell, 1981, 1984). Others have questioned the role of parasitoids more directly. In particular, Dempster (1983) failed to find convincing evidence for natural enemies being at all important, at least in the regulation of temperate Lepidoptera populations. In this paper, I shall consider first to what extent parasitoids have the *potential* to regulate their host populations, before turning to consider what evidence exists beyond this that they indeed do so and to some methodological problems in demonstrating this.

The term "regulation" is clearly central to this discussion, but unfortunately has been trapped in something of a semantic bog over the past few decades. In a purely deterministic world there is no problem with the concept of population regulation. A regulated population always tends to return to an equilibrium position following some disturbance, and this must result from some process whose net effect is broadly directly density dependent ("broadly" because some time-delayed density dependent factors can also be quite sufficient to maintain such an equilibrium). The question of locally unstable populations exhibiting limit cycles as their natural pattern falls outside this definition although, of course, this is a perfectly good mechanism for population persistence.

Observed population patterns are, however, the result of deterministic and random processes acting together, so that an idealized equilibrium gives way to a fluctuating population whose equilibrium is now represented by some average over the period of study. Such populations are still regulated if the density dependent tendency to return to the equilibrium is great enough for the probability of extinction to be small, at least over ecological time, despite the random elements acting in the system. But now the definition has lost some of its crispness and, depending on the relative strengths of the density dependent and density independent processes, there can be strongly regulated populations and weakly regulated ones with a much greater chance of extinction. Of course, density dependent processes themselves will never be perfectly precise in their action. They will appear as relationships with some degree of variance, and thus can contribute to population fluctuation as well as population stability. This, in essence, is Strong's "density vagueness" (1984), which is much the same as Southwood's density dependent relationships (1967) with sufficient variance that they can at the same time be key factors primarily responsible for population change. Despite this interplay between "precise" and "noisy" factors which affect populations, it often remains valuable to set the stochastic elements to one side and focus on the deterministic processes, as I have done in the first part of this paper in an attempt to understand how some key features of an interaction can promote population regulation.

Most studies on host-parasitoid population dynamics have focused on discrete-generation, coupled and synchronized interactions. This assumes that the parasitoids are effectively specialists and neglects the large category that is broadly polyphagous (see Chapters 8 and 9) and which has quite a different dynamical relationship with its hosts. In particular, a broad host range will tend to buffer the populations of such generalists from fluctuations in abundance of any one of their host species, giving dynamics that are largely uncoupled from that particular host population (Hassell and Comins, 1976; Southwood and Comins, 1976; Comins and Hassell, in press). In this paper I shall consider both generalists and specialists in turn to demonstrate their potential to regulate a host population, before combining them to see if substantially different dynamics emerge.

II. GENERALISTS

Two essential ingredients in modelling any prey-predator or host-parasitoid interaction are descriptions of the natural enemy's functional and numerical responses. The *functional response* defines the *per capita* ability of the parasitoids to attack hosts at different host densities, and the general form has been thoroughly explored in the literature (Holling, 1959a, 1959b; Hassell *et al.*, 1976; van Lenteren and Bakker, 1976, 1978). For the purpose of this paper, typical type II responses are assumed for both the generalist and specialist parasitoids. However, instead of describing these in the usual way on the basis of random encounters between parasitoids and hosts (Rogers, 1972; Hassell, 1978; Arditi, 1983), a negative binomial distribution of encounters is assumed, to give for G_t searching generalists

$$N_a = N_t \left[1 - \left\{ 1 + \frac{aG_t}{k(1 + aT_h N_t)} \right\}^{-k} \right]. \tag{1}$$

Here N_t is the number of hosts in generation t, N_a is the number of these attacked, a is the *per capita* searching efficiency of the parasitoids, T_h is their "handling time" and k is the parameter of the negative binomial distribution determining the degree of contagion in the frequency distribution of attacks among the N_t hosts. In contrast to the usual functional response equations based on random exploitation, Eq. (1) has the merit of recognizing, albeit in a phenomenological way, that many processes (spatial, temporal and genetic) combine to make some host individuals more susceptible to parasitism than others. The choice of the negative binomial is arbitrary, but it has been widely used in describing insect and other population distributions (Southwood, 1978), and it does provide a convenient and simple way of introducing non-random distributions into a variety of population models (May, 1978; Anderson and May, 1978; May and Hassell, 1981; Hassell and May, in press).

The *numerical response* of generalist natural enemies has attracted much less modelling effort than the functional response, despite often taking the same simple form across a wide range of predators from small mammals and birds to insects. The examples in Fig. 1 (unfortunately none from parasitoids) all show a tendency for G_t to rise monotonically with increasing prey density towards a maximum, and are conveniently described by the expression used by Kowalski (1976) and Southwood and Comins (1976):

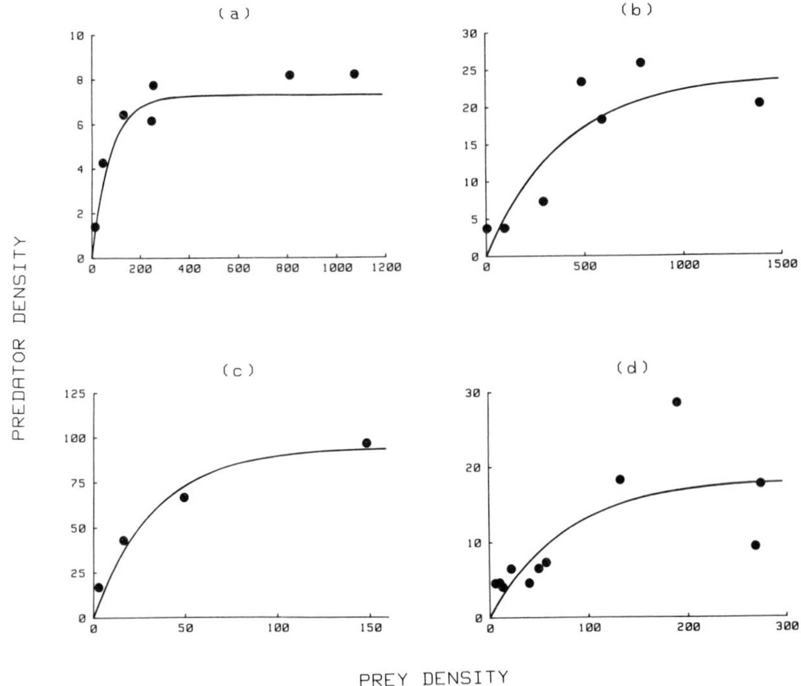

Fig. 1. Numerical responses of four generalist predators. (a) *Peromyscus maniculatus* Hoy & Kennicott and (b) *Sorex cinereus* Kerr (both as numbers per acre) in relation to the density of larch sawfly, *Neodiprion sertifer* (Geoffroy) cocoons (thousands per acre; data from Holling, 1959a). (c) The bay-breasted warbler, *Dendroica fusca* (Müller) (nesting pairs per 100 acres), in relation to third instar larvae of the spruce budworm, *Choristoneura fumiferana* (Clemens) (numbers per 10 foot^2 of foliage; data from Mook, 1963). (d) *Philonthus decorus* (Gravenhorst) (pitfall trap index) in relation to winter moth, *Operophtera brumata* (L.), larvae per m^{-2} (after Kowalski, 1976). In each case Eq. (2) has been fitted to the data using a non-linear least squares procedure with parameter values estimated as follows (±95% confidence limits): (a) $h=7.30 \pm 1.07$, $b=76.87 \pm 49.16$; (b) $h=24.12 \pm 14.02$, $b=390.58 \pm 579.81$; (c) $h=94.32 \pm 42.21$, $b=32.97 \pm 42.27$; (d) $h=18.36 \pm 9.45$, $b=77.56 \pm 106.59$ (from Hassell and May, in press).

$$G_t = h[1 - \exp(N_t/b)], \tag{2}$$

where h is the saturation number of natural enemies and b determines the typical host density at which this maximum is approached. In effect, we are assuming that these generalists have a fast numerical response in relation to changes in N_t, which might arise from a reproductive response if reproduction was rapid relative to the time scale of their hosts or, and perhaps more likely, a behavioural response involving "switching" from feeding elsewhere (Royama, 1970) or on other prey species (Murdoch, 1969).

Such numerical responses, when combined with the functional response of Eqn. (1), give a host-generalist model of the form:

$$N_{t+1}=\lambda N_t\left[1+\frac{ah[1-\exp(-N_t/b)]}{k(1+aT_hN_t)}\right]^{-k}. \qquad (3)$$

Here N_t and N_{t+1} are the host populations in successive generations t and $t+1$, and λ is the host's finite rate of increase.

Parasitism in this model is clearly directly density dependent from generation to generation, at least over a range of host densities, as shown in Fig. 2 for a range of parameter values. The transition to inverse density dependence at high host densities is due to the effects of the type II functional response as the numerical response saturates. Figure 3 shows some comparable field examples from predation of lepidopterous pupae in the soil, predominantly by generalist carabid and staphylinid beetles. That this predation does not decline at high prey densities suggests that either the functional or numerical responses (or both) would saturate only at higher prey densities than observed.

The dynamics of Eq. (3) can be summarized as follows with reference to Figs 4 and 5 (a fuller treatment of this model is given in Hassell and May, in press):

1. The "Ricker" curves in Fig. 4 show that a generalist-maintained equilibrium (N^*) may (curves C, D) or may not (curves A, B) occur. If $T_h=0$, then N^* requires

$$ah>k(\lambda^{1/k}-1). \qquad (4)$$

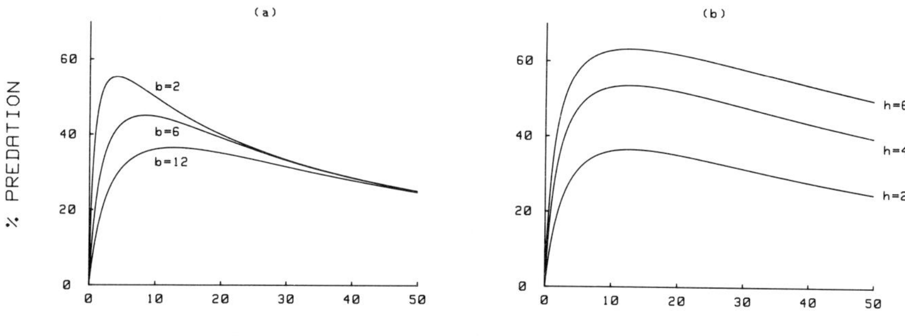

Fig. 2. Relationships between the level of predation and prey density from Eqs. (1) and (2) obtained by varying the numerical response parameters b and h. (a) $a=1$, $T_h=0.1$, $h=2$, $k=1$ and b as shown; (b) $a=1$, $T_h=0.1$, $b=12$, $k=1$ and h as shown (from Hassell and May, in press).

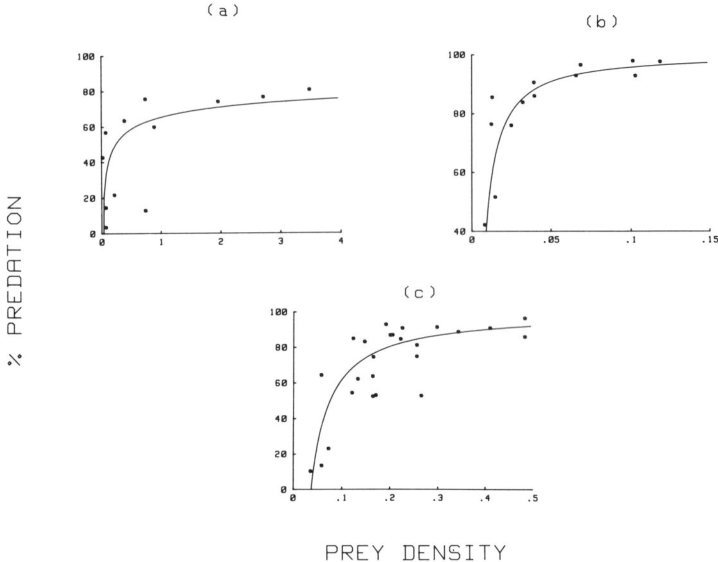

Fig. 3. Examples of density dependent pupal mortality from generation to generation for three species of soil pupating Lepidoptera, ascribed primarily to predation by carabid and staphylinid beetle. Curves fitted by transforming the original published relationships between k-values (Varley and Gradwell, 1960) and log population density (N). (a) Mortality of *Erranis defoliaria* (L.) per m^{-2}; $k = 0.46 + 0.21 \log N$ (after Ekanayake 1967). Mortality of *Pardia tripunctatana* Denis & Schiffermüller pupae per 0.18 m^{-2}; $k = 1.47 + 1.1 \log N$ (after Bauer, 1985). The original k-values were estimated between the larval and subsequent egg stages. An arbitrary 10-fold reproductive rate (λ) has thus been assumed here to avoid negative mortalities. Other assumed values of λ (if constant) will not alter the general shape of the curve. (c) Mortality of *Notocelia roborana* Denis & Schiffermüller pupae per 0.18 m^{-2}; $k = 0.39 + 0.97 \log N$ (after Bauer, 1985). $\lambda = 10$ assumed as in (c).

Broadly speaking, ah is a measure of the generalist's overall efficiency since it combines the *per capita* efficiency (a) with the asymptotic level of the numerical response (h). Thus more efficient generalists are needed to create an equilibrium as the host's rate of increase (λ) gets larger or as the clumping of attacks gets greater ($k \rightarrow 0$)

2. With finite handling time, similar cases apply but now even when an equilibrium exists, the host population, if sufficiently large, "escapes" to increase until checked by other processes. Thus the host population is only regulated at N^*, provided N_t remains below some threshold value (N_T), as shown in Fig. 4, curve C
3. The level of this host equilibrium depends on several parameters, as shown in Fig. 5. It is reduced firstly as the generalist efficiency (ah)

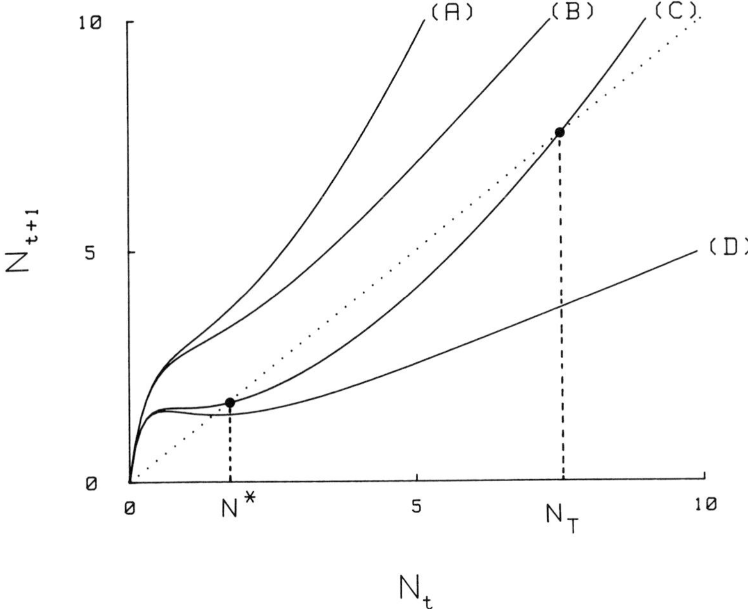

Fig. 4. Map of the prey population densities in successive generations t and $t+1$, obtained from Eq. (3). Equilibria occur when the curves intersect the dotted 45° line. Curve A: $ah=3$, $abT_h=0.04$; curve B: $ah=3$, $abT_h=0$; curve C: $ah=2$, $abT_h=0.04$. A locally stable equilibrium occurs at N^* and an unstable, "release" point at N_T. Curve D: $ah=2$, $abT_h=0$. For further details see text (after Hassell and May, in press).

increases (Fig. 5a), secondly as parsitism is distributed more evenly amongst the hosts (i.e. as $k \to \infty$, Fig. 5b) and thirdly as the host rate of increase (λ) gets smaller (Fig. 5c)

4. Finally, these equilibria need not all be locally stable, in which case the populations will show limit cycles or even chaotic behaviour (May and Oster, 1976). These cases, in which the populations will persist, but not at a steady state, are made more likely if the generalists cause severe density dependent mortality in the region of the host equilibrium. This will be promoted by large k, large ah, and intermediate λ.

In short, such generalists cause classical, directly density dependent mortality which may be sufficient to regulate a host population. They differ from specialists in that there are none of the time delays inherent in coupled host-parasitoid interactions and hence no tendency to produce typical host-parasitoid or prey-predator oscillations. In the real world such generalists are likely to abound, but in some cases, although feeding on a range of species,

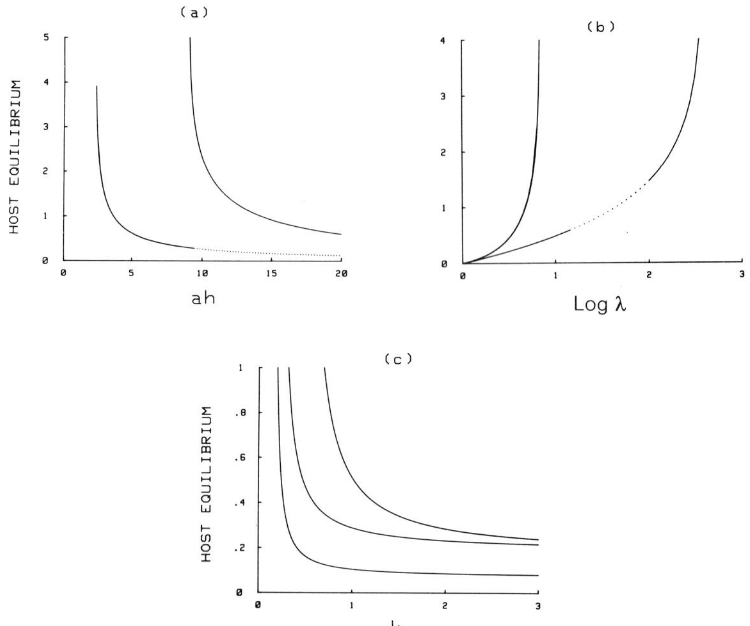

Fig. 5. The dependence of the host equilibrium N^*, from Eq. (3) on (a) the predator efficiency term, *ah*, for $\lambda = 10$ and $k \to \infty$ (left curve) and $k = 1$ (right curve): (b) the host rate of increase, λ, for $ah = 6$ and $k = 1$ (left curve) and $k \to \infty$ (right curve) and (c) the clumping term, *k*, for attacks on hosts for $\lambda = 2$ and $ah = 10$ (lower curve), $\lambda = 2$ and $ah = 4$ (middle curve) and $\lambda = 5$ and $ah = 10$ (upper curve). The dotted lines indicate locally unstable equilibria. (From Hassell and May, in press.)

the abundance of a given host or prey may have some effect on reproductive success. For example, Fig. 6 shows the density dependent predation of winter moth pupae in the soil over 18 generations, caused largely by carabid and staphylinid beetles (cf. Fig. 3). The slight but obvious tendency for an anticlockwise spiralling of the points when serially linked indicates some influence of winter moth density on predator numbers in the following generation, although this is insufficient to destroy the obvious direct density dependent relationship. Such examples fall into an intermediate category between the generalists discussed above and the specialists of the next section.

III. SPECIALISTS

The dynamics of coupled host-parasitoid interactions have been widely

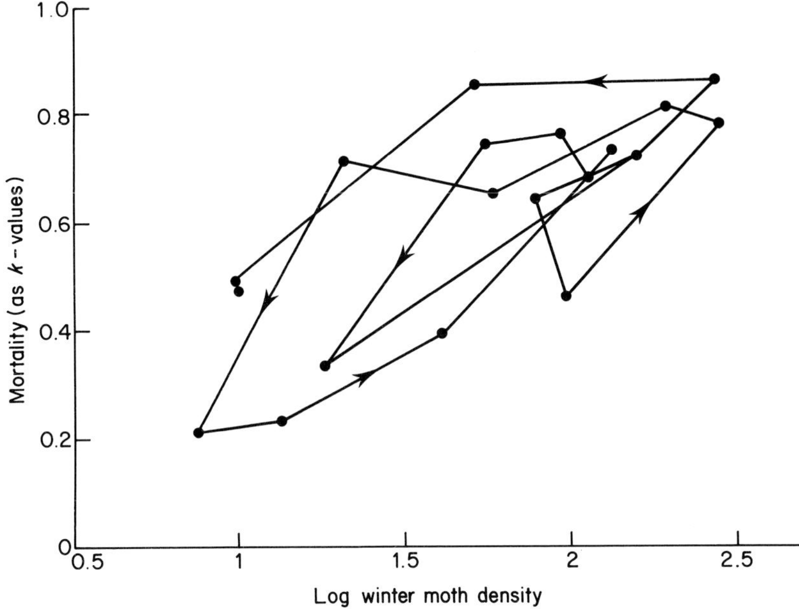

Fig. 6. Density dependent pupal mortality (expressed as *k*-values), due predominately to carabid and staphylinid beetles, in relation to winter moth density per generation. The data from 1950 to 1968 are serially linked to show the time-delayed density dependent component. (After Varley *et al.*, 1973.)

discussed in the literature, both in continuous (Murdoch and Oaten, 1975) and discrete time (Hassell, 1978). Consider the following discrete time model:

$$N_{t+1} = \lambda N_t g(N_t) + f(N_t, P_t)I \tag{5a}$$

$$P_{t+1} = cs(N_t, P_t)N_t[1 - f(N_t, P_t)]. \tag{5b}$$

Here the function f gives the survival from parasitism, the function g expresses any other density dependence affecting the host population, the function I gives the survival from all density independent processes, P are the adult female parasitoids in successive generations t and $t+1$, the function s defines the sex ratio of the parasitoid progeny and c gives the average surviving progeny per host attacked. Clearly, this model contains a vast amount of ecology (Waage and Hassell, 1982; Hassell and Waage, 1984) and this section will focus only on the effects of parasitism contained in the function f, paying particular attention to the fate of different host individuals within generations. Such "heterogeneity" includes any processes that render some individuals within the host population more susceptible to parasitism than others. This could arise, for example, from (1) non-random parasitism between patches of different host density (Hassell and May, 1974; Murdoch, 1977); (2) some hosts being protected from parasitism within a physical refuge of some sort (Maynard Smith, 1974; Hassell, 1978); (3) some degree of temporal asynchrony between the searching parasitoids and the host stage (a refuge in time; Griffiths and Holling, 1969; Hassell, 1969) and (4) some genetic or phenotypic differences between host individuals making some more able to resist parasitism (perhaps by physical repulsion or a physiological defence mechanism; Hassell and Anderson, 1984).

All of these mechanisms serve in their own way to give different probabilities of parasitism between different host individuals or classes of host individual, which in turn has a considerable effect on stabilizing coupled host-parasitoid interactions (see below). Let us consider in more detail just one from the above list—the spatial distribution of parasitism in a patchy environment. Three questions now arise:

1. How do the searching parasitoids allocate their foraging time between patches?
2. What patterns of parasitism result from this?
3. What effects do such patterns have on the dynamics of the interactions?

The first question has been well addressed in both the theoretical (Charnov, 1976; Cook and Hubbard, 1977; Hubbard and Cook, 1978; Lessells, 1985) and experimental literature (Waage, 1979; Hassell, 1982), and is discussed in detail in Chapter 2. In short, optimal foraging theory in its various forms predicts that parasitoids should tend to spend more time searching, and hence aggregate, in patches of high host density, and what field and laboratory data that exist show either some degree of such aggregation or no response at all to patch density (Waage, 1983).

The second question has been well answered by Lessells (1985) from a survey of the literature where she found 45 examples of the spatial distribution of parasitism, with 15 examples of direct density dependence, 17 examples of inverse density dependence and 13 examples showing no clear relationship.

These very different kinds of response can be explained by the same basic mechanistic explanation (Hassell, 1982; Lessells, 1985). In essence, whether the tendency is for direct, inverse or density independent patterns of parasitism to occur depends upon the balance between the tendency for parasitoids to spend more foraging time in patches of high host density (the "aggregative response") and the maximum attack rate per parasitoid within a patch set by the within-patch functional response. Thus, even if foraging time is much greater in high host density patches than low ones, inverse density dependent parasitism from patch to patch can still occur if each parasitoid is sufficiently limited in its ability to exploit a patch per unit time. Examples are given in Hassell (1982) and Hassell *et al.* (1985), and illustrated here in Fig. 7, showing how the combination of no aggregative response (Fig. 7a) and a within-patch functional response with low maximum attack rate (Fig. 7b) combine to predict satisfactorily the observed inverse density dependent distribution of parasitism (Fig. 7c).

It is now well documented that density dependent patterns of parasitism from patch to patch can contribute strongly to population stability, as shown by the numerical examples in Fig. 8a and b. Less well appreciated is that this is also perfectly true for the inverse patterns described above (Hassell, 1984). Which contributes more depends on the details of the host distribution. In general, direct density dependent patterns become increasingly important as the fraction of the total hosts in high density patches increases, and inverse patterns as more hosts occur in patches of low density. While both these patterns promote stability due to the same mechanism of differential parasitism on different segments of the host population (and the effects of this on the implicit competition between parasitoid larvae when superparasitism occurs (A. D. Taylor, personal communication), their effect on host equilibrium levels is not as symmetrical. This is illustrated by the examples in Fig. 9. The "index of density dependence" (μ) is explained by Hassell (1984) and reflects density independent parasitism across patches when $\mu = 0$, inverse density dependence when $\mu < 0$ and direct density dependence when $\mu > 0$. Note that inverse patterns tend to *increase* the host equilibrium (N^*) compared to that with $\mu = 0$, while direct patterns *decrease* N^*, at least up to some threshold μ-value, when N^* again begins to rise due to parasitism becoming increasingly confined to the high density patches, leaving the remaining hosts relatively free from parasitism.

In short, specialist parasitoids clearly have the potential to regulate their host populations by a variety of mechanisms. Some of these are within-generation processes, such as the spatial heterogeneity between patches or the other forms of heterogeneity listed at the beginning of this section (refuges, temporal asynchrony, genetic and phenotypic variability), whose effects all depend on differential probabilities of parasitism between host individuals. No one of these mechanisms generating heterogeneity is likely to predominate in the real world (Hassell and Anderson, 1984), which poses considerable problems in the design of field studies aimed at assessing the role of natural enemies (see below). As Murdoch *et al.* (1984, 1985) and Reeve and Murdoch (1985) stress, primarily in the context of biological control programmes, the spatial aggregation of parasitism in response to a patchily distributed host population is not necessary to account for stable

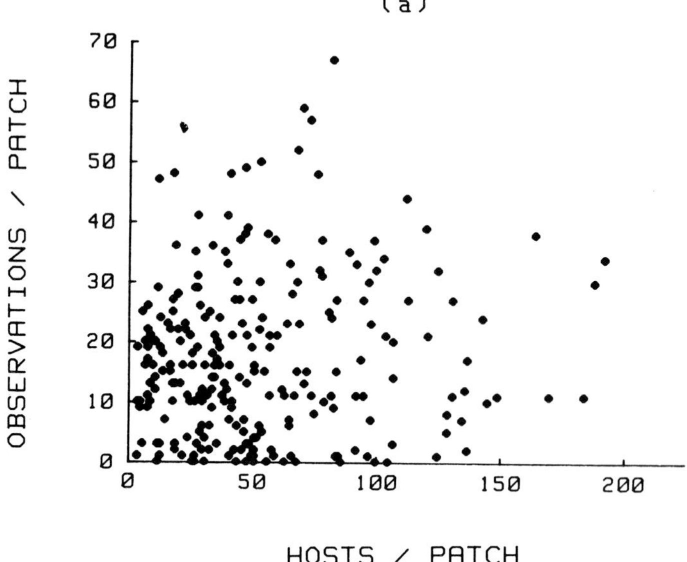
(a)
OBSERVATIONS / PATCH
70
60
50
40
30
20
10
0
0
50
100
150
200
HOSTS / PATCH

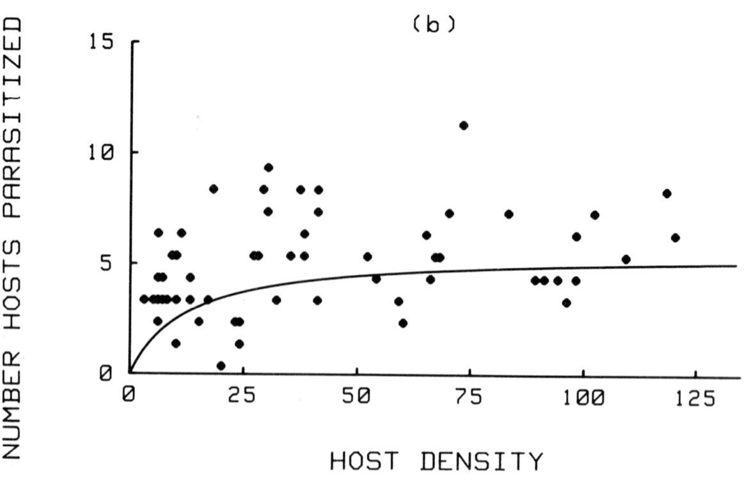
(b)
NUMBER HOSTS PARASITIZED
15
10
5
0
0
25
50
75
100
125
HOST DENSITY

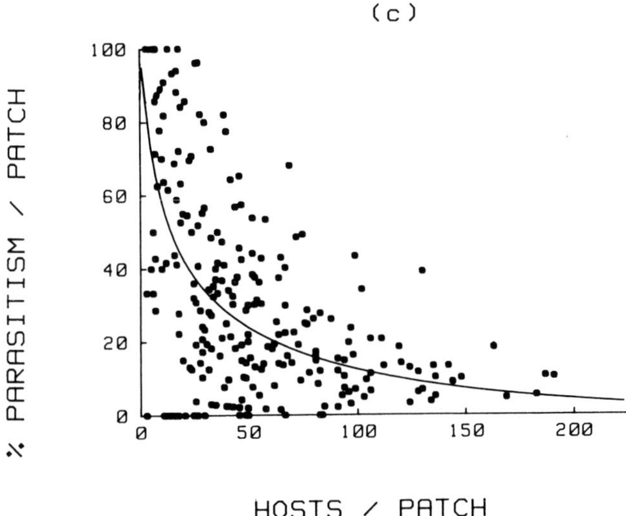

Fig. 7. Inverse density dependent parasitism of *Callosobruchus chinensis* (L.) larvae from patch to patch by the braconid parasitoid, *Heterospilus prosopidus* Viereck, in a laboratory system. (a) The relationship between the observations of parasitoids made on each patch in each of 10 replicates and host density per patch, showing the lack of any aggregative response ($r = 0.098$; $P > 0.05$). (b) The functional response obtained by confining a single female parasitoid for 24 h on a patch of varying host density. (c) The relationship between percentage parasitism and host density per patch. The 250 points represent the 25 patches in 10 replicates. The fitted curve is derived using an estimate of the average time spent per patch and the functional response parameters, a and T_h, obtained from the data in (b). (See Hassell *et al.*, 1985, for further details.)

populations at low equilibrium levels (Hassell and May, 1973; Beddington *et al.*, 1978). This is but one mechanism generating heterogeneity. In addition to these within-generation processes are others, such as mutual interference (Hassell and Varley, 1969; Royama, 1971; Beddington, 1975) and density dependent shifts in sex ratio seen in some parasitic Hymenoptera (Hassell *et al.*, 1983) which are more likely to be observed by looking at total populations from generation to generation. Whatever its source, population regulation in coupled host-parasitoid interactions with their pronounced time delays is fundamentally somewhat different from that involved in the host-generalist interactions discussed earlier, where regulation depends considerably upon a strong numerical response without a one-generation time delay. Specialist interactions are thus more likely than generalist ones to show classical host-parasitoid or prey-predator oscillations.

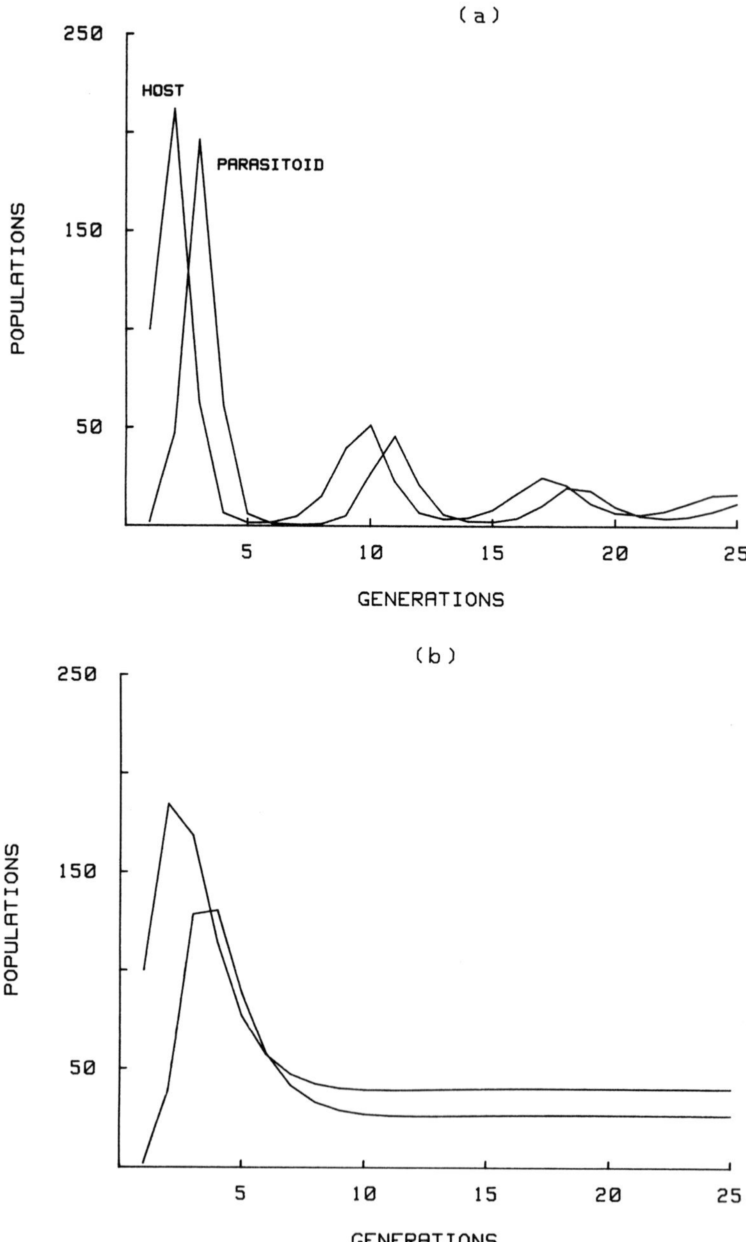

Fig. 8. Numerical examples from Eqs. (5a,b) where $g = I = c = s = 1$; $f = \sum [\alpha_i \exp(-a\beta_i P_t)]$ where n (the number of patches) = 5; $a = 0.2$; $\{\alpha_i\} = 0.5, 0.2, 0.1, 0.1, 0.1$ and $\beta_i = x\alpha_i^{\mu}$ with (a) $\mu = 2$ and (b) $\mu = 3$. (x is a normalization constant.) (See Hassell and May, 1973 for further details.)

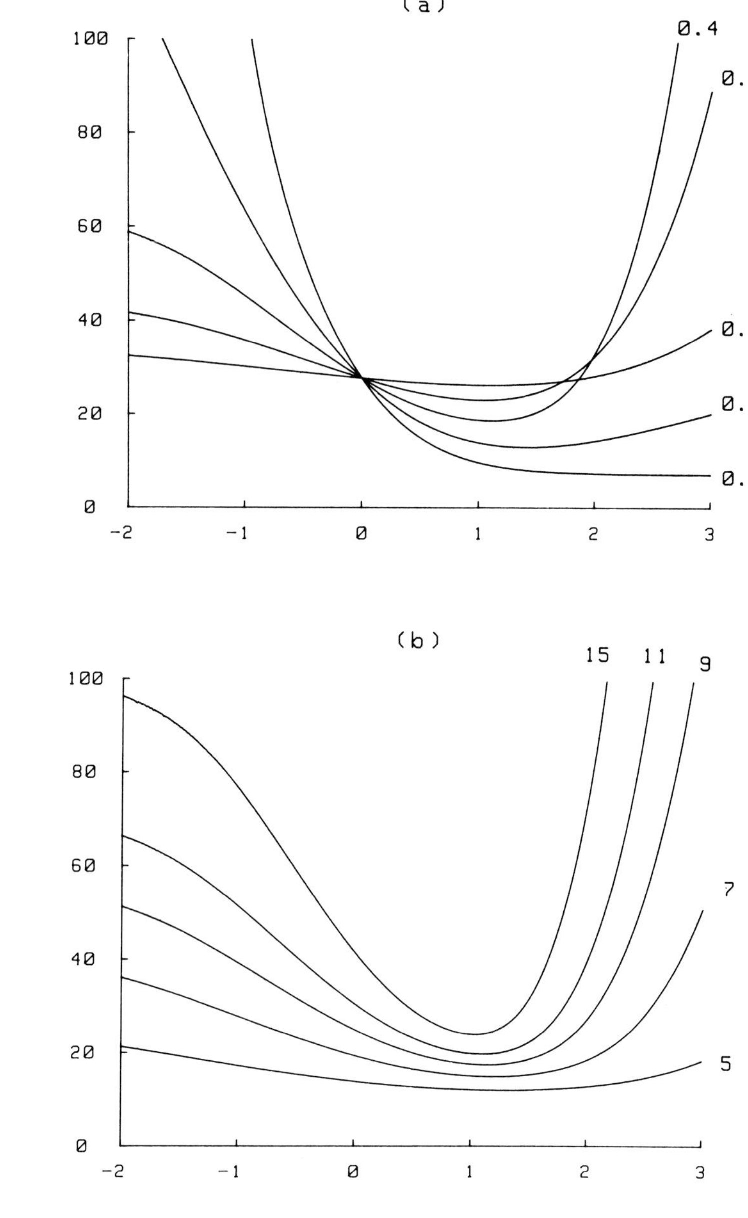

Fig. 9. Examples showing how the host equilibrium, N^*, varies with the index of density dependence from patch to patch (μ in legend to Fig. 8) in a system with a single high density patch containing a fraction α of the total hosts and $n-1$ remaining patches all with the same fraction $(1-\alpha)/(n-1)$. (a) Values of α as shown, and $a=0.5$, $n=10$ and $\lambda=2$. (b) Values of n as shown, and $a=0.05$, $\alpha=0.4$ and $\lambda=2$. (From Hassell, 1984.)

IV. SPECIALISTS AND GENERALISTS

From the dynamics of generalists and specialists acting separately on a host species the framework will now be broadened to include both types acting in concert (Hassell and May, in press). In particular, we shall examine if the action of both together introduces unexpected dynamics or whether we are just left with some kind of blending of the properties of the separate pairwise interactions.

When different mortalities act within a discrete life cycle it is necessary to be explicit on their relative timing, since this can itself affect the dynamics (Wang and Gutierrez, 1980; May *et al.*, 1981). Thus, with the specialists acting before the generalists, we have:

$$N_{t+1} = \lambda N_t f(P_t) g(N_t f) \tag{6a}$$

$$P_{t+1} = N_t[1 - f(P_t)] \tag{6b}$$

where

$$g = [1 + (ah/k)[1 - \exp(-N_t/b)]^{-k} \tag{7}$$

and

$$f = (1 + a'P_t/k')^{-k'}. \tag{8}$$

Here a' is the *per capita* searching efficiency of the specialists (cf. a for the generalists) and k' defines the degree of clumping in the distribution of parasitism by the specialist among the host population within any generation (cf. k for the generalists). Other parameters are as in Eq. (3), but now handling time for both specialist and generalist is assumed to be negligible compared with the total time available for searching.

However, if the generalists act first, we now have

$$N_{t+1} = \lambda N_t g(N_t) f(P_t) \tag{9a}$$

$$P_{t+1} = N_t g(N_t)[1 - f(P_t)]. \tag{9b}$$

The following conclusions stand out from an analysis of these models:

1. A specialist can invade and coexist more easily if acting before the generalist in the host's life cycle
2. A generalist can exclude the specialist, making a persistent three-species interaction impossible. This occurs if λ is too low (not enough host

reproduction to support both species) as shown by the broken lines in Fig. 10), or if the generalist's overall searching efficiency (ah) is too high relative to that of the specialist

3. With less efficient generalists and/or a higher host rate of increase, the specialist can invade and a persistent three-species state can occur (solid lines in Fig. 10)
4. This three-species state can be simple, in which case there is just one equilibrium point (or stable cycle), or much more complex in which there are a variety of possible alternative stable states. Thus, there may be two alternative persistent states, one with only a generalist and the other with both generalist and specialist present (curve B, Fig. 10) or two alternative three-species states in which the interaction may "flip" between high and low levels if sufficiently perturbed (curve C, Fig. 10)
5. A three-species stable system can readily exist when one or both of the two-species interactions alone would be unstable, as shown by the example in Fig. 11.

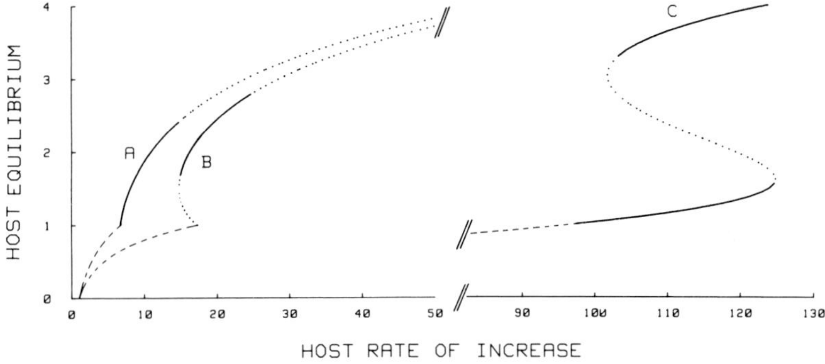

Fig. 10. Three examples of the dependence of the host equilibrium, N^*, on the host rate of increase, λ, from Eqs. (6–8). Curve A: $h=20$ and $b=10$; curve B: $h=30$ and $b=10$ and curve C: $h=5$ and $b=0.4$, with $a=a'=1$ and $k=k'\to\infty$ throughout. The broken lines indicate host-generalist interactions, the solid lines locally stable host-generalist-specialist interactions and the dotted lines locally unstable ones. (From Hassell and May, in press.)

In short, unlike the two-species host-generalist or host-specialist interactions, each of which has rather straightforward dynamics, the combined three-species system presents a rather wider, and in some respects unexpected, range of properties. Thus, while it is clearly important to understand the properties of individual pairwise interactions, this may well be insufficient for a complete picture of how the dynamics of more complicated systems with combinations of species interactions can influence community structure.

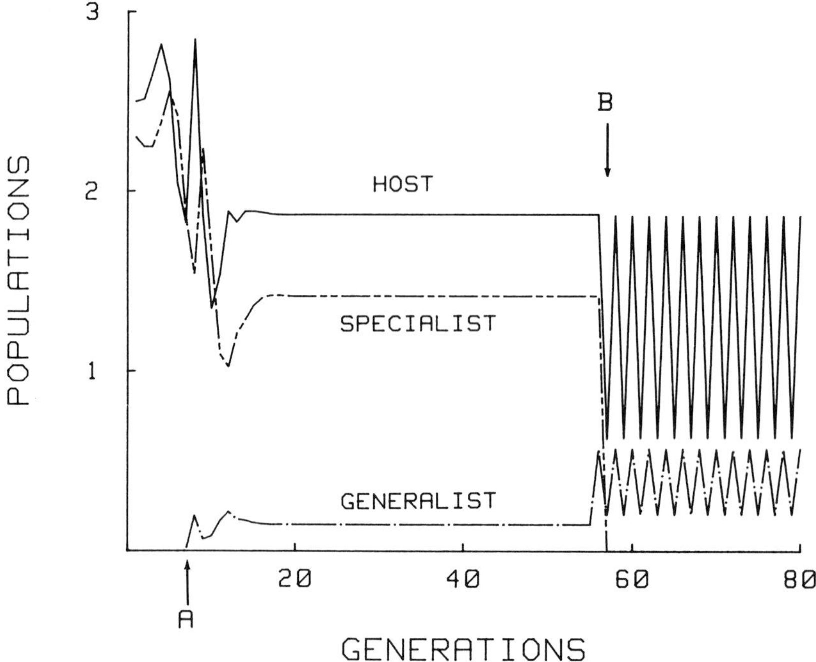

Fig. 11. Numerical example from Eqs. (6–8) illustrating a stable three-species interaction where each of the two-species interactions are locally unstable. The generalist (drawn at 1/6th scale) is introduced at point A and the specialist is removed at B. Parameter values: $\lambda = 10$, $a = a' = 1$, $k = k' \rightarrow \infty$, $h = 20$, $c = 10$. (From Hassell and May, in press.)

V. DO PARASITOIDS REGULATE THEIR HOST POPULATIONS?

Moving on from the question of whether or not parasitoids *can* regulate their hosts (and they clearly can in several different ways), this section considers briefly what evidence exists that they actually do so. And here the picture is far less clear, primarily because there are still very few field studies in which sufficient suitable data have been collected. Added to this is the problem of detecting some of the regulatory processes discussed above, particularly from the body of existing life table data.

The best evidence for insect generalist natural enemies being central to the regulation of their hosts or prey comes from soil-pupating Lepidoptera where there are several examples of complexes of pupal predators (mainly carabids and staphylinids) causing major density dependent mortality from generation to generation (see Fig. 3). Furthermore, in each of these cases, specific population models were developed showing the observed density dependence to be sufficient for the regulation of the prey population. Such natural enemies, albeit predators and not parasitoids, are precisely those envisaged earlier in considering the dynamics of generalists.

The case for specialist parasitoids being important to the persistence of their hosts is less direct. There are many biological control successes where introductions of such parasitoids have been instrumental in dramatically reducing the level of their host populations which have then persisted at low levels in a seemingly stable interaction. Very few of these, however, have been documented to the point where density dependent relationships can be identified and shown to be sufficient for regulation. The winter moth, *Operophtera brumata* (L.), in Nova Scotia is one case where empirical models have been developed and predict a stable equilibrium caused by the introduced parasitoids (Hassell, 1980). Similarly, Ives (1976) identified *Olesicampe benefactor* Hinz as the only density dependent factor affecting the larch sawfly, *Pristiphora erichsonii* (Hartig), and demonstrated, using a simple population model, its ability to regulate the host population.

The evidence however is sparse and it is only the unambiguous conclusions from theoretical studies on the *potential* of natural enemies to regulate their hosts that has led many workers to accept that insect natural enemies are indeed important for this. But this is by no means a universal point of view, and recently Dempster (1983) questioned whether they are at all important, at least in the regulation of temperate Lepidoptera populations. This was based on a review of 24 life tables on various Lepidoptera with more-or-less discrete generations, from which he concluded:

> "In eight studies no density dependent relationships could be identified, and in a further 13 the only density dependence demonstrated was due to intraspecific competition for resources. In the remaining three studies, natural enemies are thought to be acting in a density dependent manner, but their ability to regulate the population is questioned."

Clearly, in drawing such conclusions, the methods of testing for density dependent relationships in such field studies are critically important. Conventionally, average mortality per generation, usually as percentages or as k-values (Varley and Gradwell, 1960), is plotted against the average population density per generation on which the mortality acts: a positive correlation, if statistically satisfactory, indicating a density dependent relationship. To illustrate some difficulties with this technique, let us consider the example of a stable host-specialist interaction in which parasitism is the sole regulatory factor, arising from one of the mechanisms of non-random parasitism within each generation described above. We now consider the extent to which density dependence acting spatially within each generation will shine through from a routine generation-to-generation analysis based on mean populations per generation. In a perfectly deterministic situation there may be no problem as shown in Fig. 12a and c, although even here the time delayed density dependence in such coupled systems may tend to obscure this relationship, as shown in Fig. 12b and d where the populations are showing damped oscillations towards an equilibrium. Let us now add a small amount of random variation in turn to some of the elements of the stable interaction in Fig. 12a. The populations are still regulated by the parasitoids in just the same manner as previously, but any hope of a perfect equilibrium has disappeared (Fig. 13 a–c). When these data are examined as if collected as part of a life table study, and mortality over the 50 generations is plotted against population density per generation, the obvious density dependent

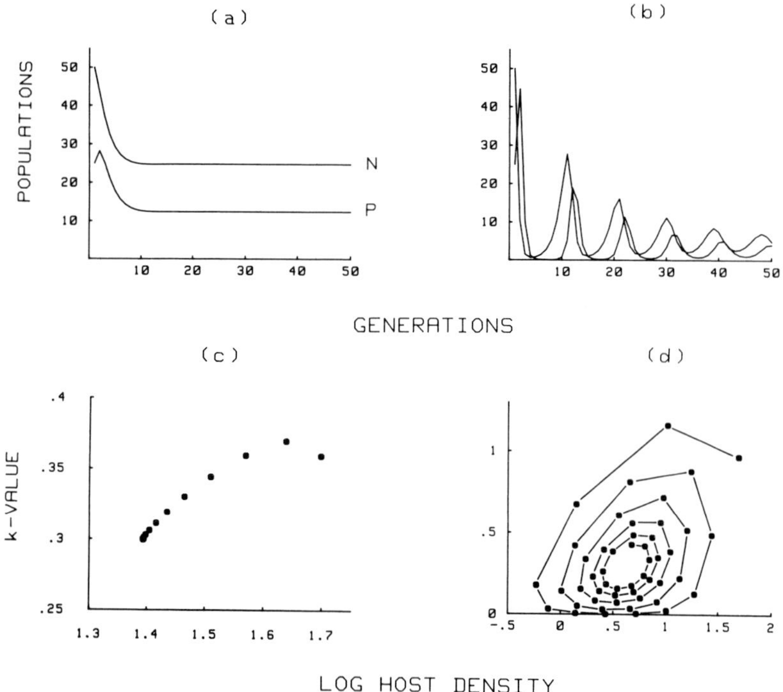

Fig. 12. (a) and (b) Simulations using Eqs. (5a,b) with $g=c=s=1$ and f defined by Eq. (8) for two levels of clumping in the distribution of parasitism: (a) $k=0.2$, (b) $k=0.8$. Other parameters: $\lambda=4$, $I=0.5$, $a=0.5$. (c) and (d) The corresponding relationships between the k-value for parasitism per generation and $\log_{10}$ host density for the 50 generations. The points in (d) are linked serially to emphasize the delayed density dependent nature of the parasitism. (From Hassell, 1985.)

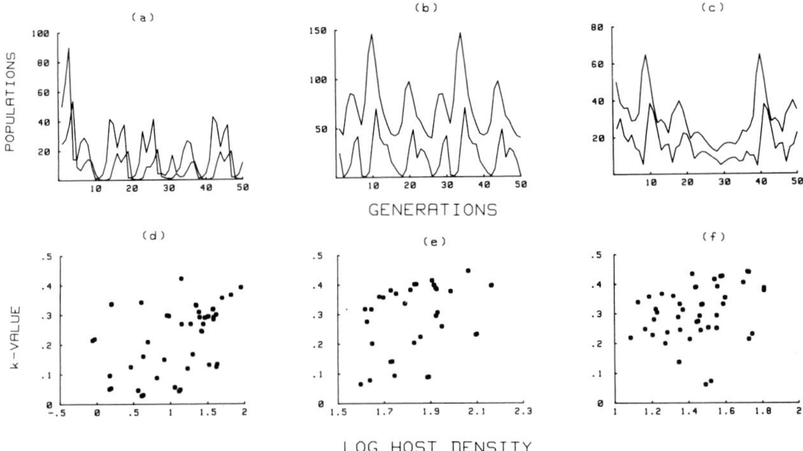

Fig. 13. (a)–(c) Simulations as in Fig. 12a except for one component now treated as a normally distributed random variable: (a) $I = 0.5 \pm 0.5$ (s.d.); (b) $c = 0.5 \pm 0.5$ and (c) $a = 0.5 \pm 0.5$. (d)–(f) The corresponding relationships between the k-value for parasitism and $\log_{10}$ host density, with regression statistics as follows: (d) $Y = 0.128 + 0.084X$; $r^2 = 0.156$, $P < 0.05$. (e) $Y = -0.188 + 0.255X$; $r^2 = 0.121$, $P < 0.05$. (f) $Y = 0.135 + 0.119X$; $r^2 = 0.056$, NS.

relationship of Fig. 12c has been largely obscured (although the correlation is statistically significant in the case of Fig. 13d and e). The combination of the random noise and time delays in the system make the detection of the regulatory ability of the parasitoids extremely difficult without more detailed information. Similar problems apply to host-generalist interactions (Hassell, 1985) and to detecting density dependence in single species populations (Hassell, Southwood and Reader, in press).

Thus, major problems remain in evaluating whether or not natural enemies are generally important in the regulation of their insect prey. In the first place, we have seen that within-generation effects can have a marked impact on population dynamics and yet can easily be masked in an analysis using average populations per generation. Secondly, it will often be difficult to detect and quantify some of the less obvious forms of within-generation heterogeneity (e.g. genetic or phenotypic variability in susceptibility) which can have just as much an effect on dynamics. The solution to these problems would appear to lie in much more detailed field studies, still in the form of life tables over as many generations as possible, but with much more emphasis placed on how natality and mortality are distributed within the populations of each generation, and coupled wherever possible with manipulation experiments. Such data should provide every chance to evaluate properly the factors affecting the dynamics of insect populations.

Acknowledgements

I am very grateful to Mick Crawley, Vicky Taylor and Jeff Waage for their helpful comments on the manuscript, and to David Hassell for help in presenting the figures.

REFERENCES

Anderson, R. M., and May, R. M. (1978). Regulation and stability of host-parasite population interactions. 1. Regulatory processes. *Journal of Animal Ecology* **47,** 219–247.

Arditi, R. (1983). A unified model of the functional response of predators and parasitoids. *Journal of Animal Ecology* **52,** 293–303.

Bauer, G. (1985). Population ecology of *Pardia tripunctatana* Schiff. and *Notocelia roborana* Den. and Schiff. (Lepidoptera, Tortricidae)—an example of "equilibrium species". *Oecologia* **65,** 437–441.

Beddington, J. R. (1975). Mutual interference between parasites or predators and its effect on searching efficiency. *Journal of Animal Ecology* **44,** 331–340.

Beddington, J. R., Free, C. A., and Lawton, J. H. (1978). Characteristics of successful natural enemies in models of biological control of insect pests. *Nature* (*London*) **273,** 513–519.

Charnov, E. L. (1976). Optimal foraging: the marginal theorem. *Theoretical Population Biology* **9,** 126–136.

Comins, H. N., and Hassell, M. P. (in press). The dynamics of predation and competition in patchy environments. *Theoretical Population Biology.*

Cook, R. M., and Hubbard, S. F. (1977). Adaptive searching strategies in insect parasites. *Journal of Animal Ecology* **46,** 115–125.

Dempster, J. P. (1983). The natural control of populations of butterflies and moths. *Biological Reviews* **58,** 461–481.

Ekanayake, U. B. M. (1967). "Parasitism of Four Species of *Erannis*". DPhil. Thesis, University of Oxford.

Griffiths, K. J., and Holling, C. S. (1969). A competition sub model for parasites and predators. *Canadian Entomologist* **101,** 785–818.

Hassell, M. P. (1969). A study of the mortality factors acting upon *Cyzenis albicans* (Fall.), a tachinid parasite of the winter moth *Operophtera brumata* (L.). *Journal of Animal Ecology* **38,** 329–339.

Hassell, M. P. (1978). "The Dynamics of Arthropod Predator-prey Systems." Princeton University Press, Princeton, New Jersey.

Hassell, M. P. (1980). Foraging strategies, population models and biological control: a case study. *Journal of Animal Ecology* **49,** 603–628.

Hassell, M. P. (1982). Patterns of parasitism by insect parasites in patchy environments. *Ecological Entomology* **7,** 365–377.

Hassell, M. P. (1984). Parasitism in patchy environments: inverse density dependence can be stabilizing. *IMA Journal of Mathematics Applied in Medicine and Biology* **1,** 123–133.

Hassell, M. P. (1985). Insect natural enemies as regulating factors. *Journal of Animal Ecology* **54,** 323–334.

Hassell, M. P., and Anderson, R. M. (1984). Host susceptibility as a component in host-parasitoid systems. *Journal of Animal Ecology* **53,** 611–621.

Hassell, M. P., and Comins, H. N. (1976). Discrete time models for two-species competition. *Theoretical Population Biology* **9,** 202–221.

Hassell, M. P., and May, R. M. (1973). Stability in insect host-parasite models. *Journal of Animal Ecology* **42,** 693–726.

Hassell, M. P., and May, R. M. (1974). Aggregation in predators and insect parasites and its effect on stability. *Journal of Animal Ecology* **43,** 567–594.

Hassell, M. P., and May, R. M. (in press). Generalist and specialist natural enemies in insect predator-prey interactions. *Journal of Animal Ecology*.

Hassell, M. P., and Varley, G. C. (1969). New inductive population model for insect parasites and its bearing on biological control. *Nature* (*London*) **223,** 1133–1137.

Hassell, M. P., and Waage, J. K. (1984). Host-parasitoid population interactions. *Annual Review of Entomology* **29,** 89–114.

Hassell, M. P., Lawton, J. H., and Beddington, J. R. (1976). The components of arthropod predation. I. The prey death-rate. *Journal of Animal Ecology* **46,** 249–262.

Hassell, M. P., Waage, J. K., and May, R. M. (1983). Variable parasitoid sex ratios and their effect on host parasitoid dynamics. *Journal of Animal Ecology* **52,** 889–904.

Hassell, M. P., Lessells, C. M., and McGavin, G. C. (1985). Inverse density dependent parasitism in a patchy environment: a laboratory system. *Ecological Entomology* **10,** 393–402.

Hassell, M. P., Southwood, T. R. E., and Reader, P. M. (in press). The dynamics of the viburnum whitefly (*Aleurotrachelus jelinekii* (Frauenf.)): a case study in population regulation. *Journal of Animal Ecology*.

Holling, C. S. (1959a). The components of predation as revealed by a study of small mammal predation of the European pine sawfly. *Canadian Entomologist* **91,** 293–320.

Holling, C. S. (1959b). Some characteristics of simple types of predation and parasitism. *Canadian Entomologist* **91,** 385–398.

Hubbard, S. F., and Cook, R. M. (1978). Optimal foraging by parasitoid wasps. *Journal of Animal Ecology* **47,** 593–604.

Ives, W. G. H. (1976). The dynamics of larch sawfly (Hymenoptera: Tenthredinidae) populations in southeastern Manitoba. *Canadian Entomologist* **108,** 701–730.

Kowalski, R. (1976). *Philonthus decorus* (Gr.) (Coleoptera: Staphylinidae). Its biology in relation to its action as a predator of winter moth pupae (*Operophtera brumata*) (Lepidoptera: Geometridae). *Pediobiologia* **16,** 233–242.

Lawton, J. H., and Hassell, M. P. (1981). Asymmetrical competition in insects. *Nature* (*London*) **289,** 793–795.

Lawton, J. H., and Hassell, M. P. (1984). Interspecific competition in insects. *In* "Ecological Entomology." (C. B. Huffaker and R. L. Rabb, eds.), pp. 451–495. John Wiley, New York.

Lawton, J. H., and Strong, D. R. (1981). Community patterns and competition in folivorous insects. *American Naturalist* **118,** 317–338.

van Lenteren, J. C., and Bakker, K. (1976). Functional responses in invertebrates. *Netherlands Journal of Zoology* **26,** 567–572.

van Lenteren, J. C., and Bakker, K. (1978). Behavioural aspects of the functional responses of a parasite (*Pseudeucoila bochei* Weld) to its host *Drosophila melanogaster*. *Netherlands Journal of Zoology* **28,** 213–233.

Lessells, C. M. (1985). Parasitoid foraging: should parasitism be density dependent? *Journal of Animal Ecology* **54,** 27–41.

Lewin, R. (1983a). Santa Rosalia was a goat. *Science* **221,** 636–639.

Lewin, R. (1983b). Predators and hurricanes change ecology. *Science* **221,** 737–740.

May, R. M. (1978). Host-parasitoid systems in patchy environments: a phenomenological model. *Journal of Animal Ecology* **47,** 833–843.

May, R. M., and Hassell, M. P. (1981). The dynamics of multiparasitoid-host interactions. *American Naturalist* **117,** 234–261.

May, R. M., and Oster, G. F. (1976). Bifurcations and dynamic complexity in simple ecological models. *American Naturalist* **110,** 573–599.

May, R. M., Hassell, M. P., Anderson, R. M., and Tonkyn, D. W. (1981). Density dependence in host parasitoid models. *Journal of Animal Ecology* **50**, 855–865.

Maynard Smith, J. (1974). "Models in Ecology." Cambridge University Press, Cambridge.

Mook, L. J. (1963). Birds and the spruce budworm. *In* "The Dynamics of Epidemic Spruce Budworm Populations" (R. F. Morris, ed.), pp. 268–271. *Memoirs of the Entomological Society of Canada*, 31.

Murdoch, W. W. (1969). Switching in general predators: experiments on predator and stability of prey populations. *Ecological Monographs* **39,** 355–354.

Murdoch, W. W. (1977). Stabilizing effects of spatial heterogeneity in predator-prey systems. *Theoretical Population Biology* **11,** 252–273.

Murdoch, W. W., and Oaten, A. (1975). Predation and population stability. *Advances in Ecological Research* **9,** 2–131.

Murdoch, W. W., Reeve, J. D., Huffaker, C. B., and Kennett, C. E. (1984). Biological control of olive scale and its relevance to ecological theory. *American Naturalist* **123,** 371–392.

Murdoch, W. W., Chesson, J., and Chesson, P. L. (1985). Biological control in theory and practice. *American Naturalist* **125,** 344–366.

Reeve, J. D., and Murdoch, W. W. (1985). Aggregation by parasitoids in the successful control of the California red scale: a test of theory. *Journal of Animal Ecology* **54,** 797–816.

Rogers, D. J. (1972). Random search and insect population models. *Journal of Animal Ecology* **41,** 369–383.

Royama, T. (1970). Factors governing the hunting behaviour and selection of food by the great tit (*Parus major* L.). *Journal of Animal Ecology* **39,** 619–668.

Royama, T. (1971). A comparative study of models of predation and parasitism. *Researches in Population Ecology* **1** (suppl), 1–91.

Southwood, T. R. E. (1967). The interpretation of population change. *Journal of Animal Ecology* **36,** 519–529.

Southwood, T. R. E. (1978). "Ecological Methods with Particular Reference to the Study of Insect Populations." Chapman and Hall, London.

Southwood, T. R. E., and Comins, H. N. (1976). A synoptic population model. *Journal of Animal Ecology* **45,** 949–965.

Strong, D. R. (1984). Density-vague ecology and liberal population regulation in insects. *In* "A New Ecology" (P. W. Price, C. N. Slobodchikoff and W. S. Gaud, eds.), pp. 313–327. John Wiley, New York.

Varley, G. C., and Gradwell, G. R. (1960). Key factors in population studies. *Journal of Animal Ecology* **29,** 399–401.

Varley, G. C., Gradwell, G. R., and Hassell, M. P. (1973). "Insect Population Ecology." Blackwell Scientific Publications, Oxford.

Waage, J. K. (1979). Foraging for patchily distributed hosts by the parasitoid, *Nemeritis canescens*. *Journal of Animal Ecology* **48,** 353–371.

Waage, J. K. (1983). Aggregation in field parasitoid populations: foraging time allocation by a population of *Diadegma* (Hymenoptera, Ichneumonidae). *Ecological Entomology* **8,** 447–453.

Waage, J. K., and Hassell, M. P. (1982). Parasitoids as biological control agents—a fundamental approach. *Parasitology* **84,** 241–268.

Wang, Y. H., and Gutierrez, A. P. (1980). An assessment of the use of stability analyses in population ecology. *Journal of Animal Ecology* **49,** 435–452.

8

Parasitoid Communities: their Size, Structure and Development

R. R. ASKEW AND M. R. SHAW

I. INTRODUCTION

Most insects may have their lives curtailed by insect parasitoids, especially parasitic Hymenoptera. In Britain, the only pterygote orders apparently free from attack are Ephemeroptera, Plecoptera, Phthiraptera and Strepsiptera. However, although parasitoids attack a very broad range of host types, their distribution is uneven. Some insect species are host to a complex of 20 or more parasitoid species, but a large number are attacked by just one or two.

A community is a group of species having a high degree of spatial and temporal concordance, and in which the member species mutually interact to a greater or lesser extent. Because biologically analogous hosts, such as gall wasps on oak or leaf-miners on deciduous trees, often share parasitoid species, their parasitoids can be regarded as constituting parasitoid communities. It is useful to employ the more restricted, host-oriented term "parasitoid complex" to the parasitoids attacking a particular taxon. Both terms should be understood to include secondary as well as primary parasitoids. Parasitoid communities can seldom be sharply defined because many parasitoid species regularly or occasionally enter the parasitoid complexes of more than one host, some of which we may not recognize as belonging to the same community.

In this chapter we first consider the literature on host-parasitoid associations and its limitations. We then review ideas that have been applied to the study of parasitoid communities. Communities that have been well investigated form a very limited group: they concern, almost exclusively, endophytic hosts. We have a much more comprehensive knowledge of host ranges of parasitoids within communities such as those of galls and leaf-mines than we have for those in the much less easily studied exophytic communities. Because of this shortcoming, we develop a system of categorizing parasitoids, based on a fundamental dichotomy in their biology, which allows us to make predictions about host range. The size of parasitoid complexes and differences in composition between those of taxonomically related hosts can then be discussed in terms of host life history and parasitoid biology.

II. USE OF THE LITERATURE

The best way to try to understand the host ranges of parasitoids and factors influencing the extent of parasitoid complexes might seem to be from a collation of the considerable available literature on parasitism of particular hosts. However, the literature is of such limited value that a discussion of its limitations is itself worthwhile.

Firstly, much of what has been written is very unrepresentative. Most research, until quite recently, was in the context of biological control. The hosts were pest species, and levels of parasitism and the composition of parasitoid complexes were often recorded in a situation of monoculture and from huge host populations that had expanded beyond their normal limits. Even the sampling of hosts of no economic significance has often been in an outbreak context when the host has outgrown its normal population levels and controls (Benson, 1950). Data obtained in such atypical circumstances are likely to be a poor reflection of the "natural" situation in which evolutionary adaptations have occurred.

Secondly, and another consequence of the pest control background to so much of our knowledge, is the tendency to assume that parasitoids are far more host-specific than they really are. Relatively very few are strictly monophagous and many are substantially polyphagous, but the vital and interesting problem of elucidating the host ranges of parasitoids is difficult to tackle. It is easy to investigate the parasitoids of a particular host, but quite another matter to discover all the hosts of a particular parasitoid. Attempts to collate records in a way that sums up host ranges, usually in the form of host/parasite catalogues, have at best only limited success and at worst are positively misleading (Shaw, 1981a). Our lack of accurate knowledge of the host ranges of parasitoids has three main consequences:

1. An important factor—the extent to which a parasitoid population is dispersed outside the complex or community under consideration—will be missing from any analysis. Many species, although always more or less niche-specialized, are generalists to the extent of entering the parasitoid complex of a given host species only facultatively, and they can be responsible for considerable local variation
2. Models of single host/single parasitoid interactions assume effective isolation of the system (see Chapter 7). In a varied ecosystem parasitoid populations are probably very rarely so constrained and the validity of applying such models to the natural situation is questionable
3. The behaviour of a parasitoid in the laboratory in relation to just one species of host must similarly be interpreted with care. Its performance in the field may be mediated by the availability of alternative hosts.

A third problem in trying to inter-relate published studies on parasitoid complexes of given hosts is that it entails comparing data from disparate and often incompatible sources. For example, parasitoid complexes of arboreal hosts, though broadly similar, are often richer in ancient and undisturbed habitats than in altered ones (personal observation), as would be expected at the high trophic levels concerned (Pimm and Lawton, 1977). In contrast, Miller (1980) found more parasitoid species attacking *Spodoptera* in disrupted habitats than in non-disrupted ones. Studies at a single site are likely to give very misleading impressions. For example, two studies of *Ladoga camilla* (L.) in Britain showed it to be practically unparasitized at the edge of its recently expanded range (Pollard, 1979) whilst several specialized parasitoids were of considerable importance at sites where the butterfly has been continuously present over a long period (Shaw, 1981b). Similarly, Randall

(1982) found that the parasitoid complex of *Coleophora alticolella* Zeller diminished steadily with increasing altitude from seven species at 15 m to one at 395 m and none above 400 m.

Temporal changes may also occur in a parasitoid complex, causing samples taken only at a particular time to be of limited value. Depending on exactly when a given host is sampled, not only will differing levels of parasitism be found but also, very often, different parasitoid species. In the case of endophytic host populations it is usually possible to discover, from remains in the larval feeding site, the cause, sequence and magnitude of past mortalities by sampling at the generation endpoint. This cannot be done for exposed, mobile hosts. A striking illustration of how sampling in a less than continuous way could be misleading is provided by two collections of *Xestia xanthographa* (Denis and Schiffermüller) larvae of the same generation made near Reading, England on different days in the spring of 1979 (Table I). The parasitoids reared from the sample collected in early March had all overwintered in young host larvae which they killed before the end of April. *Ophion scutellaris* Thomson emerges from its cocoon in late March or early April and attacks older larvae. These two samples, differing only by the 10 weeks separating their dates of collection, yielded respectively eight and one entirely different species of parasitoids.

TABLE I. Parasitism of larvae of *Xestia xanthographa* (Denis & Schiffermüller) collected near Reading on two dates during spring, 1979. The figures refer to larval parasitism only and take no account of larval/pupal parasitoids such as *Amblyteles armatorius* (Forster)

		Sampling dates	
		March 2	May 12
Numbers of unparasitized larvae		188	50
Numbers of larvae parasitized by			
Tachinidae:	*Periscepsia spathulata* (Fallén)	24	0
	Pales pavida (Meigen)	1	0
Braconidae:	*Apanteles fulvipes* (Haliday)	8	0
	A. hyphantriae Riley	1	0
	Meteorus gyrator (Thunberg)	1	0
	Aleiodes sp. A	46	0
	Aleiodes sp. B	10	0
Ichneumonidae:	*Hyposoter* sp.	1	0
	Ophion scutellaris Thomson	0	6
Total larvae sampled		280	56
Percentage parasitism		33	11

In addition to the need for care in within-generation sampling, it is necessary to be aware of sometimes profound and regular differences between generations of multivoltine hosts in temperate climates. These differences are chiefly quantitative: there is often an overall greater level of parasitism in late summer host generations (Askew and Shaw, 1974, 1979a) and some parasitoid species increase in numbers more than others towards the end of the season, thus affecting the relative composition of communities. Exceptionally, however, parasitism differs qualitatively according to season. A clear example is provided by cynipid oak galls where the difference is correlated with an alternation in gall site and form, the alternate generations of the same host species being treated as equivalent to two completely different species by the parasitoid community (Askew, 1961a).

In the absence of such profound seasonal differences in host behaviour, qualitative changes involving the regular entry of a parasitoid species into the parasitoid complex of a multivoltine host only at particular times of year are rarely noted. However, this situation occurs in the rich parasitoid complex associated with the multivoltine choreutid moth *Anthophila fabriciana* (L.) in Britain. Larvae live under a slight web on *Urtica* and overwinter to complete feeding in spring. The species reappears as a larva from July in one or two overlapping summer generations. One parasitoid, *Lissonota stigmator* Aubert, which in Britain seems to be exclusively monophagous, emerges from its cocoon in February or early March and oviposits into post-hibernation larvae of the overwintering host generation. During May, this species kills its prepupal host and develops rapidly to the pharate adult stage, in which it remains for up to 10 months before emerging early in the following spring, missing the summer generations of its host. Another parasitoid, *Clinocentrus gracilipes* (Thomson), attacks late instar host larvae and does not emerge from its overwintering cocoon until around midsummer, when the overwintering host generation is virtually past the larval stage. Parasitoids associated with *A. fabriciana* have been the subject of a long-term study and samples have been collected from all over Britain (Shaw, unpublished). A striking finding is that each of the two parasitoid species mentioned above is dominant in its generation and present at almost all sites.

III. PARASITOID COMMUNITY STRUCTURE

A. Parasitoid Categories and the Structure of Communities

Early work on parasitoid complexes listed and described species reared from particular hosts, and it soon became evident that some hosts supported a very large number of species. Gahan's monumental study (1933) of the hessian fly, *Mayetiola destructor* (Say) associated no fewer than 41 parasitoid species with this host. Quantitative as well as qualitative accounts of the composition of parasitoid complexes soon followed, often including descriptions of immature stages. Investigations by Salt (1931) of the wheat-stem sawfly, *Cephus pygmaeus* (L.), by Taylor (1937) of the coconut beetle, *Promecotheca caeruleipennis* Blanchard, by Cameron (1939) of the holly leaf-miner, *Phytomyza ilicis* Curtis, and by Carleton (1939) of the bean gall sawfly, *Pontania proxima* (Lepeletier), were all motivated, at least in part, by economic considerations. One of the principal aims of these studies was to evaluate the general level of host mortality inflicted by each parasitoid species and the probability of its success as a possible biological control agent.

About the middle of this century ecological thinking about parasitoid communities centred upon two related problems. The first was how the sizes of host and parasitoid populations are regulated. Varley (1947) used a non-economic host, the knapweed gallfly, *Urophora jaceana* (Hering) and its parasitoid complex, for a field analysis of population dynamics and demonstrated that whilst most host mortality could be attributed to factors that were probably not density dependent, a parasitoid species was responsible for a small but significant proportion of delayed density dependent mortality which could control the population of *Urophora*. The enormous volume of literature subsequently devoted to host-parasitoid population dynamics has centred upon quantifying and modelling the balance between reproductive capacity (including searching efficiency of parasitoids) and mortality (see Chapter 7).

The second problem was how a single host species can support a large number of parasitoid species and, in particular, how several often closely related parasitoid species are able to coexist within the same parasitoid complex. Blair (1944), in his study of parasitoids associated with the bedeguar gall wasp *Diplolepis rosae* (L.), pioneered research into the very complex and subtle trophic relationships existing in some parasitoid complexes. Askew (1961a) followed Blair's approach in examining the large but seemingly structured parasitoid community associated with a range of cynipid oak galls. Differences in distribution and trophic relationships between parasitoid species were interpreted as reflecting the outcome of evolutionary pressure to minimize interspecific competition: gall diversity provided sufficient environmental heterogeneity to accommodate 60 parasitoid and inquiline species in a community of 24 gall-forming cynipid species.

Whether community composition is deterministic or stochastic, the extent to which species in a community interacted (competed) in the past to influence the pattern of resource partitioning which we see today (Connell,

1980) and the extent to which random colonization processes contribute to the pattern (Dean and Ricklefs, 1979; Shorrocks *et al.*, 1984), has recently attracted much attention (Bouton *et al.*, 1980; Strong *et al.*, 1984; Southwood, 1985). Interspecific competition between parasitoids provides a very probable explanation of instances of the displacement from a complex of one species by another, although all examples concern biological control situations and introduced parasitoid species (Bess *et al.*, 1961; DeBach and Sundby, 1963; Flanders, 1966; Orphanides, 1984; Tremblay, 1984). Similarly, the orderly interdigitation in large communities of parasitoid species which have an array of trophic relationships, phenologies and other biological features is seen by many (Askew, 1961b, 1975; Zwölfer, 1971; Hawkins and Goeden, 1984; Ehler, 1985) as an indication that interspecific competition between parasitoids has been a potent structuring force.

The coexistence of a number of parasitoid species attacking the same host species is explained by Zwölfer (1971) and Schröder (1974) on the basis of "counter-balanced competition". This operates between well-adapted and highly synchronized species tending towards monophagy which are, however, intrinsically poor competitors when multiparasitism occurs, and intrinsically superior but less host-specialized species. It is clear that parasitoids tend to polarize about contrasting biologies, but uncertainty about the most useful way of defining these is reflected in the number of different proposals that have been made. The old division into ectoparasitoids and endoparasitoids, although readily made, carries too few ecological correlates to be of general value in clarifying their position in communities. It was an important step forward when Force (1972) and Price (1973) analysed parasitoid complexes associated with, respectively, the gall midge *Rhopalomyia californica* Felt and the sawfly *Neodiprion swainei* Middleton in terms of MacArthur and Wilson's *r*-selection and *K*-selection (1967). Force assigned *r* values to parasitoids and found an inverse relationship between these and their competitive abilities; *r*-selected species died in already parasitized hosts and females normally avoided oviposition in such hosts. Price found that the total number of parasitoid species in the complex increased with plant succession and different parasitoid adaptations were favoured at different points of the succession. Parasitoids of larvae, with high fecundity and adult mobility (*r*-selected), predominated in early successional stages (or in peripheral locations), whilst pupal parasitoids, with low fecundity but often able to act as facultative secondary parasitoids (better competitive ability; *K*-selected) tended to displace *r*-strategists from long-established host populations. The wisdom of extending *r*-*K* terminology to a taxonomically restricted group of species at predominantly one trophic level is questionable, as Force (1975) himself acknowledges. Nevertheless, recognition of a diversification in parasitoid biology based upon relatively high or low reproductive potential was very important because reproductive potential could be associated with other attributes which helped to explain a species' position in the community.

Askew (1975) did not use *r-K* terminology in analysing parasitoid communities associated with cynipid oak galls and lepidopterous leaf-mines. Instead, two sets of parasitoids were defined: those which attack the host early in its development and those which attack it late. These approximated to, but in some features differed from, the *r* and *K*-selected categories of Force and Price. For each set, a suite of associated characteristics was recognized. "Early" parasitoids attack host larvae at a time when they are abundant, making a high reproductive capacity (many eggs, thelytoky, etc.) advantageous. Individual hosts tend to be small when attacked so that endoparasitism, which is most compatible with appreciable host development, is favoured. Endoparasitism, because it is believed to demand closer adjustment to the host's biology, is often associated with a narrow host range. "Late" parasitoids, in contrast, face a situation of relative scarcity of primary hosts which may increase the cost (searching time) of each egg laid. The host is killed when, or very soon after, it is attacked as it is sufficiently large to sustain a parasitoid without further growth. Ectoparasitism is favoured over endoparasitism, probably to avoid the host's physiological defence system, and the host range can be broad (usually permitting facultative secondary parasitism).

A fuller understanding of the pattern of parasitoid community structure encouraged some tentative evolutionary speculation. The narrower specializations required of "early" parasitoids, and their tendency to be taxonomically more isolated, led Askew (1975) to suggest that they have been longer established in communities and that "late" parasitoids are more recent acquisitions. This supports the contention of Price (1973) and Force (1975) that *K*-selected parasitoids may displace *r*-selected species. These predictions, formulated by basically similar arguments but based upon somewhat different categorizations, highlight the desirability of having a biological classification of parasitoids that is simple, all-embracing and based upon an absolute rather than a relative character.

B. Koinobionts and Idiobionts: a Useful Categorization

For the purpose of developing an ecological theory about parasitoids, the recognition of specialists and generalists—those with respectively narrow and broad host ranges—is crucial (as revealed in Chapters 7 and 9). However, to label a parasitoid as either a specialist or a generalist implies that we have considerable information about its host range. Measuring this dimension of parasitoid biology demands very extensive study and has been achieved only in certain endophyte communities. In these, parasitoids are known from a number of host species, and host range within the community can be assessed. For parasitoids in many less sharply-defined communities however, especially those based upon exophytic hosts, information on host ranges is at present very poor and it is premature to differentiate between specialists and generalists. What is required is a practical means of categorizing parasitoids, based on minimum biological information, which correlates with their usually unknown host ranges. The simplest classification might be based upon endoparasitism as opposed to ectoparasitism, the hypothesis being that endoparasitoids have a narrower host range than ectoparasitoids. However, there are many exceptions to this pattern, the major ones being ectoparasitoids which allow their hosts to continue to develop and endoparasitoids which kill the host at the stage attacked. A more useful hypothesis was presented by Haeselbarth (1979), who divided parasitoids into two groups on the basis of whether or not they permit the host to grow or metamorphose beyond the stage attacked. Those that do he termed "koinophytes", and those that do not "idiophytes". We prefer to employ the terms "koinobionts" (Greek *koinos*, in common; *bios*, life) and "idiobionts" (Greek *idios*, individual).

1. Koinobionts

Koinobionts include most endoparasitoids of larvae and of adult insects, some groups of specialized ectoparasitoids of mobile hosts (e.g. Tryphoninae), and almost all groups of obligatory secondary parasitoids (e.g. Mesochorinae, Eucerotinae, Trigonalidae, Perilampidae).

The parasitized host continues to be mobile and able to defend itself; larval hosts are often not killed until they have prepared cryptic pupation retreats. The host may not live very long after parasitism, but the cardinal point is that the koinobiont benefits from the continued life of its host.

2. Idiobionts

Idiobionts include the many ectoparasitoids which permanently paralyse or kill the host before the egg hatches; egg parasitoids (but not, of course, egg-larval parasitoids) and pupal endoparasitoids (e.g. Pimplini). The host is consumed in the location and state it is in when attacked. Many idiobionts can function as facultative secondary parasitoids.

This categorization is tested here as a correlate of host range with respect to the arboreal leaf-miner community in which the biology of each species has been ascertained (Table II). Host range is measured not as the number of

TABLE II. The numbers of host families attacked by chalcid parasitoids in the arboreal leaf-miner community. Host range is more closely related to the koinobiotic/idiobiotic categorization of parasitoids than to the endoparasitic/ectoparasitic division. Data are from Askew and Shaw (1974) and are restricted to parasitoid species for which more than five records were available

Parasitoid species	Number of host families	Endoparasitoid (*) or ectoparasitoid	Koinobiont (*) or idiobiont
Elachertus inunctus (Nees)	4		*
Cirrospilus vittatus Walker	5		
C. diallus Walker	7		
C. pictus (Nees)	5		
C. lyncus Walker	2		
Diglyphus minoeus (Walker)	1		
Sympiesis sericeicornis (Nees)	5		
S. grahami Erdös	1		
S. gordius (Walker)	1		
Pnigalio pectinicornis (L.)	8		
P. agraules (Walker)	6		
P. longulus (Zetterstedt)	6		
P. soemius (Walker)	6		
Pediobius alcaeus (Walker)	1	*	*
P. saulius (Walker)	2	*	*
Chrysocharis laomedon (Walker)	4	*	
C. nephereus (Walker)	7	*	
C. nitetis (Walker)	6	*	
C. pentheus (Walker)	4	*	
C. phryne (Walker)	1	*	*
C. prodice (Walker)	2	*	*
C. gemma (Walker)	1	*	*
C. melaenis (Walker)	1	*	*
Entedon punctiscapus Thomson	1	*	*
Achrysocharoides cilla (Walker)	1	*	*
A. acerianus (Askew & Ruse)	1	*	*
A. latreillei (Curtis)	1	*	*
A. splendens (Delucchi)	1	*	*
A. zwoelferi (Delucchi)	1	*	*
A. niveipes (Thomson)	1	*	*
A. suprafolius (Askew & Ruse)	1	*	*
A. atys (Walker)	1	*	*
Chrysonotomyia chlorogaster (Erdös)	4	*	*
Closterocerus trifasciatus Westwood	6	*	
Tetrastichus miser (Nees)	2	*	
T. ecus (Walker)	3		
Pteromalus dolichurus (Thomson)	2		
Trichomalus inscitus (Walker)	1		

Mean number of families in host range of: endoparasitoids = 2.3; ectoparasitoids = 3.9; koinobionts = 1.5; idiobionts = 4.2.

host species but the number of higher taxa (families) utilized (Pschorn-Walcher, 1957), and the koinobiont/idiobiont categorization is shown to be much more closely correlated than endoparasitism/ectoparasitism with the breadth of host range. Ideally, the correlation should be tested further on data involving a much broader taxonomic spectrum of parasitoids, but we know of no data complete enough to be used for this.

The conclusion that koinobiosis should be superior to endoparasitism as a correlate of a limited host range accords with what we know of the demands on a parasitoid associated with a still-living host, and also the specializations needed to benefit from this mode of life. These requirements are not imposed in idiobiosis where the parasitoid does not depend upon further host development and where the host is arrested and usually killed very rapidly. Rapid host death is certainly a usual consequence of ectoparasitism by idiobionts and in at least some cases of endoparasitism by idiobionts (Führer and Kilincer, 1972; see Chapter 4). Vinson and Iwantsch (1980a, 1980b) review factors which determine the suitability of a host. Of these, the evasion of host defences (in particular, encapsulatory responses; Salt, 1970) by regulation of the host's physiology, or by precise placement of progeny within the host, are applicable exclusively to endoparasitic koinobionts, whilst both endoparasitic and ectoparasitic koinobionts may have the capacity for regulating host development or behaviour in a favourable manner, possibly in some cases by the production of specialized venoms (Vinson, 1975; Shaw, 1981c). It seems reasonable, on this evidence, to hypothesize that the host ranges of koinobionts are limited by their need to interact with the living host. Haeselbarth (1979) outlined the advantages of synchronization and searching efficiency that accrue from exploitation of a narrow host range.

Evidence from artificial rearing suggests that idiobionts have basically broad nutritional requirements. Many can be reared on surrogate hosts which are not at all closely related to those attacked in nature (Simmonds, 1944; Fedde *et al.*, 1984). For example, *Rhyssa persuasoria* (L.) can develop satisfactorily on *Apis* larvae (Spradbery, 1968) and *Pimpla turionellae* (L.) on *Tenebrio* pupae (Sandlan, 1980). Even a completely artificial medium can be used to rear *Itoplectis conquisitor* (Say) (House, 1978). Nutritional flexibility is also seen in the many idiobionts that function as facultative secondary parasitoids, which usually involves hosts of at least two orders.

The specializations available to and demanded of koinobionts but not idiobionts seem very likely to impose a greater restriction on host ranges of koinobionts. In particular within a given community the koinobionts are in general expected to have the narrower host ranges.

We stress that we believe the separation of parasitoids into koinobionts and idiobionts to be the most practical indication of host range that can be used at present. It has the advantage that parasitoids can be categorized simply from knowledge of whether or not the host continues its life after parasitism, and also that, in many groups of parasitoids (but, in general, not Chalcidoidea), it matches divisions between higher taxa (usually subfamilies or tribes in Ichneumonoidea, for example) which makes it possible to categorize a good proportion of species in non-reared samples. The hypothesis that a primary division into koinobionts and idiobionts is a good indicator of specialization will be refined and qualified as more biological information becomes available. For example, Scelionidae are idiobionts but they exhibit a very high degree of host specialization through their searching behaviour (see Chapter 4).

Ecological factors are a major determinant of the host ranges of parasitoids (Vinson, 1976). Every parasitoid, whether idiobiont or koinobiont, has evolved a productive searching behaviour; it searches environments in which, over evolutionary time, suitable hosts have been found (see Chapter 2). Host ranges are therefore a function of the searching environment, and both idiobionts and koinobionts can recruit to their host ranges potential hosts encountered. Recruitment appears to be easiest for idiobionts, free of constraints imposed by the need to adapt to a living host, and they can be more opportunistic and accommodate themselves to hosts which are only irregularly encountered [e.g. *Mesopolobus gramnium* (Hårdh) (Graham, 1969), *Pachyneuron concolor* (Förster) (Rosen and Kfir, 1983)]. It is often simpler to define an idiobiont's host range in terms of searching environment (host habitat) than host taxonomy; different *Scambus* species can be very accurately described as parasitoids of hosts in grass stems, in seed pods or in composite flower-heads even though hosts belonging to several families or even orders comprise each host range (Shaw, personal observation). Even idiobionts specialized to a very narrow searching environment may widen their host range by virtue of their capacity to be facultative secondary parasitoids, for example *Rhyssa* (Hanson, 1939) and *Perithous* (Danks, 1971).

In contrast, it appears that koinobionts do not attack hosts facultatively and we suggest that more prolonged contact with a potential host, in evolutionary time, is required for its incorporation to the host range (Shaw, 1983). We can give no instance of a koinobiont behaving as a facultative secondary parasitoid. In koinobiont braconids of the genus *Aleiodes*, each species uses a number of lepidopterous hosts found in its general searching environment (such as low-growing plants or deciduous trees) but, although readily attacked in the laboratory, closely allied species of Lepidoptera from the "wrong" environment will encapsulate the *Aleiodes* eggs (Shaw, 1983). Prolonged contact in evolutionary time probably accounts for instances of relatively unrelated hosts being included in the host ranges of koinobionts; *A. bicolor* (Spinola) parasitizes *Polyommatus* and *Zygaena*, larvae of both of which feed on *Lotus*.

In general, host ranges of koinobionts can be described in terms of searching environment and also taxonomy since they encompass a much smaller number of host taxa, or host taxa of lower rank, than those of idiobionts (this statement is based chiefly on our extensive rearings of hymenopterous parasitoids). In the wide sense in which many idiobionts can be said to be generalists, it seems that koinobionts (excluding Tachinidae; see discussion in Section IV. B.) normally cannot.

Many workers will know of a particular koinobiont that appears to have an exceptionally broad host range, and we suggest that such species would repay further study. Some may turn out to be species aggregates, but others may provide biological insights that will lead to useful refinements of the koinobiont/idiobiont host range hypothesis.

IV. GEOGRAPHICAL VARIATION IN PARASITOID COMMUNITIES

A. Stability in the Composition of Complexes

The parasitoid complex of a given host appears to be reasonably constant in composition over much of the host's range, and similar parasitoid complexes attack taxonomically and ecologically related hosts in widely separate (but climatically similar) areas.

Lampe (1984) found a remarkably constant parasitoid complex associated with *Coleophora alticolella* from Norway, through Germany, to Austria and Switzerland, with regional differences occurring only on a microgeographical scale associated with ecologically extreme habitats.

The complex of inhabitants of *Diplolepis rosae* (L.) galls is concluded by Schröder (1967) to be "very much the same all over Europe". Almost everywhere it is dominated by the inquiline cynipid *Periclistus brandti* (Ratzeburg) and by four parasitoids, the ichneumonid *Orthopelma mediator* (Thunberg) and the chalcids *Glyphomerus stigma* (Fabricius), *Torymus bedeguaris* (L.) and *Pteromalus bedeguaris* Thomson. In all national samples (Schröder, 1967; Askew, 1984 and unpublished), these five species together comprise at least 90% of the parasitoid fauna, in total at least 13 species. In North America, Shorthouse (1973) found that *Diplolepis polita* (Ashmead), which forms unilocular galls on *Rosa*, supported a community dominated in hierarchical order by *Periclistus pirata* (Osten Sacken), *Eurytoma longavena* Bugbee, *Pteromalus* sp. indet., *Glyphomerus stigma* (Fabricius) and *Torymus bedeguaris* (L.). Thus there are close and obvious parallels between the parasitoid communities associated with the palaearctic *D. rosae* and the nearctic *D. polita*.

Deyrup (1984) drew similar parallels at the generic level between parasitoids associated with palaearctic and nearctic species of *Xiphydria*.

B. Temperate/Tropical Variation in Community Structure

Tropical vegetation climaxes are characterized by a far greater diversity of plants and insect herbivores than is seen in temperate climaxes (Janzen, 1970). In recent years there has been considerable speculation on the nature of the fauna of parasitic Hymenoptera in the tropics compared with temperate regions.

Owen and Owen (1974) found evidence that Ichneumonidae in tropical Africa are neither more abundant nor more diverse than in Britain, and suggested that no more niches are available to them in the tropics than in temperate areas, despite the greater species-richness of major host groups in the tropics. Janzen and Pond (1975) produced similar evidence by sampling from secondary vegetation in the Americas. This departure from the usual trend in insect groups of species diversity increasing with decreasing latitude to a maximum in the tropics has been explained in terms of resource fragmentation (Janzen and Pond, 1975; Janzen, 1981); potential phytophagous hosts exist in the highly diverse tropical vegetation as small, scattered populations which demand of their parasitoids impossibly high degrees of polyphagy or host location efficiency.

Morrison *et al.* (1979) and Hespenheide (1979) contended that the sampling methods were biased towards large insects active in exposed situations, and small Hymenoptera were probably seriously underestimated. Hespenheide suggested that Chalcidoidea may not show the trend evident in Ichneumonidae, and Morrison *et al.* argued that relative host scarcity in tropical habitats will pressurize parasitoids into attacking eggs, the most abundant host life stage.

We suggest that the division of parasitoids into koinobionts and idiobionts correlates sufficiently well with the potential for respectively narrow and broad host ranges to provide a useful test of the resource fragmentation hypothesis. Predominantly idiobiotic groups such as Braconinae, Pimplinae and Phygadeuontinae should be less constrained by resource fragmentation than koinobiotic groups such as Microgasterinae, Banchinae and Campopleginae, and therefore, in tropical climax vegetation, koinobionts would be expected to suffer a greater depression in species-richness than idiobionts.

Janzen (1981), in demonstrating that the peak of species-richness in North American Ichneumonidae lies between 38° and 42°N and declines towards the tropics, published the numbers of species, in each of eight subfamilies, known to occur in 2.5° latitudinal bands. Janzen's data (excluding the insufficiently known Acaenitinae) are here re-examined to test whether or not the decline in species-richness with falling latitude is greater for koinobiotic than for idiobiotic groups. The number of species per 10^6 km^2 in the peak latitudinal band and the number in the band 12.5° to the south (this analysis eliminates the latitudinal bias of particular groups) is given for each subfamily (Table III). From the ratios of these figures it is seen that there is indeed a much sharper decline in species numbers towards the tropics in the major groups of koinobionts compared with idiobionts, with only the two smallest taxa, Xoridinae and Diplazontinae (both of which have narrow host or habitat requirements), falling slightly out of sequence. Gauld (1986; see Chapter 1) indicates a similar trend in the Australian ichneumonid fauna, in which the ratio of larval parasitoids (presumed to be the most host-specific) to pupal and prepupal parasitoids declines with latitude.

TABLE III. Decline in species richness of some subfamilies of North American Ichneumonidae with falling latitude (after Janzen, 1981)

Taxon	Category	Peak latitude °N	Spp. per 10^6 km^2 at peak latitude	Spp. per 10^6 km^2 12.5° south of peak	Decline ratio
Phygadeuontinae (= Gelinae)	I (almost all)	40.0–42.4	243.4	127.5	1.91
Pimplinae (= Ephialtinae)	I (almost all)	37.5–39.9	120.3	48.4	2.49
Diplazontinae	K	45.0–47.4	69.6	23.2	3.00
Xoridinae	I	37.5–39.9	36.3	11.5	3.16
Mesochorinae	K	37.5–39.9	72.7	20.8	3.50
Metopiinae	K	40.0–42.4	91.4	22.2	4.12
Banchinae	K	37.5–39.9	275.1	62.3	4.42

I = idiobiotic; K = koinobiotic subfamilies.

The expectation that koinobionts will be proportionally more poorly represented than idiobionts in the tropics requires qualification. Idiobiotic groups that have encountered sufficient resource stability in temperate ecosystems to speciate with a relatively high degree of host specialization (e.g. many Pteromalinae; Graham, 1969) may be relatively less well represented in tropical ecosystems than other idiobionts. Conversely, host ranges of species belonging to particular koinobiont taxa in temperate ecosystems are known to be uncharacteristically broad. In Britain, many species of Tachinidae in the subfamilies Tachininae, Goniinae and Macquartiinae parasitize several families or even orders of insects (van Emden, 1954). This is probably connected with external egg placement, the ability of many Tachinidae to divert the host's encapsulatory response to provide them with a respiratory funnel, and their respiration via the host's tracheal system or the atmosphere directly (Shaw and Askew, 1976a). Janzen (1981) suggested that Tachinidae would suffer particularly severe resource depletion in the tropics because so many potential hosts are inaccessible in endophytic situations. Such differences in resource availability may indeed be limiting, but we suggest that Tachinidae may be expected to achieve a relatively higher representation among the koinobionts attacking exposed hosts in the tropics than in temperate regions.

V. EFFECT OF HOST PLANT ON PARASITOID COMMUNITIES

A. Direct Effects of the Host Plant

Vinson (1976) outlined specific mechanisms whereby parasitoids may be able to utilize oligophagous hosts on only a part of their food plant range. These may include greater toxicity of the plant to the parasitoid than to the host (Flanders, 1942; Mueller, 1983), unsuitable surface properties of the host plant for the searching parasitoid (Rabb and Bradley, 1968) and perhaps a lack of correspondence in orientation cues used by host and parasitoid. Such mechanisms may lead to a virtual exclusion of a parasitoid from a part of the host population.

Edmunds (1976) suggested that low predation pressure on aposematic hosts may encourage the development of a parasitoid complex. Some larval *Zygaena* contain cyanides derived directly or indirectly from the food plant (reviewed by Tremewan, 1985). These toxins appear to protect them from attack by generalist, polyphagous parasitoids, but they have accumulated a large and taxonomically diverse assemblage of specialized idiobiotic as well as koinobiotic ichneumonoid and tachinid primary parasitoids which have apparently developed the ability to detoxify cyanides (Jones *et al.*, 1962). Once these primary parasitoids have consumed and detoxified the host tissues, however, their cocoons are heavily attacked by a wide range of common, polyphagous idiobionts including some, such as *Itoplectis*, which in other parasitoid communities would function, at least partly, as primary parasitoids (Shaw, unpublished).

In general surveys of the parasitoids of arboreal leaf-miners (Askew and Shaw, 1974; Shaw and Askew, 1976b) we found subtle, quantitative differences between parasitoid representations across *Phyllonorycter* complexes on a range of host trees. Qualitative differences (presence or absence of species) are predominantly due to the distribution of koinobionts; for example, 11 of 12 idiobiont species but only one of five koinobionts are common to both oak and birch at the same site (Askew and Shaw, 1979b). This we explain on the basis of the specialization of koinobionts to a narrower host range.

However, the idiobiont species, although mostly broadly represented in many *Phyllonorycter* complexes on a large number of trees, account for a significant part of the quantitative variation in the parasitoid complexes (Table IV). Askew and Shaw (1974) demonstrated that complexes are most similar on taxonomically related trees and this is, to a large extent, due to similarity in idiobiont representation. It is concluded that the environment provided by a particular tree species has a strong influence on the extent to which idiobionts search for hosts on it.

TABLE IV. Numbers of 11 species of idiobiotic parasitoids of *Phyllonorycter* associated with hosts on both oak and birch at one site in Cheshire. The distribution is highly heterogeneous ($\chi^2_{10}=254.4$, $P<0.001$). Data from Askew and Shaw (1979b)

	Oak	*Birch*
Colastes braconius Haliday	9	38
Cirrospilus diallus Walker	120	39
C. lyncus Walker	14	12
C. vittatus Walker	91	29
Sympiesis sericeicornis (Nees)	43	17
S. gordius (Walker)	18	43
Pnigalio longulus (Zetterstedt)	8	92
P. pectinicornis (L.)	45	71
P. soemius (Walker)	1	40
Chrysocharis nephereus (Walker)	36	14
C. laomedon (Walker)	13	48

We draw attention here to a different influence that host plants may exert on the parasitoid complex of an oligophagous host when the timing of development of endoparasitized hosts differs crucially from that of unparasitized hosts. Intermittent sampling of larvae of *Leucoma salicis* (L.) on the sand dunes of the Lancashire coast (Shaw, unpublished) showed that the two principal endoparasitoids, the solitary braconids *Aleiodes pallidator* (Thunberg) and *Cotesia* (*Apanteles sensu lato, partim*) *melanoscelus* (Ratzeburg), achieved substantially different representation in the parasitoid complex on adjacent stands of the low-growing *Salix repens* L. and mixed arborescent *Populus* species (*nigra* L. predominating). While both species occurred on *Populus*, with *C. melanoscelus* the more common (177 *C. melanoscelus* and 22 *A. pallidator* reared from 566 host larvae), only *A. pallidator* could be found associated with *S. repens*, on which it was very common. *C. melanoscelus* could, however, be reared from *L. salicis* larvae collected from *Populus* but fed on *S. repens*.

Both *A. pallidator* and *C. melanoscelus* overwinter in diapausing young host larvae, and the latter become active again early in spring at about the time of bud burst. Parker (1935) has shown that *L. salicis* harbouring *C. melanoscelus* (=*solitarius*) break diapause in advance of unparasitized control larvae. Our own experiments with *A. pallidator* have shown, in contrast, that under outdoor conditions a cohort of host larvae ($n=57$) parasitized by *A. pallidator* came out of their hibernacula a mean of 9.2 days later than controls from the same egg batch ($n=66$; $t=9.18$; $P<0.001$). In most years, buds burst earlier on *Populus nigra* than on *Salix repens*, and the interval, although variable, may be two or more weeks.

We interpret these data as indicating that the timing of bud burst on *Salix repens* is critical to the parasitoids of *L. salicis*. *Leucoma* harbouring *C. melanoscelus* would normally be searching for food on *S. repens* before it is available early in spring. In some years even unparasitized hosts seem at risk from starvation. The delayed appearance of hosts parasitized by *A. pallidator*, however, probably ensures that a much higher proportion of these can find food on *S. repens*. This implies that the level of parasitism due to *A. pallidator* in larvae successfully feeding in spring could be higher than that actually parasitized the previous autumn. Absence of *C. melanoscelus* from *Leucoma* on *S. repens* is probably not because *S. repens* is outside its usual searching environment (*C. melanoscelus* has been reared from *Orgyia* on this plant at the same site), but more likely because *Salix repens* is an unsatisfactory food plant for *Leucoma* parasitized by *C. melanoscelus*.

With regard to the secondary parasitoid component of the *L. salicis* complex, the data presented in Table V show particularly how generalized the hyperparasitoid fauna is with respect to both food plants and primary hosts. Most of the commoner species develop in cocoons of both *Cotesia* and *Aleiodes*. This is to be expected of idiobionts. Koinobiont *Mesochorus* species, all of which are obligatory secondary parasitoids, appear to be able to develop in a relatively broad range of primary parasitoids (including in some cases both ichneumonoids and tachinids). It would be interesting to know whether this is because parasitoids have not been under sustained evolutionary pressure to evolve a potent physiological defence system, or because Mesochorinae have a broadly effective means of overcoming immunological defence systems.

The only idiobionts to partition the host resource according to food plant are the three *Gelis* species and this is readily correlated with morphological differences: *G. areator* (Panzer) and *Gelis sulcata* (Blunck) are fully-winged and associated only with the arboreal hosts, while apterous females of *G. ?instabilis* (Förster) search only among low plants.

TABLE V. Hyperparasitoids of *Leucoma salicis* (L.) reared from cocoons of *Aleiodes pallidator* (Thunberg) and *Cotesia melanoscelus* (Ratzeburg) at Ainsdale, Lancashire, 1973–1977

	Aleiodes pallidator/ *Salix repens* L.	*Aleiodes pallidator*/ *Populus*	*Cotesia melanoscelus*/ *Populus*
Total collected	78	105	135
Koinobionts			
Mesochorus tachypus Holmgren	0	9	1
Mesochorus tetricus Holmgren	0	2	0
Mesochorus sp. A	1	5	4
Mesochorus sp. B	0	0	1
Idiobionts			
Itoplectis maculator (F.)	0	6	0
Gelis ? instabilis (Förster)	5	0	0
Gelis areator (Panzer)	0	5	9
Gelis sulcata (Blunck)	0	0	1
Pteromalus chrysos Walker	3	12	30
Pteromalus semotus (Walker)	4	9	21
Dibrachys cavus (Walker)	0	2	4
Dibrachys boarmiae (Walker)	0	0	1
Trichomalopsis sp. A	0	1	0

B. Host Plant Apparency

Because they represent a prominent resource that is extensive in both space and time, trees in temperate forests are considered to be highly apparent plants (Feeny, 1976 and many subsequent authors). It has been postulated that plant species of high apparency are subject to the highest levels of colonization by herbivorous insects, and it would follow that parasitoids should find the greatest range of hosts on such plants and likewise be concentrated upon them. The arboreal concentration of herbivorous insects may be only a correlate of the greater structural complexity ("architecture") of highly apparent plants (Lawton, 1983; but see also Fowler, 1985), since plant architecture provides for greater niche diversification on trees than on low-growing plants. Some niches, however, correspond on low plants and trees. Insects forming leaf-mines, for example, occupy narrow niches and the additional niches available on trees will be quite irrelevant to their niche-specialized parasitoids. Parasitoid complexes of equivalent niche-specialized hosts should then be comparable in terms of host plant apparency freed from the complication of differences in plant architecture.

Askew (1980) found higher species diversity of parasitoids of Cynipinae and *Phyllonorycter* associated with hosts on trees than with hosts on herbaceous plants and shrubs. On the same grounds, Lampe (1984) suggested that *Coleophora* species feeding on early successional plants in temporary habitats are less apparent to parasitoids than are species living on more "predictable" plants (trees, shrubs, *Juncus*).

Apparency theory predicts that the resource stability of apparent plants should promote specialization by insect colonists. Three studies on *Coleophora* species in Britain record data on parasitism as follows:

C. serratella (L.) on *Betula* and *Ulmus*: two koinobiont and six idiobiont parasitoid species, 70% parasitism; Coshan (1974).
C. alticolella on *Juncus squarrosus* L.: one koinobiont and six idiobiont parasitoid species, up to 61% parasitism at low altitude; Randall (1982).
C. discordella Zeller on *Lotus corniculatus* L.: three idiobiont parasitoid species, 76% parasitism; Compton (1981).

The small number of parasitoid species on the *Lotus*-feeding host accords with the suggested relatively low apparency of the food plant. Also, phytophagous hosts specialized to feed on unapparent plants in heterogeneous environments might be expected to be attacked least by host-specialized koinobiont parasitoids which require, it is suggested, regular discovery of a host over evolutionary time to incorporate it in the host range. The high percentage parasitism on *Lotus* is due entirely to fairly generalized idiobiotic parasitoids. The complex on *Juncus* is also dominated by idiobionts, although some are specialists (reflecting resource stability; Lampe, 1984), while koinobionts feature more prominently in the complex of the arboreal host, and in fact achieve numerical dominance at most sites. Lampe (1984) compares more extensively sampled parasitoid complexes of several *Coleophora* species in Europe and finds similar trends.

A further illustration of the preponderance of parasitoid species on hosts feeding on late successional, or in this case climax, vegetation is provided by a comparison of two leaf-mining sawflies, *Profenusa pygmaea* (Klug) on *Quercus* which supports a very large parasitoid complex (two koinobionts, 11 idiobionts; Table VI), and *Fenella nigrita* Westwood on early successional *Potentilla* from large samples of which we have reared only the idiobiont *Xenarcha lustrator* (Haliday).

VI. EFFECT OF HOST BIOLOGY ON COMMUNITY SIZE

We have shown that high spatio-temporal prominence (apparency) of a host is probably conducive to colonization by parasitoids and is needed particularly by koinobionts. Host food plant apparency, however, is just one of several factors which may affect the size of a parasitoid complex. These factors pertain to what we may describe as the discoverability of a host, a measure of how readily and predictably it can be found.

1. Host habitat

In a general account of parasitism of British Diptera, Shaw and Askew (1979) stressed the importance of host habitat and how discoverable a given host species was likely to be as a determinant of whether or not it would be heavily, or regularly, parasitized. Predictability of discovery must be a major factor in determining how many parasitoid species colonize a particular host. Furthermore, habitat and predictability of discovery are strongly linked. Hosts dispersed within a relatively homogeneous environment, such as mud or water, are probably difficult to locate, but hosts restricted to a particular element within a more heterogeneous environment will be more reliably discoverable. Both a sufficient and stable host resource and the existence of suitable environmental cues are necessary for the development of a parasitoid complex. Larval Syrphidae develop in a variety of habitats, but the only group to have attracted an appreciable parasitoid complex (Weems, 1954) is the aphidophagous Syrphinae which, living sometimes at high density on leaf surfaces among rather static, honeydew producing aphid colonies, are probably the most accessible and detectable to parasitoids (Shaw and Askew, 1979; Rotheray, 1981).

Other hosts which are strongly localized, and therefore similarly predictable, for example those feeding on flower- and seed-heads, often seem subject to high levels of parasitism, a penalty which is incurred by feeding on these rich food sources.

2. Immobility

Hosts completely confined within plant tissue, as in galls and leaf-mines, cannot escape the propinquity of signals generated during their lives and possibly used by parasitoids in host location. The generally high levels of parasitism of these endophytic hosts have already been mentioned. Other factors, not necessarily related to discoverability, which may encourage parasitism of endophytic hosts, are their inability to escape by wriggling or dropping from aerial vegetation, or to employ repellent secretions, and their low levels of predation, infection by disease and mortality due to abiotic factors such as weather.

3. Longevity of resource

The longer a host stage exists, the more available for discovery it will be. This might be reflected in a high level of parasitism simply in relation to the increased time for discovery of individual host specimens, or it might also be reflected in a larger number of parasitoid species that have, over evolutionary time, become associated with it as a result of its high temporal availability. Unambiguous evidence for this is hard to find. Very long-lived insects such as larval cicadas and soil-dwelling larval Elateridae, in fact, appear to suffer little parasitism, but this is likely to be because a subterranean existence is cryptic in other respects. Conversely, those hosts mentioned above which feed on a rich food source are able to develop rapidly, yet suffer much parasitism as a result of their high discoverability. Many wood-boring insects are long-lived and subject to quite high levels of parasitism, but this could also be due to enhanced discoverability through often living in association with fungal activity (Crowson, 1981).

To look for differential longevity in otherwise comparable hosts, we again turn to leaf-mining Lepidoptera. In Nepticulidae, *Phyllonorycter*, *Caloptilia* and *Parornix*, the duration of occupancy of larval feeding sites varies, being longest in *Phyllonorycter* and shortest in *Caloptilia* and *Parornix*. Parasitoid complexes of these leaf-mining Lepidoptera (Table VII) are considered in detail later, and it is sufficient to note here that, on the same food plants, *Caloptilia* and *Parornix* have the smallest parasitoid complexes and *Phyllonorycter* the largest.

4. Host taxonomy

There is close similarity between the parasitoid complexes of different *Phyllonorycter* species on the same tree species and indeed, as far as idiobionts are concerned, between the parasitoid complexes of different higher taxa of leaf-miners (Askew and Shaw, 1974). Idiobiotic parasitoids generalize between host species (and higher taxa) so that a rare host augments the potential host group and is as much at risk as an abundant host. Idiobiotic parasitoids are unlikely to be affected by host taxonomic isolation provided that it is not linked with an exceptional biology. Koinobiotic parasitoids, on the other hand, do discriminate to a greater or lesser extent between higher taxa. It might be expected, therefore, that taxonomically isolated hosts would be less easily colonized by koinobionts than hosts in large taxa. Data needed to test this hypothesis are not presently available.

VII. HOST LIFE HISTORY EFFECTS ON COMMUNITY STRUCTURE

We have considered factors which might mould the host ranges of koinobionts and idiobionts, emphasizing the importance of the environment in which they search, and discussed some features of hosts which may influence the probability of their colonization by parasitoids. We will now examine, in these terms, larval parasitism of arboreal Tenthredinidae and parasitism of some groups of leaf-mining Lepidoptera.

A. Larval Tenthredinid Sawflies

This sawfly family is selected because it incorporates a variety of larval feeding habits (exophytic, gall-inhabiting, leaf-mining), and because it is relatively well studied. By considering only arboreal hosts, we should minimize differences due to differential host plant apparency.

Data on larval parasitism of Tenthredinidae (Table VI) are from several sources, and therefore caution must be exercised in making comparisons, but a number of general points can safely be made because the pattern of parasitism within each of the three larval feeding categories is so consistent.

Parasitoids are tabulated under families or subfamilies because here division into koinobionts and idiobionts follows, to a large extent, these higher taxa. However, this is not always the case and currently, recognized subfamilies of Chalcidoidea, in particular, tend to include a mixture of koinobiotic and idiobiotic species.

1. Exophytic hosts

Koinobiotic endoparasitism is favoured over ectoparasitism in exposed hosts. Apart from the ichneumonid subfamily Adelognathinae, the biology of which is discussed by Fitton *et al.* (1982) and which is rare and unrepresented in the surveys included in Table VI, the only larval ectoparasitoid representation in the complex is by Tryphoninae. Tryphonines attach their eggs firmly to the integument of growing tenthredinid larvae. As the parasitoid larvae feed only a little before killing the host in its cocooned prepupal stage, Tryphoninae have to be considered to be koinobionts, as indeed are all other parasitoids of exophytic Tenthredinidae included in Table VI. Many Chalcidoidea, and probably all of those attacking Tenthredinidae, are idiobionts and in Europe Chalcidoidea do not appear to parasitize exposed tenthredinid larvae. On the other hand, it is only the exophytic category that is available to Tachinidae because these parasitoids,

TABLE VI. Categories and numbers of parasitoid species, in families and subfamilies, associated with larval Tenthredinidae on deciduous trees and shrubs. Parasitic inquilines (Curculionidae, Tephritidae) in sawfly galls and their parasitoids are excluded.
Information sources: 1 Zinnert, 1969; 2 Carleton, 1939; 3 Kopelke, 1983, 1985; 4 Liston, 1982; 5 Eichhorn and Pschorn-Walcher, 1973; 6 Altenhofer, 1980; 7 Hanapi, 1980; 8 our own data.

		Koinobionts				*Idiobionts*									
		Endo			*Ecto*	*Endo*	*Ecto*								
Host Tenthredinid	Host plant	Ichneumonidae[a]	Braconidae[b]	Tachinidae	Ichneumonidae[c]	Eulophidae[d]	Eulophidae[e]	Pteromalidae	Eurytomidae	Torymidae	Eupelmidae	Ichneumonidae[f]	Braconidae[g]	Total spp.	Source
Exophytic															
Croesus latipes (Villaret)	*Betula*	6	0	2	1	0	0	0	0	0	0	0	0	9	1
C. septentrionalis (L.)	Betulaceae	7	0	2	1	0	0	0	0	0	0	0	0	10	1
Trichiocampa viminalis (Fallén)	*Populus*	1	0	0	1	0	0	0	0	0	0	0	0	2	1
Pristiphora moesta (Zaddach)	*Malus*	3	1	0	1	0	0	0	0	0	0	0	0	5	1
P. geniculata (Hartig)	*Sorbus*	4	0	2	1	0	0	0	0	0	0	0	0	7	1
P. testacea (Jurine)	*Betula*	2	0	2	1	0	0	0	0	0	0	0	0	5	1
Hemichroa crocea (Fourcroy)	Betulaceae	4	1	4	2	0	0	0	0	0	0	0	0	11	1
Nematus pavidus Lepeletier	*Salix*	7	1	2	2	0	0	0	0	0	0	0	0	12	1
N. melanaspis Hartig	Salicaceae	5	0	2	2	0	0	0	0	0	0	0	0	9	1
N. capreae (L.)	*Salix*	1	0	1	0	0	0	0	0	0	0	0	0	2	1
N. leucotrochus Hartig	*Ribes*	3	1	0	1	0	0	0	0	0	0	0	0	5	1
N. melanocephalus Hartig	*Corylus*	1	0	1	0	0	0	0	0	0	0	0	0	2	1

Gallicolous															
Pontania proxima (Lepeletier)	*Salix*	2	0	0	1	0	1	1	0	0	1	1	2	9	2,3,7,8
P. collactanea (Förster)	*Salix*	0	0	0	0	0	0	1	2	0	1	1	2	7	7,8
P. viminalis (L.)	*Salix*	0	0	0	0	0	1	4	1	0	0	1	2	9	3,7,8
P. pedunculi (Hartig)	*Salix*	0	1	0	0	0	1	2	1	0	0	1	1	7	7,8
P. vesicator (Bremi-Wolf)	*Salix*	0	0	0	0	0	0	3	1	0	1	1	1	7	3,8
P. bridgmani (Cameron)	*Salix*	1	0	0	0	0	1	0	0	0	1	1	1	5	7,8
Euura mucronata (Hartig)	*Salix*	0	0	0	0	0	1	1	1	0	0	1	1	5	7
E. amerinae (L.)	*Salix*	1	0	0	0	0	0	0	0	1	0	1	0	3	4
Leaf-mining															
Heterarthrus aceris (Kaltenbach)	*Acer*	0	0	0	0	2	3	0	0	0	0	0	1	6	8
H. vagans (Fallén)	*Alnus*	1	0	0	0	5	9	0	0	0	0	0	1	16	6, 8
Fenusa pusilla (Lepeletier)	*Betula*	2	0	0	2	4	9	0	0	0	0	0	1	18	5,6,8
F. ulmi (Sundewall)	*Ulmus*	0	0	0	0	4	5	1	0	0	0	0	2	12	6,8
Profenusa pygmaea (Klug)	*Quercus*	1	0	0	1	5	5	0	0	0	0	0	1	13	6,8
Messa hortulana (Klug)	*Populus*	1	0	0	1	2	6	1	0	0	0	0	1	12	6,8
M. nana (Klug)	*Betula*	1	0	0	0	4	2	1	0	0	0	0	1	9	6,8
Scolioneura betuleti (Klug)	*Betula*	1	0	0	0	3	5	0	0	0	0	1	1	11	6,8

[a]subfamilies Campopleginae, Ctenopelmatinae, Mesochorinae; [b]subfamily Ichneutinae; [c]subfamily Tryphoninae; [d]subfamily Entedontinae; [e]subfamilies Eulophinae, Tetrastichinae; [f]subfamily Pimplinae; [g]subfamilies Braconinae (gallicolous hosts), Rogadinae (leaf-mining hosts).

lacking piercing ovipositors, cannot gain access to hosts fully enclosed in plant tissue.

2. Gallicolous hosts

Species of *Pontania* and *Euura* are parasitized mainly by idiobionts with just a few koinobionts represented in the complex. This correlates with the protection afforded by the gall favouring ectoparasitism. Idiobiotic Ichneumonoidea and especially Chalcidoidea are well represented, the latter by five families. Gallicolous species are the only tenthredinids to be parasitized by Eurytomidae, Eupelmidae and Torymidae, three chalcid families which include a large number of species associated with insect galls. The idiobionts, whilst mostly attacking only the inhabitants of sawfly galls on *Salix*, are relatively polyphagous within this host range. *Eurytoma aciculata* Ratzeburg kills and devours the young sawfly larva and completes its development by feeding on gall tissue (Hanapi, 1980).

3. Leaf-mining hosts

The parasitoid faunas, with strong representation of idiobionts, much more closely resemble those of gallicolous than of exophytic sawflies. Leaf-miner complexes differ from gallicolous complexes chiefly in the almost complete absence of idiobiotic Ichneumonidae (probably because they provide too small a food resource) and in the greater numbers of Chalcidoidea. The chalcids, however, are of only two families with Eulophidae very much the dominant group.

4. Comparisons between the feeding categories

The major division of the tenthredinid larval parasitoid complexes is between the exophytic and endophytic hosts and can be explained in terms of host exposure or concealment, the former condition favouring endoparasitism and koinobiosis. Many of the koinobiotic endoparasitoid Ichneumonidae have very narrow host ranges, either restricted to a single host species or to related hosts on the same or a group of allied plants (Zinnert, 1969), and this can be attributed to the cost of developing a capacity to overcome defence systems in a range of hosts. Only *Mesochorus globulator* (Thunberg) has a relatively broad host range, developing as an obligatory secondary endoparasitoid of other endoparasitic Ichneumonidae in five of the tabulated exophytic sawflies. Possible explanations of the broad host ranges of mesochorines have been suggested earlier.

The Tachinidae recorded by Zinnert (1969) as parasitoids of exophytic tenthredinids are polyphagous species, at least one of which, *Bessa selecta* (Meigen), is also known to attack several lepidopterous families (van Emden, 1954). We have already discussed possible mechanisms permitting broad host ranges in Tachinidae.

The ectoparasitic Tryphoninae show an even higher degree of host-specificity than the endoparasitic ichneumonid subfamilies, which initially may seem surprising. However, the temporal and behavioural specializations required by a tryphonine to oviposit with precision on the thoracic region of a well-grown, hostile host larva may provide an explanation.

Both categories of endophytic sawflies are considered to be relatively discoverable, but the gallicolous species may be less so than the leaf-miners. *Pontania* and *Euura* are virtually restricted to *Salix* but each tenthredinid is usually more or less plant-specific so that no single *Salix* species accommodates a large range of possible host species. Furthermore, a quite high proportion of sawfly galls are empty. A sawfly gall, unlike a cynipid gall, will develop fully without containing a living and growing sawfly larva; the stimulus for cecidogenesis comes from the ovipositing female and, even if the egg or small larva dies, gall growth continues (Magnus, 1914). Galls in which such mortality has occurred may decoy parasitoids and lower the predictability of discovery of the host population. Temporal availability of gallicolous sawfly larvae may also be lowered by the usual habit of the fully grown larva of vacating the gall. A comparable situation in leaf-mining Lepidoptera is shown below to be associated with a reduction in parasitism, but in sawflies there is no evidence that this is the case. Parasitism of *Euura amerinae* (L.), which pupates in the gall, does not differ to any marked extent from that of *E. mucronata* (Hartig) which leaves the gall, and parasitism of the American stem-galling species *E. lasiolepis* Smith, which also pupates in the gall, is very low indeed (two parasitoid species, 12% parasitism; Price and Craig, 1984). The suggested lower discoverability of gallicolous sawflies compared with leaf-mining species may, however, contribute to their overall smaller number of associated parasitoids.

Leaf-mining tenthredinids are highly vulnerable to parasitism. The basic discoverability of endophytic life is augmented by the fact that it has been adopted by many different groups of insects and most tree species have a large and varied leaf-mining fauna. We would expect in such a situation that polyphagous idiobionts would dominate the parasitoid community. This is indeed the case. Many of the Eulophidae are polyphagous to the extent of attacking leaf-miners of different orders on a wide range of host plants (Askew and Shaw, 1974; see also Table II). This is also true for most idiobiotic endoparasitoids, and only one, *Chrysocharis eurynota* Graham which we have associated only with *Profenusa pygmaea*, shows any indication of being host-specific: further study of *C. eurynota* may, in fact, reveal it to be a koinobiont. The large parasitoid complexes and generally high levels of parasitism in leaf-mines give scope for secondary parasitism and *Chrysocharis purpurea* Bukowski, *Closterocerus trifasciatus* Westwood and the pteromalid *Mokrzeckia obscura* Graham are chiefly, if not obligatorily, secondary parasitoids. In contrast, secondary parasitism in tenthredinid galls is never more than facultative.

It may be noted that *Heterarthrus* species, unlike the other tabulated leaf-mining tenthredinids, pupate inside the mines, but again there is no evidence that this greater availability has increased the size of the parasitoid complex. However, it may be significant that the only British species to mature early in the year [*H. aceris* (Kaltenbach)] forms a deciduous disc, and so escapes the searching environment of arboreal idiobionts, whilst other species make their fixed discs only shortly before leaf-fall.

B. Leaf-Mining Lepidoptera

An opportunity to examine more closely the influence of particular variations in host life histories is provided by Lepidoptera which feed entirely or partly within leaf-mines. We will compare our own data on the parasitoids of Nepticulidae and two groups of Gracillariidae (*Phyllonorycter*, and the genera *Caloptilia* and *Parornix* taken together). These differ considerably in the duration of their occupancy of the larval feeding site and in whether or not they leave it to pupate; differences which favour some categories of parasitoid and are disadvantageous to others. Another factor influencing the parasitoid complexes is the respective body size of the primary host: Nepticulidae are smaller than *Phyllonorycter*, which are smaller than *Parornix* and (especially) *Caloptilia.* To avoid complications arising from varying host plant apparencies, we will confine our attention to just those species feeding on four arboreal plant genera. The duration of occupancy of essentially similar feeding sites equates with different spatio-temporal availability of the hosts themselves and, assuming any other differences to be unimportant, this allows us to compare them in terms of apparency. Data on which our discussion is based are presented in Table VII.

1. *Phyllonorycter*

All larval stages and also the pupal stage are spent in a single mine. The pre-adult stages are therefore highly localized and, because a mine has a high probability of still being occupied by a larval or pupal host, particularly apparent.

High apparency accords with the large number of parasitoid species associated with *Phyllonorycter*. The majority of these are relatively polyphagous idiobionts, mostly Eulophidae, which very often function as facultative hyperparasitoids. The prolonged occupancy of *Phyllonorycter* mines, and the many species mining leaves in a generally similar way on apparent plants, together create conditions that are well-suited for parasitoid exploitation. An important benefit often inherent in koinobiosis, that of allowing the host to live long enough to move away from the vulnerable larval feeding site, is not enjoyed by koinobiont parasitoids of *Phyllonorycter*, which are therefore under severe pressure from facultatively hyperparasitic idiobionts. In fact, many of the koinobiont genera and species we have reared from *Phyllonorycter* occur in the complex only occasionally and are much better represented in (and probably primarily adapted to) the parasitoid communities of hosts other than *Phyllonorycter*. At the generic level only the majority of *Achrysocharoides* species, and the unevenly occurring, polyembryonic *Holcothorax*, have specialized as koinobionts of *Phyllonorycter*, although several species of *Pholetesor* (*Apanteles sensu lato, partim*) depend seasonally or sometimes entirely on *Phyllonorycter* species. These braconids spin unusual cocoons which are slung hammock-like in the tentiform mine of the host, and the chalcids are gregarious and individually small. Both may be adaptations limiting attack by idiobiont eulophids.

2. *Nepticulidae*

The entire larval life is spent in a single mine, but the fully-fed larva leaves to spin a cocoon and pupate in a different situation (with the exception of *Fomoria* and some *Trifurcula* species which are on unapparent plants and do not concern us here). Larval development is relatively rapid so that mines are occupied rather briefly. This diminishes the apparency of Nepticulidae compared with *Phyllonorycter*, as do the smaller number of arboreal species (many are on less apparent plants) and possibly the greater variety of mine forms (Emmet, 1976).

Parasitoid complexes of Nepticulidae are smaller overall than those of *Phyllonorycter*, as we would predict from the lesser degree of apparency of their mining stage. Parasitoids which rely on the hosts' feeding sites as searching cues may respond positively to vacated mines, and in so doing waste time and energy. An idiobiont able to develop as a facultative hyperparasitoid would therefore benefit by being attracted preferentially to other feeding sites which are most often full (for example, those of *Phyllonorycter*). The idiobionts in the nepticulid communities are mainly *Chrysocharis* subgenus *Nesomyia* and the smaller *Cirrospilus* species which are specialized to the extent of concentrating their attack mainly on smaller, and hence more often occupied, mines. *Sympiesis*, *Pnigalio* and *Colastes*, which attack mainly larger mines and late feeding stages, especially of *Phyllonorycter*, are

TABLE VII. Parasitoid complexes associated with species of *Phyllonorycter*, Nepticulidae, *Caloptilia* and *Parornix* feeding on *Quercus* (*robur* L, *petraea* (Maruschka) Lieblein), *Betula* (*pendula* Roth, *pubescens* Ehrhart), *Alnus* (*glutinosa* (L.) Gaertner, *incana* (L.) Moench) and *Crataegus* (*monogyna* Jacquin, *oxyacanthoides* Thruillier) in Britain. Parasitoids are listed as either idiobionts or koinobionts, with the total number of species in each taxon associated with the tabulated hosts. Numbers are host mortalities attributed to each parasitoid taxon, and the figure in parentheses is the number of parasitoid species which have contributed to that mortality

Name of species	No. of species	*Phyllonorycter*				Nepticulidae			*Caloptilia*						*Parornix*			
		9	4	4	2	7	7	2	2		1		1		1		1	
		Quercus	*Betula*	*Alnus*	*Crataegus*	*Quercus*	*Betula*	*Crataegus*	*Quercus*		*Betula*		*Alnus*		*Betula*		*Crataegus*	
									Cones	Cocoons	Cones	Cocoons	Cones	Cocoons	Mines	Folds	Mines	Folds
Idiobionts																		
Cirrospilus	4	603(4)	132(4)	31(4)	17(4)	32(3)	14(2)	34(2)	0	0	0	1	0	0	0	0	1	1(2)
Sympiesis	6	615(2)	182(3)	148(3)	147(2)	0	0	1	11(3)	0	55(4)	1	18(3)	0	4(2)	3(1)	0	2(1)
Pnigalio	4	340(4)	636(4)	175(3)	61(2)	9(3)	12(3)	6(2)	16(1)	0	6(2)	3(2)	0	0	5(1)	1	1	4(2)
Hemiptarsenus dropion (Walker)		0	1	0	0	0	2	0	0	0	0	0	0	0	1	0	0	0
Dimmockia brevicornis (Erdös)		0	0	0	0	0	0	0	4	0	0	0	0	0	0	0	0	0
Chrysocharis (*Nesomyia*)	5	717(3)	400(3)	137(3)	154(3)	9(3)	29(3)	26(4)	0	0	0	0	0	0	0	0	1	2(1)
Chrysonotomyia chlorogaster (Erdös)		1	0	0	1	3	6	13	0	0	0	0	0	0	0	0	0	0
Closterocerus trifasciatus Westwood		123	2	0	2	1	1	0	0	0	0	0	0	0	0	0	0	0
Pediobius	2	528(2)	91(1)	316(2)	1	0	0	0	1	0	0	0	0	0	0	0	0	0
Tetrastichus	4	100(2)	8(2)	5(1)	0	2(1)	1	0	2(1)	0	0	0	0	0	0	0	0	0
Pteromalus	2	1	0	0	0	0	0	0	0	0	2(1)	0	0	0	0	0	0	0
Colastes braconius Haliday		468	193	196	29	9	0	4	0	0	0	0	0	0	0	0	0	1
Scambus annulatus (Kiss)		7	3	1	2	0	0	0	46	0	25	0	3	0	0	6	0	1
Itoplectis alternans (Gravenhorst)		2	0	0	0	0	0	0	0	0	0	1	0	2	0	0	0	0
Encrateola laevigata (Ratzeburg)		0	4	1	0	0	0	0	0	1	1	0	1	0	0	0	0	0
Diaglyptidea conformis (Gmelin)		0	0	0	0	0	0	0	0	6	0	9	0	3	0	0	0	0
Gelis	2	2(1)	0	0	0	0	0	0	0	1	1	10(1)	0	3(1)	0	0	0	0
Bathythrix	2	0	0	0	0	0	0	0	0	0	0	2(2)	0	2(1)	0	0	0	0
Total of idiobiont species		24	22	19	16	13	12	11	8	3	10	9	5	4	5		8	
Total idiobiont mortalities		3507	1652	1010	414	65	65	84	80	8	90	27	22	10	20		14	

Koinobionts																
Elachertus	3	5(1)	11(2)	2(1)	13(1)	0	0	0	17(1)	0	47(3)	0	12(2)	0	2(1) 3(2)	3(1) 39(1)
Chrysocharis (*Nesomyia*)	5	0	0	0	0	45(4)	7(2)	24(1)	0	0	0	0	0	0	0	0
Chrysocharis s.str.	3	258(2)	4(1)	0	0	0	1	0	0	0	0	0	0	0	0	0
Derostenus	2	0	0	0	0	12(1)	2(1)	1	0	0	0	0	0	0	0	0
Achrysocharoides	6(?)	683(3)	1335(2)	14(1)	603(2)	0	0	0	73(1)	0	0	0	0	0	0	7(1)
Chrysonotomyia formosa (Westwood)		0	0	0	0	0	0	0	0	0	0	0	0	0	19	0
Seladerma	2	0	0	0	0	0	1	2(1)	0	0	0	0	0	0	0	0
Holcothorax	2	71(1)	0	18(1)	0	0	0	0	0	0	0	0	0	0	0	0
Gnaptodon pumilio (Nees)		0	0	0	0	0	6	0	0	0	0	0	0	0	0	0
Acoelius	3	0	0	0	0	23(1)	1	8(1)	0	0	0	0	0	0	0	0
Mirax	1	0	0	0	0	0	2	0	0	0	0	0	0	0	0	0
Microgasterinae	8	56(4)	36(3)	13(3)	6(2)	0	0	0	32(1)	0	99(4)	0	56(4)	0	7(2)	16(2)
Rhysipolis	3	8(1)	4(2)	0	0	0	0	0	188(1)	0	6(2)	0	21(1)	0	20(1)	0
Campopleginae	7	1	1	3(1)	20(1?)	0	0	0	35(3)	0	48(2)	0	29(3)	0	27(1)	25(3)
Mesochorinae	2	0	0	0	0	0	0	0	1	0	0	0	1	0	0	0
Total of koinobiont species		13	11	7	6	6	8	4	8		11		11		7	7
Total koinobiont mortalities		1082	1391	50	642	80	20	35	346		200		119		78	90

scarce. This may be attributed either to the frequent emptiness of full-size nepticulid mines or to the inadequate size attained by most fully grown nepticulid larvae. The latter factor may well account for the absence of ichneumonid parasitoids, both idiobionts and koinobionts, from the parasitoid complexes of Nepticulidae.

Vacation of the mine after termination of a relatively brief larval feeding period should favour more synchronized, specialized koinobionts which allow the host to transport them to a safer pupation site. In fact, the koinobiont component of the parasitoid complex, although proportionally not very different in size from that of *Phyllonorycter*, is much more specialized. Three subfamilies of Braconidae, all primary parasitoids, are exclusively associated with Nepticulidae, and we interpret this as an indication of a very long-standing relationship. The other major group of nepticulid koinobionts comprises species of *Chrysocharis* subgenus *Nesomyia* and the allied eulophid genus *Derostenus*. Some *Nesomyia* species are idiobionts in both *Phyllonorycter* and nepticulid complexes, as well as in complexes associated with other groups of leaf-mining hosts, but the koinobionts in nepticulid complexes are very specialized, mostly monophagous to relatively narrowly oligophagous and often with a short, presumably synchronized, flight period. Both koinobiotic *Nesomyia* and the Braconidae take advantage of their host's continuing life to be transported away from the larval mine (and its associated cues to searching, potentially hyperparasitic idiobionts) to a cryptic situation protected by the host's cocoon.

3. *Parornix and Caloptilia*

Larvae start as miners but then vacate the mine to make a series of one, two or three folds or cones at the leaf margin in which feeding continues: even the last feeding site is vacated fairly quickly, however, as pupation takes place in yet another site. The mine of *Parornix* is very like that of a small *Phyllonorycter*, and feeding continues in a fold at the leaf edge. Pupation is in a small, cryptic cocoon sometimes made away from the tree, and this stage has not been well enough sampled to be included in Table VII. The mine of all the included *Caloptilia* species is minute when vacated. It is attacked very little by idiobiotic parasitoids, but endoparasitic koinobionts attacking *Caloptilia* (and also *Parornix*) no doubt oviposit into hosts mainly during their mining stages. Parasitism by koinobionts is, however, most conveniently assessed from the cone stage onwards (Table VII), in which the mortality occurs. Unlike *Parornix*, *Caloptilia* pupates in a conspicuous cocoon, usually on a fresh leaf, which superficially resembles a *Phyllonorycter* mine.

4. Comparisons between the feeding categories

The brief duration of occupancy of each feeding site should make *Parornix* and *Caloptilia* larvae relatively unapparent and less subject to parasitism by idiobionts. An increased proportion of koinobiotic parasitoids, compared with the parasitoid complexes of *Phyllonorycter* or Nepticulidae, should therefore be evident.

Indeed, there is a pronounced increase in the proportion of koinobionts (Table VII). A high proportion of the small number of records of idiobionts describe them functioning as facultative hyperparasitoids of the numerous koinobionts or as parasitoids attacking the host in its cocoon. The major groups of koinobionts are Ichneumonidae and Braconidae, and of particular significance is the importance of *Rhysipolis* which, like *Colastes* (idiobiotic), is in the subfamily Rogadinae. *Colastes* features prominently in *Phyllonorycter* parasitoid complexes, is scarce in nepticulid complexes, and almost absent from *Caloptilia* and *Parornix*. Conversely, *Rhysipolis* is scarce as a parasitoid of *Phyllonorycter* but features prominently as a parasitoid of *Caloptilia* and *Parornix*. *Rhysipolis* is unusual among koinobionts in being ectoparasitic and it has a highly specialized biology that makes the most of the hosts' mobility (Shaw, 1983). As in Tryphoninae, the small larva feeds little before the host spins its cocoon away from the previous feeding site, but the hosts of *Rhysipolis* spin up an instar earlier than usual if the host is attacked early. *Elachertus*, too, is ectoparasitic and is classed as a koinobiont because (usually but not invariably) it stimulates and exploits the capacity of the host to make a fresh, kairomone-free retreat after being attacked. All of these specialized koinobiont ectoparasitoids have a moderately high degree of host specificity (they are never facultative hyperparasitoids), and certainly some produce a specialized venom which controls the expiry of the host rather than being immediately debilitating or fatal as is usual with idiobiont ectoparasitoids (Shaw, 1983; see also *Eulophus*, Shaw, 1981c).

Lampe (1984) found fewer polyphagous idiobiont parasitoids of *Coleophora alticolella* in the middle of large, uniform stands of rushes, and suggested that these poorly synchronized species required alternative hosts to survive (see also Haeselbarth, 1979). In the leaf-miner community examined here, moderately polyphagous idiobionts are again sources of considerable local variation, and the local status of alternative hosts may again be important. Indeed, the idiobionts may behave differently in different parasitoid complexes. Overall, we have reared only males of *Scambus annulatus* (Kiss) as primary parasitoids of *Phyllonorycter* (19 specimens) or as hyperparasitoids in *Caloptilia* cones (31 specimens), while both sexes (35 males and 23 females) developed as primary parasitoids of *Caloptilia*. Similarly, almost all of the *Colastes braconius* Haliday reared from the smaller Nepticulidae were males (32 males and two females; but also two males and 12 females from the relatively large *Nepticula sorbi* Stainton) while a more even sex ratio (595 males and 445 females) resulted from *Phyllonorycter* mines. These data show that sex ratios, and also the behaviour of *S. annulatus* as a secondary parasitoid, can vary from one complex to another. In the local absence of hosts suitable for female development, these species would presumably fail to enter the parasitoid complexes of their smaller hosts. Like *C. braconius* (Shaw, 1983), some of the idiobiotic eulophids (e.g. *Sympiesis*, *Pnigalio*) which have so successfully colonized *Phyllonorycter* in the leaf-miner community may have entered the parasitoid complex of Nepticulidae only through resource augmentation, their evolution as parasitoids of leaf-miners having been dependent on more suitable host groups such as *Phyllonorycter*. Such speculations are difficult to test but they suggest ways by which the appearance of a novel host resource could influence established parasitoid complexes attacking other hosts. This aspect of community structure can be investigated only with thorough knowledge of the real host ranges of the parasitoids concerned.

VIII. CONCLUDING REMARKS

In this chapter we have followed Haeselbarth (1979) in suggesting that a basic life history division of parasitoids into idiobionts and koinobionts provides a workable categorization that correlates well with the *potential* for broad or narrowly specialized host associations, through which community structure can usefully be analysed. Exceptions will of course occur. In particular, resource security often allows idiobionts to be specialists, and there is certainly an element of expansion through recruitment from the searching environment in the host ranges of many koinobionts. We expect that a good deal of useful refinement will become possible as more is found out about the idiosyncrasies of particular higher taxa of parasitoids.

We have certainly found it helpful to categorize parasitoids into idiobionts and koinobionts in the communities we have investigated: however, there is so little accurate information on host ranges of parasitoids in exophytic host communities, in which koinobionts are most plentiful, that we have been unable to test adequately the broader applicability of the koinobiont/idiobiont categorization as an indicator of host range. Most of the available data on host associations of parasitoids come from host-oriented studies and, until more research is specifically aimed at examining the host spectra of particular parasitoids, we will have to depend upon hypothesized correlates, as we have done here, simply because we know so little about actual host ranges.

As the mechanisms underlying host selection and host suitability become better understood, and the ecology of parasitoids becomes more accurately modelled, it suddenly matters that the host ranges of parasitoids have been largely ignored: we are faced with a lack of the empirical data needed to assess these theoretical and mechanistic advances. To challenge Thompson and Parker's assertion (1927) that "the laws underlying host relations are not capable of expression in scientific terms nor discoverable by scientific methods", we first have to gather the information they say we cannot predict. Strong *et al.* (1984) and Southwood (1985) urge us to continue to study the idiosyncrasies of insects. Let us see host range as the fundamental idiosyncrasy of a parasitoid.

Acknowledgements

We are very much indebted to all those entomologists, too numerous to mention individually, who have supplied us with field data to supplement our own on the biology of parasitoids. We would like to thank Gordon Blower, Charles Godfray and Graham Rotheray for their comments on earlier drafts of this paper, and Jeff Waage for his most helpful criticism of our ideas. Views expressed in this paper are our own and do not necessarily concur with those of any of these people.

REFERENCES

Altenhofer, E. (1980). Zur Systematik und Ökologie der Larvenparasiten (Hym., Ichneumonidae, Braconidae, Eulophidae) der minierenden Blattwespen (Hym., Tenthredinidae). *Zeitschrift für Angewandte Entomologie* **89,** 250–259.

Askew, R. R. (1961a). On the biology of the inhabitants of oak galls of Cynipidae (Hymenoptera) in Britain. *Transactions of the Society for British Entomology* **14,** 237–268.

Askew, R. R. (1961b). A study of the biology of species of the genus *Mesopolobus* Westwood (Hymenoptera: Pteromalidae) associated with cynipid galls on oak. *Transactions of the Royal Entomological Society of London* **113,** 155–173.

Askew, R. R. (1975). The organisation of chalcid-dominated parasitoid communities centred upon endophytic hosts. *In* "Evolutionary Strategies of Parasitic Insects and Mites" (P. W. Price, ed.), pp. 130–153. Plenum, London.

Askew, R. R. (1980). The diversity of insect communities in leaf-mines and plant galls. *Journal of Animal Ecology* **49,** 817–829.

Askew, R. R. (1984). The biology of gall wasps. *In* "Biology of Gall Insects" (T. N. Ananthakrishnan, ed.), pp. 223–271. Oxford & IBH Publishing Co., New Delhi.

Askew, R. R., and Shaw, M. R. (1974). An account of the Chalcidoidea (Hymenoptera) parasitising leaf-mining insects of deciduous trees in Britain. *Biological Journal of the Linnean Society* **6,** 289–335.

Askew, R. R., and Shaw, M. R. (1979a). Mortality factors affecting the leaf-mining stages of

Phyllonorycter (Lepidoptera: Gracillariidae) on oak and birch. 1. Analysis of the mortality factors. *Zoological Journal of the Linnean Society* **67,** 31–49.

Askew, R. R., and Shaw, M. R. (1979b). Mortality factors affecting the leaf-mining stages of *Phyllonorycter* (Lepidoptera: Gracillariidae) on oak and birch. 2. Biology of the parasite species. *Zoological Journal of the Linnean Society* **67,** 51–64.

Benson, R. B. (1950). An introduction to the natural history of British sawflies (Hymenoptera Symphyta). *Transactions of the Society for British Entomology* **10,** 45–142.

Bess, H. A., van den Bosch, R., and Haramoto, F. H. (1961). Fruit fly parasites and their activities in Hawaii. *Proceedings of the Hawaiian Entomological Society* **17,** 367–378.

Blair, K. G. (1944). A note on the economy of the rose bedeguar gall, *Rhodites rosae* L. *Proceedings of the South London Entomological and Natural History Society*, 55–59.

Bouton, C. E., McPheron, B. A., and Weis, A. E. (1980). Parasitoids and competition. *American Naturalist* **116,** 876–881.

Cameron, E. (1939). The holly leaf-miner (*Phytomyza ilicis* Curt.) and its parasites. *Bulletin of Entomological Research* **30,** 173–208.

Carleton, M. (1939). The biology of *Pontania proxima* Lep., the bean gall sawfly of willows. *Journal of the Linnean Society (Zoology)* **40,** 575–624.

Compton, S. (1981). Observations on the biology and parasites of *Trifurcula cryptella* (Stainton) and *Coleophora discordella* Zeller (Lepidoptera). *Entomologist's Gazette* **32,** 169–173.

Connell, J. H. (1980). Diversity and the coevolution of competitors, or the ghost of competition past. *Oikos* **35,** 131–138.

Coshan, P. F. (1974). The biology of *Coleophora serratella* (L.) (Lepidoptera: Coleophoridae). *Transactions of the Royal Entomological Society of London* **126,** 169–188.

Crowson, R. A. (1981). "The Biology of the Coleoptera." Academic Press, London.

Danks, H. V. (1971). Biology of some stem-nesting aculeate Hymenoptera. *Transactions of the Royal Entomological Society of London* **122,** 323–399.

Dean, J. M., and Ricklefs, R. E. (1979). Do parasites of Lepidoptera larvae compete for hosts? No! *American Naturalist* **113,** 302–306.

DeBach, P., and Sundby, R. A. (1963). Competitive displacement between ecological homologues. *Hilgardia* **34,** 105–166.

Deyrup, M. A. (1984). A maple wood wasp, *Xiphydria maculata*, and its insect enemies (Hymenoptera: Xiphydriidae). *Great Lakes Entomologist* **17,** 17–28.

Edmunds, M. (1976). Larval mortality and population regulation in the butterfly *Danaus chrysippus* in Ghana. *Zoological Journal of the Linnean Society* **58,** 129–145.

Ehler, L. E. (1985). Species-dependent mortality in a parasite guild and its relevance to biological control. *Environmental Entomology* **14,** 1–6.

Eichhorn, O., and Pschorn-Walcher, H. (1973). The parasites of the birch leaf-mining sawfly (*Fenusa pusilla* (Lep.), Hym.: Tenthredinidae) in Central Europe. *Commonwealth Institute of Biological Control Technical Bulletin* **16,** 79–104.

van Emden, F. I. (1954). Diptera Cyclorrhapha, Calyptrata (1) Section (a) Tachinidae and Calliphoridae. *Handbook for the Identification of British Insects* **10,** 4a.

Emmet, A. M. (1976). Nepticulidae. In "The Moths and Butterflies of Great Britain and Ireland" Vol. 1. (J. Heath, ed.), pp. 171–267. Blackwell, Oxford.

Fedde, V. H., Fedde, G. F., and Drooz, A. T. (1984). Factitious hosts in insect parasitoid rearings. *Entomophaga* **27** (1982), 379–386.

Feeny, P. (1976). Plant apparency and chemical defence. *Recent Advances in Phytochemistry* **10,** 1–40.

Fitton, M. G., Gauld, I. D., and Shaw, M. R. (1982). The taxonomy and biology of the British *Adelognathinae* (Hymenoptera: Ichneumonidae). *Journal of Natural History* **16,** 275–283.

Flanders, S. E. (1942). Abortive development in parasitic Hymenoptera, induced by the food-plant of the insect host. *Journal of Economic Entomology* **35,** 834–835.

Flanders, S. E. (1966). The circumstances of species replacement among parasitic Hymenoptera. *Canadian Entomologist* **98,** 1009–1024.

Force, D. C. (1972). *r*- and *K*-strategists in endemic host-parasitoid communities. *Bulletin of the Entomological Society of America* **18,** 135–137.

Force, D. C. (1975). Succession of *r*- and *K*-strategies in parasitoids. *In* "Evolutionary Strategies of Parasitic Insects and Mites" (P. W. Price, ed.), pp. 112–129. Plenum, London.

Fowler, S. V. (1985). Differences in insect species richness and faunal composition of birch seedlings, saplings and trees: the importance of plant architecture. *Ecological Entomology* **10,** 159–169.

Führer, E., and Kilincer, N. (1972). Die motorische Aktivität der endoparasitischen Larven von *Pimpla turionellae* L. und *Pimpla flavicoxis* Ths. in der Wirtspuppe. *Entomophaga* **17,** 149–165.

Gahan, A. B. (1933). The serphoid and chalcidoid parasites of the hessian fly. *United States Department of Agriculture Miscellaneous Publication* **174.**

Gauld, I. D. (1986). Latitudinal gradients in ichneumonid species-richness in Australia. *Ecological Entomology* **11,** 155–161.

Graham, M. W. R. de V. (1969). The Pteromalidae of north-western Europe (Hymenoptera: Chalcidoidea). *Bulletin of the British Museum (Natural History), Entomology Supplement* **16.**

Haeselbarth, E. (1979). Zur Parasitierung der Puppen von Forleule (*Panolis flammea* [Schiff.]), Kiefernspanner (*Bupalus piniarius* [L.]) und Heidelbeerspanner (*Boarmia bistortana* [Goeze]) in bayerischen Kiefernwäldern. *Zeitschrift für Angewandte Entomologie* **87,** 186–202; 311–322.

Hanapi, S. B. (1980). "A Study of Insect Communities Associated with Galls of Sawflies and Other Insects on Willows." PhD Thesis, University of Manchester.

Hanson, H. S. (1939). Ecological notes on the *Sirex* wood wasps and their parasites. *Bulletin of Entomological Research* **30,** 27–76.

Hawkins, B. A., and Goeden, R. D. (1984). Organization of a parasitoid community associated with a complex of galls on *Atriplex* spp. in Southern California. *Ecological Entomology* **9,** 271–292.

Hespenheide, H. A. (1979). Are there fewer parasitoids in the tropics? *American Naturalist* **113,** 766–769.

House, H. L. (1978). An artificial host: encapsulated synthetic medium for *in vitro* oviposition and rearing the endoparasitoid *Itoplectis conquisitor*. *Canadian Entomologist* **110,** 331–333.

Janzen, D. H. (1970). Herbivores and the number of tree species in tropical forests. *American Naturalist* **104,** 501–528.

Janzen, D. H. (1981). The peak in North American ichneumonid species richness lies between 38° and 42°N. *Ecology* **62,** 532–537.

Janzen, D. H., and Pond, C. M. (1975). A comparison, by sweep sampling, of the arthropod fauna of secondary vegetation in Michigan, England and Costa Rica. *Transactions of the Royal Entomological Society of London* **127,** 33–50.

Jones, D. A., Parsons, J., and Rothschild, M. (1962). Release of hydrocyanic acid from crushed tissues of all stages in the life-cycle of species of the Zygaeninae (Lepidoptera). *Nature* **193,** 52–53.

Kopelke, J.-P. (1983). Die gallenbildenden *Pontania*-Arten—ihre Sonderstellung unter den Blattwespen. Teil II: Konkurrenz und Koexistenz, Parasitismus. *Natur und Museum* **113,** 11–23.

Kopelke, J.-P. (1985). Biologie und Parasiten der gallenbildenden Blattwespe *Pontania proxima* (Lepeletier 1823) (Insecta: Hymenoptera: Tenthredinidae). *Senkenbergiana Biologica* **65** (1984), 215–239.

Lampe, K.-H. (1984). Struktur und Dynamik des Parasitenkomplexes der Binsensackträgermotte *Coleophora alticolella* Zeller (Lep.: Coleophoridae) in Mitteleuropa. *Zoologische Jahrbücher Systematik, Ökologie und Geographie der Tiere* **111,** 449–492.

Lawton, J. H. (1983). Plant architecture and the diversity of phytophagous insects. *Annual Review of Entomology* **28,** 23–39.

Liston, A. D. (1982). Aspects of the biology of *Euura amerinae* (Linnaeus) (Hymenoptera, Tenthredinidae). *Zeitschrift für Angewandte Entomologie* **94,** 56–61.

Mac Arthur, R. H., and Wilson, E. O. (1967). "The Theory of Island Biogeography." Princeton University Press, Princeton, New Jersey.

Magnus, W. (1914). "Die Enstehung der Pflanzengallen verursacht durch Hymenopteren." Jena.

Miller, J. C. (1980). Niche relationships among parasitic insects occurring in a temporary habitat. *Ecology* **61,** 270–275.

Morrison, G., Auerbach, M., and McCoy, E. D. (1979). Anomalous diversity of tropical parasitoids: a general phenomenon? *American Naturalist* **114,** 303–307.

Mueller, T. F. (1983). The effect of plants on the host relations of a specialist parasitoid of *Heliothis* larvae. *Entomologia Experimentalis et Applicata* **34,** 78–84.

Orphanides, G. M. (1984). Competitive displacement between *Aphytis* spp. (Hym. Aphelinidae) parasites of the California red scale in Cyprus. *Entomophaga* **29,** 275–281.

Owen, D. F., and Owen, J. (1974). Species diversity in temperate and tropical Ichneumonidae. *Nature* **249,** 583–584.

Parker, D. L. (1935). *Apanteles solitarius* (Ratzeburg), an introduced braconid parasite of the satin moth. *United States Department of Agriculture Bulletin* **477.**

Pimm, S. L., and Lawton, J. H. (1977). Number of trophic levels in ecological communities. *Nature* **268,** 329–331.

Pollard, E. (1979). Population ecology and change in range of the white admiral butterfly *Ladoga camilla* L. in England. *Ecological Entomology* **4,** 61–74.

Price, P. W. (1973). Parasitoid strategies and community organisation. *Environmental Entomology* **2,** 623–626.

Price, P. W., and Craig, T. P. (1984). Life history, phenology, and survivorship of a stem-galling sawfly, *Euura lasiolepis* (Hymenoptera: Tenthredinidae), on the Arroyo Willow, *Salix lasiolepis*, in Northern Arizona. *Annals of the Entomological Society of America* **77,** 712–719.

Pschorn-Walcher, H. (1957). Probleme der Wirtswahl parasitischer Insekten. *Bericht über die 8. Wanderversammlung Deutscher Entomologen* **11,** 79–85.

Rabb, R. L., and Bradley, J. R. (1968). The influence of host plants on parasitism of eggs of the tobacco hornworm. *Journal of Economic Entomology* **61,** 1249–1252.

Randall, M. G. M. (1982). The ectoparasitism of *Coleophora alticolella* (Lepidoptera) in relation to its altitudinal distribution. *Ecological Entomology* **7,** 177–185.

Rosen, D., and Kfir, R. (1983). A hyperparasite of coccids develops as a primary parasite of fly puparia. *Entomophaga* **28,** 83–88.

Rotheray, G. E. (1981). Host searching and oviposition behaviour of some parasitoids of aphidophagous Syrphidae. *Ecological Entomology* **6,** 79–87.

Salt, G. (1931). Parasites of the wheat-stem sawfly, *Cephus pygmaeus*, Linnaeus, in England. *Bulletin of Entomological Research* **22,** 479–545.

Salt, G. (1970). "The Cellular Defence Reactions of Insects." Cambridge University Press, Cambridge.

Sandlan, K. (1980). Host location by *Coccygomimus turionellae* (Hymenoptera: Ichneumonidae). *Entomologia Experimentalis et Applicata* **27,** 233–245.

Schröder, D. (1967). *Diplolepis* (= *Rhodites*) *rosae* (L.) (Hym: Cynipidae) and a review of its parasite complex in Europe. *Commonwealth Institute of Biological Control Technical Bulletin* **9,** 93–131.

Schröder, D. (1974). A study of the interactions between the internal larval parasites of *Rhyacionia buoliana* (Lepidoptera: Olethreutidae). *Entomophaga* **19,** 145–171.

Shaw, M. R. (1981a). Parasitic control: section A, general information. *In* "Large White Butterfly; the Biology, Biochemistry and Physiology of *Pieris brassicae* (Linnaeus)" (J. Feltwell, ed.), pp. 401–407. W. Junk, The Hague.

Shaw, M. R. (1981b). Parasitism by Hymenoptera of larvae of the white admiral butterfly, *Ladoga camilla* (L.), in England. *Ecological Entomology* **6,** 333–335.

Shaw, M. R. (1981c). Delayed inhibition of host development by the nonparalyzing venoms of parasitic wasps. *Journal of Invertebrate Pathology* **37,** 215–221.

Shaw, M. R. (1983). On[e] evolution of endoparasitism: the biology of some genera of Rogadinae (Braconidae). *Contributions of the American Entomological Institute* **20,** 307–328.

Shaw, M. R., and Askew, R. R. (1976a). Parasites. *In* "The Moths and Butterflies of Great Britain and Ireland" (J. Heath, ed.), pp. 24–56. Blackwell, Oxford.

Shaw, M. R., and Askew, R. R. (1976b). Ichneumonoidea (Hymenoptera) parasitic upon leaf-mining insects of the orders Lepidoptera, Hymenoptera and Coleoptera. *Ecological Entomology* **1,** 127–133.

Shaw, M. R., and Askew, R. R. (1979). Hymenopterous parasites of Diptera (Hymenoptera Parasitica). *In* "A Dipterist's Handbook" (A. E. Stubbs, and P. J. Chandler, eds.), pp. 164–171. Amateur Entomologist's Society, Hanworth.

Shorrocks, B., Rosewell, J., Edwards, K., and Atkinson, W. (1984). Interspecific competition is not a major organizing force in many insect communities. *Nature* **310,** 310–312.

Shorthouse, J. D. (1973). The insect community associated with rose galls of *Diplolepis polita* (Cynipidae, Hymenoptera). *Quaestiones Entomologicae* **9,** 55–98.

Simmonds, F. J. (1944). The propagation of insect parasites on unnatural hosts. *Bulletin of Entomological Research* **35,** 219–226.

Southwood, T. R. E. (1985). Insect communities. *Antenna* **9,** 108–116.

Spradbery, J. P. (1968). A technique for artificially culturing ichneumonid parasites of woodwasps (Hymenoptera: Siricidae). *Entomologia Experimentalis et Applicata* **11,** 257–260.

Strong, D. R., Simberloff, D., Abele, L. G., and Thistle, A. B. (1984). "Ecological Communities: Conceptual Issues and the Evidence." Princeton University Press, Princeton, New Jersey.

Taylor, T. H. C. (1937). "The Biological Control of an Insect in Fiji. An Account of the Coconut Leaf-mining Beetle and its Parasite Complex." Richard Clay, Bungay, Suffolk.

Thompson, W. R., and Parker, H. L. (1927). The problem of host relations with special reference to entomophagous parasites. *Parasitology* **19,** 1–34.

Tremblay, E. (1984). The parasitoid complex (Hym.: Ichneumonoidea) of *Toxoptera aurantii* (Hom.: Aphidoidea) in the Mediterranean area. *Entomophaga* **29,** 203–209.

Tremewan, W. G. (1985). Zygaenidae. *In* "The Moths and Butterflies of Great Britain and Ireland (J. Heath and A. M. Emmet, eds.), pp. 74–123. Harley Books, Colchester.

Varley, G. C. (1947). The natural control of population balance in the knapweed gall-fly (*Urophora jaceana*). *Journal of Animal Ecology* **16,** 139–187.

Vinson, S. B. (1975). Biochemical coevolution between parasitoids and their hosts. *In* "Evolutionary Strategies of Parasitic Insects and Mites" (Price, P. W., ed.), pp. 14–48. Plenum, London.

Vinson, S. B. (1976). Host selection by insect parasitoids. *Annual Review of Entomology* **21,** 109–133.

Vinson, S. B., and Iwantsch, G. F. (1980a). Host suitability for insect parasitoids. *Annual Review of Entomology* **25,** 397–419.

Vinson, S. B., and Iwantsch, G. F. (1980b). Host regulation by insect parasitoids. *Quarterly Review of Biology* **55,** 143–165.

Weems, H. V. (1954). Natural enemies and insecticides that are detrimental to beneficial Syrphidae. *Ohio Journal of Science* **54,** 45–54.

Zinnert, K.-D. (1969). Vergleichende Untersuchungen zur Morphologie und Biologie der Larvenparasiten (Hymenoptera: Ichneumonidae und Braconidae) mitteleuropäischer Blattwespen aus der Subfamilie Nematinae (Hymenoptera: Tenthredinidae). *Zeitschrift für Angewandte Entomologie* **64,** 180–217; 277–306.

Zwölfer, H. (1971). The structure and effect of parasite complexes attacking phytophagous host insects. *In* "Dynamics of Populations" (P. J. den Boer and G. R. Gradwell, eds.), pp. 405–418. Centre for Agricultural Publishing and Documentation, Wageningen.

9

The Effect of Parasitoids on Phytophagous Insect Communities

J. H. LAWTON

I. INTRODUCTION

This chapter, in contrast with others in this volume, views the parasitoid-host interaction from the host's perspective. Its theme is simple and has two parts. First, the niches of phytophagous insects (insects feeding on the living tissues of higher plants; Strong *et al.*, 1984) must often be constrained and moulded by parasitoids; second, indirect interspecific competition between phytophagous insects can occur via shared polyphagous parasitoids, with potentially important consequences for the structure of phytophagous insect communities, and particularly the number of coexisting species. ("Community" is used here in its simplest sense to mean a group of co-occurring species; I have sometimes employed the more neutral term "assemblage".)

The idea that niches are maintained by interactions with enemies is not new; nor is that of species competing via shared enemies. The concept of competition for "enemy-free space" has been discovered and rediscovered repeatedly in the ecological literature in many contexts (palaeontology, freshwater biology and parasitology, to name but three; Jeffries and Lawton, 1984). Insect ecologists are no exception (Brower, 1958; Askew, 1961; Klomp, 1961; Root, 1973; Hebert *et al.*, 1974; Gilbert and Singer, 1975; Ricklefs and O'Rourke, 1975; Zwölfer, 1975; Orians and Solbrig, 1977; Otte and Joern, 1977; Schultz *et al.*, 1977; Strong, 1977; Lawton, 1978; Waage, 1979; Price *et al.*, 1980; Atsatt, 1981; Lawton and Strong, 1981; Cornell, 1983; Gilbert, 1984; Strong *et al.*, 1984). Holt (1977, 1984) calls the problem "apparent competition" and explores it theoretically.

Despite this long pedigree, many ecologists continue to think and write as though traditional interspecific interactions for "resources", particularly food, are the only or main determinants of niche structure and interspecific competition. There is therefore justification for developing alternative arguments at some length for a particular system, in this case phytophagous insects and parasitoids. Some of the material discussed here is also touched on by Freeland (1983) who includes parasitoid examples in his general review of parasites and the coexistence of host species.

This chapter has four main parts. First there are brief remarks about host specificity in parasitoids. Population biology has been particularly concerned with host-specific parasitoids (Hassell, 1978, and Chapter 7). Since most of my arguments hinge on polyphagous parasitoids, it is important to remind ourselves of their existence. There then follow more substantial sections on niche differences and resource partitioning by phytophagous insects, and arguments that niches in this group of insects must often be constrained and influenced by parasitoids. Niche differences between coexisting phytophages also appear to be important in determining relative vulnerabilities to the local parasitoid complex. These ideas are used in the final section to show how the number of coexisting species of phytophagous insects in a community could, to a degree, be determined by competition for enemy-free space via shared polyphagous parasitoids.

Clearly, the concept of the niche figures prominently in this chapter. "Niche" can be defined in several ways (Elton, 1927; Maguire, 1973; Whittaker *et al.*, 1973; Hutchinson, 1978). For present purposes differences between these definitions are unimportant. Ecologists use various interpretations of the niche concept to discuss a number of questions. For example, how and why do species differ in their use of resources, i.e. what is the significance of niche differences between species? Are there limits to the number of coexisting species in communities, set by limits to niche space and niche overlap? I address these and other related questions here.

There is probably no such thing as absolute enemy-free niche space. Safety from enemies is almost always relative, not absolute, not least because ways of escaping one enemy may increase the risk from others. Nor are parasitoids the only important enemies confronting phytophagous insects. In a symposium on parasitoids it is reasonable but simplistic to single them out for special treatment.

In concentrating on parasitoids, I have also been parsimonious. The literature on insect host-parasitoid systems is vast, and my choice of examples is illustrative, not exhaustive.

II. POLYPHAGOUS PARASITOIDS

Much of what follows hinges on parasitoids which attack more than one species of insect host; that is, on oligophagous or polyphagous parasitoids. (To avoid long-windedness, I shall subsequently refer to all species that are not monophagous as polyphagous.) As Zwölfer (1971) notes: "Strictly specific (monophagous) parasites are rare and in many natural parasite complexes they do not occur at all." In one survey the average number of aphid species attacked by a single species of parasitoid was 7.8 (Starý and Rejmánek, 1981). Some species of Aphidiidae in that study had remarkably broad host ranges; over 20% of *Ephedrus* species attack between 50 and 100 species of aphids.

Unfortunately, such data do not tell us how many species of insect hosts are utilized by particular parasitoids within local communities. Here the degree of polyphagy will inevitably be less than it appears to be from broad geographic surveys (Starý and Rejmánek, 1981). Other formidable problems in estimating host range are discussed in Chapter 8. There is no doubt, however, that polyphagous parasitoids are sufficiently common in most natural host assemblages to spread a web of interactions across a range of victim species (Zwölfer and Kraus, 1957; Askew and Shaw, 1974; Askew, 1975, 1980; Jervis, 1980; Hawkins and Goeden, 1984; see Chapter 8). But the web of interactions is not infinitely wide. It is very unusual, for example, for polyphagous chalcids to attack both leaf galls and leaf-mines (Askew and Shaw, 1974). Usually we expect attacks by polyphages to be constrained to a greater or lesser extent by host life history stage, taxonomy and mode of exploiting the food-plant (for a general discussion of these ideas in related contexts, see Freeland, 1983).

I will return these points at several places in this chapter.

III. RESOURCE PARTITIONING AND NICHE DIFFERENCES IN PHYTOPHAGOUS ASSEMBLAGES

Examples of resource partitioning (niche differences) between species are easy to find in any assemblage of phytophagous insects, be they between close relatives on the same or different species of host plants, or between unrelated species sharing one host. Examples are to be found in papers by Lawton and Strong (1981) and Gilbert (1984). In one recent study, Cornell (1983) notes: "The most closely related Cynipinae (gall-forming wasps) often have widely divergent gall structures, locations on the host tree, and sometimes altogether different host species."

Normally, ecologists interpret niche differences as a way of reducing interspecific competition for resources (usually food) over ecological and evolutionary time (Connell, 1980). The problem with phytophagous insects is that interspecific competition for food seems generally too rare and feeble to explain most niche differences. There are, of course, exceptions (Zwölfer, 1979; see also Strong *et al.*, 1984). Evidence that most species of phytophagous insects do not, apparently, compete for food most of the time is summarized by Lawton and Strong (1981), Lawton and Hassell (1984) and Strong *et al.* (1984). *Chirosia* spp. and *Dasineura* spp., dipteran leaf-miners and gall-formers on bracken (*Pteridium aquilinum*) (L.) Kuhn, are good examples of closely related species with distinct larval niche partitioning in space and time, but with no evidence of significant food limitation or interspecific competition for food (Lawton, 1984a, 1984b).

Species' niches may differ for several reasons (Rohde, 1979; Lawton and Strong, 1981; Price, 1984):

1. Niches diverge to reduce interspecific competition for some, limiting resource, usually food, but also perhaps shelter, oviposition sites, etc. This is the traditional view. Some of the niche differentiation seen in the insect communities of flower-heads is consistent with this explanation (Zwölfer, 1979, 1983)
2. Niche differences between species reflect no more than accidents of independent evolutionary histories and have nothing to do with competition, past or present (Howard and Harrison, 1984)
3. Niche specialization may facilitate mate location, particularly in rare

species, and reduce the risk of hybridization (Doolan and MacNally, 1981; Booij, 1982; Gilbert, 1984)

4. Niche differentiation may be driven by natural enemies, broadly defined to include predators, parasites, parasitoids and diseases.

It is this last possibility that I wish to develop here. The general arguments apply with equal facility to all enemies, not just parasitoids.

IV. THE ROLE OF PARASITOIDS IN MOULDING NICHES

A. General Considerations

Attack by enemies imposes selection on victims to reduce vulnerability; counter-selection works on enemies to improve efficiency. The evolutionary consequences have been likened to the Red Queen of "Alice Through the Looking Glass", who had to run as fast as she could just to remain standing where she was (van Valen, 1973; Maynard Smith, 1976).

There are two ways in which selection by enemies may mould niche differentiation in victims. First, imagine two closely related victims with very similar ecologies, but sharing no enemies. (The victims could, for example, be sympatric but attacked only by strict monophages, or allopatric with different enemy complexes in different places.) Because it is unlikely that identical mutations will arise in both groups of enemies and victims, the expectation is that the two victim species will diverge in such characteristics as appearance, morphology, and feeding sites under selection from different enemies. This is the enemy-driven version of category (2) in Section III, and involves no competition (direct or indirect) between the victim species.

Alternatively, sympatric victim species may share one or more polyphagous enemies. Again, selection will generally favour any characteristic that reduces the vulnerability of either species to their shared attackers, but there is now the potential for victim species to interact via the common enemy. Niches may again diverge as victims compete for enemy-free space.

Predicting patterns in the evolution of niche differentiation between species that share enemies is very hard. For example, does niche divergence make coexistence more or less likely? How much divergence is to be expected? Some preliminary answers are given by Holt (1977), Levin and Segel (1982) and Levin (1983), but none of these models seem sufficiently realistic to permit tests with field data, and none refer to parasitoids and hosts. The problem deserves more theoretical attention.

As far as I know, selection operates for reduced vulnerability to enemies, either shared or specific, whether enemy mortality is density dependent or density independent on the victim population. Again, theoretical studies under a variety of assumptions about victim and enemy population dynamics, and enemy foraging behaviour would be useful. Gilbert (1984) made a valuable start by developing a verbal model of the way in which parasitoids select for partitioning of different *Passiflora* host plant species by *Heliconius* butterflies.

Selection operates on a hierarchy of levels to reduce vulnerability to enemies (Sih, 1985). Victims may:

1. Avoid enemies spatially or temporally by occupying habitats or seasons when enemies are scarce or absent
2. Adopt behaviours that reduce recognition and detection by enemies
3. Develop abilities to escape once detected
4. Resist attack.

Roughly mirroring this classification, parasitoid success is usually attributed to: (a) host habitat location; (b) host location; (c) host acceptance and (d) host suitability (Doutt, 1959; Vinson, 1981, 1984a; see Chapter 2). In other words, parasitoids clearly have the potential to influence niches of phytophagous insects, principally via categories 1–3 and (a)–(c).

B. Some Evidence

The necessary conditions exist for evolutionary effects on phytophage niches whenever parasitoid-imposed mortality differs by location, time or host morphological and behavioural phenotype. Sufficient conditions involve a plethora of complexities: the existence of genetic variation for appropriate host traits, counter-balancing selection and so on. Not least of the problems of counter-balancing selection is the fact that absolute safety from enemies is virtually impossible (Jeffries and Lawton, 1984). Species usually evolve from the frying pan into marginally safer fires. Here, I merely wish to show that the potential exists for parasitoids to mould many aspects of the niches of victim species, by demonstrating that intra- and interspecific variations in patterns of parasitoid attack fulfil the necessary conditions for selection to operate, all things being equal.

An alternative way of thinking about these same data, important for developing later arguments, is that levels of parasitoid attack, and whether or not coexisting species of phytophagous insects share parasitoids, depend upon the degree of niche differentiation shown by the phytophages. "Being different" reduces the risk of attack from your neighbours' enemies.

Evidence is gathered under six headings.

1. *The Same, or Closely Related, Species of Phytophagous Insects on Different Parts of the Same Host Plant Suffer Different Levels of Attack from the Same Species of Parasitoid*

Parasitoids often confine their searching to particular parts of plants (reviews in Vinson, 1976, 1981; Price *et al.*, 1980), resulting in different levels of parasitism among individuals of one victim species collected from different positions on a plant (Starý, 1970; Münster-Swendsen, 1980; Price *et al.*, 1980; Vinson, 1981; Gross and Fritz, 1982; Alghali, 1984) or among species occupying different microhabitats (Herrebout, 1969; Claridge and Reynolds, 1972). By pupating off the host, *Zeiraphera rufimitrana* (Herrich-Schäffer) (= *Eucosma rufimitrana*) appears to escape most of the pupal parasitoids which attack related tortricids pupating on leaves (Zwölfer and Kraus, 1957).

Aphidius rhopalosiphi De Stephani preferentially searches leaves of wheat; in consequence under field conditions aphids on leaves (mainly *Metopolophium* spp.) are much more heavily parasitized than aphids on ears (predominantly *Sitobion avenae* (Fabricius)). *S. avenae* placed experimentally on leaves are more heavily parasitized than individuals on ears. Finally, *A. rhopalosiphi* has no "innate preference" for either *S. avenae* or *M. dirhodum* (Walker) when the two species are equally accessible in the laboratory (Gardner and Dixon, 1985). Thus, site-specific searching by *A. rhopalosiphi* creates within- and between-species differences in parasitoid-imposed mortality on different parts of the same host plant.

2. *The Same, or Closely Related, Species of Phytophagous Insects With Different Morphologies Suffer Different Levels of Attack from the Same Species of Parasitoid*

Several reviews show that differences in host size, shape, colour and texture influence vulnerability to parasitoids (Zwölfer and Kraus, 1957; Herrebout 1969; Arthur, 1981; Vinson, 1984a). Although not usually regarded as components of a species' niche, these attributes of insects may indirectly influence more traditional niche axes, for example feeding sites or hiding places.

Among gall-formers, differences in gall morphology may profoundly influence levels of parasitism. *Ephedrus persicae* Froggatt attacks gall and leaf-curling aphids in Europe. Species normally immune to attack [e.g. *Aphis pomi* De Geer, *Rhopalosiphum insertum* (Walker)] fall victim when populations become large enough to curl leaves (Starý, 1970). Askew (1961) argued that differences in the form and growing times of different species of *Cynips* galls on oak were important in determining relative vulnerabilities to a complex of parasitoids. Consistent with this view are data showing that different sizes of galls of one species are differentially attacked by particular parasitoids (Weis, 1982; Weis *et al.*, 1985; Zwölfer, 1979). Weis and Abrahamson (1985) have evidence for genetic effects on gall size both in the host plant (*Solidago altissima* L.) and the gall-former *Eurosta solidaginis* (Fitch) (Tephritidae), providing the raw material for selection by parasitoids to affect gall size. Selection for large galls, immune to attack by the parasitoid *Eurytoma gigantea* Hedicke may, however, be constrained by counterbalancing selection in the form of woodpeckers and chickadees which prefer to prey on large *Eurosta solidaginis* galls (Weis *et al.*, 1985).

3. There is Seasonal Variation in Vulnerability to Parasitoids

Many aspects of species' temporal niches are undoubtedly influenced by variable risks of parasitoid attack at different times of the year. Mortality due to parasitoids often differs between early- and late-hatching individuals of one species (Münster-Swendsen, 1980) or in different generations of multivoltine species (Bryan, 1983; Dempster, 1984), although as West (1985) and Holliday (1985) have shown, parasitoids alone are unlikely to be the sole determinants of seasonal phenologies. Myers (1981) has, however, argued convincingly that heavy parasitism of late-developing caterpillars by tachinids is the reason why *Malacosoma californicum* Packard feeds early in the season, when leaves are actually less suitable for caterpillar growth and survival.

It follows from these arguments that temporal niche separation of close relatives could be driven by shared parasitoids; a possible example is *Aporia crataegi* (L.) and *Pieris brassicae* (L.) attacked by *Apanteles glomeratus* (L.), discussed by Gilbert and Singer (1975).

4. The Same Species of Phytophagous Insects on Different Host Plants Experience Different Levels of Attack from Parasitoids, Because of Habitat and Host Plant Selection by Parasitoids

Such effects are now well documented and undoubtedly play a part in host plant choice, and niche segregation by food-plant, in closely related species of phytophagous insects. The idea that enemies, including parasitoids, are important in host plant choice is well developed by Gilbert and Singer (1975), and Gilbert (1984); see also comments on lycaenid butterflies in Section V. B.

Many parasitoids appear to locate suitable victims by first locating habitats and host plants (Doutt, 1959; Herrebout, 1969). Numerous examples are reviewed by DeBach (1964), Starý (1970), Price *et al.* (1980), Price (1981) and Vinson (1981, 1984b). As a result, the same species of insect often experiences markedly different levels of parasitism from the same species of parasitoid on different species of host plants (reviews in DeBach, 1964; Price *et al.*, 1980; Price, 1981; Vinson, 1981). Richards (1940) provided an early example, noting that *Artogeia rapae* (L.) (= *Pieris rapae*) larvae living on plants other than cabbages are seldom attacked by parasitoids.

Some indication of the balance of selective forces which maintain host plant choice in the face of intense differences in levels of parasitism on different species of plants is provided by Gibson and Mani (1984). *Danaus chrysippus* L. caterpillars are heavily parasitized when feeding on milkweeds high in alkaloids, but are ignored by birds. The reverse is true on milkweeds lacking alkaloids.

Habitat and host plant selection by parasitoids has the interesting corollary that many polyphagous parasitoids are more likely to attack taxonomically unrelated insects found on one plant species or group of species, or in one habitat, than they are to attack taxonomically related insects occupying different host plants or habitats (Vinson, 1981, 1984b). Such parasitoids are essentially "niche-specific" in their choice of victims, rather than "taxon-specific".

Ichneumons show the phenomenon very clearly (Zwölfer and Kraus, 1957; Shaw and Askew, 1976; Porter, 1977; Shaw, 1977; Owen *et al.*, 1981), as do parasitoids in a number of other taxa (Starý, 1970; Askew, 1971; Askew and Shaw, 1974; Jervis, 1980). A number of these polyphagous parasitoids regularly kill victims in more than one order of insects, providing they occur in similar locations, for example, leaf- and stem-mining Lepidoptera, Diptera, Coleoptera and Hymenoptera: Tenthredinidae (e.g. *Scambus pterophori* (Ashmead) (Chalcididae) (Cushman, 1926); *Sympiesis sericeicornis* (Nees) (Eulophidae) (Askew, 1971); *Chrysocharis* spp. (Chalcididae) (Askew and Shaw, 1974); *Colastes braconius* Haliday (= *Exothecus braconius*) (Braconidae) (Shaw and Askew, 1976)). The great majority of species (17 out of 19) of parasitoids reared from leaf-mining weevils also attacked leaf-mining Lepidoptera and sawflies (Askew and Shaw, 1974). Leaf-mining agromyzids (Diptera) had a more distinct guild of parasitoids, but again eight out of 16 species of chalcids also occurred in other orders of leaf-mining insects (Askew and Shaw, 1974).

As already noted, there are limits to the degree of polyphagy shown by niche-specific parasitoids. And not all polyphagous parasitoids are niche-specific rather than taxon-specific in their selection of victims (Pschorn-Walcher, 1977). Some attack closely related victims in very different habitats (Griffiths, 1964, 1967; Jervis, 1980). The fact is, however, that niche-specific polyphagous parasitoids are common.

5. *Related Insect Species on the Same Host Plant Generally Share Fewer Parasitoids the More Dissimilar their Niches*

This is a logical corollary of previous arguments. The idea was clearly developed by Askew (1961) in his work on cynipid oak-galls: "The parasite complement of similar galls of species belonging to different genera are more alike than those of galls of congeneric species when these galls differ in position or form". In contrast, most *Phyllonorycter* (Lepidoptera) mines look alike and "as a rule there is no discrimination by polyphagous parasites between those occurring on the same plant" (Askew, 1980).

Although this idea is important for community structure in phytophagous insects, and follows logically from much that has been said above, there have been surprisingly few attempts to test it on other systems. One recent study provides little support. Hawkins and Goeden (1984) studied the gall complex of *Atriplex* spp. in southern California and found that species of gall-forming Diptera shared parasitoids in complex ways. There were some obvious host groups, for example, woody and smooth stem-galls had distinct parasitoid faunas, but, to a human observer at least, gall morphology *did not* appear to provide the key to patterns of host-insect utilization by most parasitoids. The problem deserves more attention.

6. *An Experiment*

The data and evidence assembled so far are based almost entirely on observations, not manipulative experiments. If parasitoids are effective selective agents on where, when and how victim species live, it follows that experimental manipulation of a species' niche should expose the manipulated victim to entirely different risks of parasitism. Zwölfer and Kraus (1957) reported an experiment involving the polyphagous ichneumon *Apechthis rufatus* (Gmelin) and pupae of the fir budworm, *Choristoneura murinana* Hübner, in the French Vosges. The normal hosts of *A. rufata* in this area include pupae of two leaf-rolling tortricids, *Archips xylosteana* (L.) and *Tortrix viridana* (L.), found on oak, *Quercus petraea* (Matuschka) Liebl. The host plant of *C. murinana*, silver fir, *Abies alba* Miller, grew "side by side" with oaks. In 1955 and 1956, Zwölfer and Kraus reared parasitoids from over 5000 *C. murinana* pupae collected from fir; none yielded *Apechthis rufatus*. They then transplanted unparasitized pupae of *C. murinana* into artificially formed leaf-rolls on oaks, made to look like those of *T. viridana* and *Archips xylosteana*. Thirteen out of 158 (8.5%) of these transplanted pupae yielded *Apechthis rufatus*.

This experiment is remarkable in many ways, and as far as I know is unique in the entomological literature. It shows that the risk of parasitoid attack depends to a great extent upon the exact position of a species in a habitat, and on its appearance. Similar experiments with other systems would be valuable.

C. Conclusions

Taken together, these many and varied examples leave little room for doubt that the risk of parasitoid attack probably influences many aspects of the niches of phytophagous insects. Also, whether or not polyphagous parasitoids are shared depends to a considerable extent upon similarities and differences in the niches of co-occurring phytophagous insects. These are hardly dramatic or novel conclusions. But they do have potentially important consequences for the structure of phytophagous insect communities.

V. IMPLICATIONS FOR COMMUNITY STRUCTURE

A. Species Invasion in Theory

Niche differences between phytophagous insects, and the effects of such differences on parasitoids, underpin answers to the question: can polyphagous parasitoids limit the number of species of coexisting phytophagous insects in a community? Consider, for example, the question of whether a new species of phytophage can invade an established assemblage of phytophagous insects and their resident parasitoids. For a potential invader with discrete generations a simple model of the invasion process in Nicholson-Bailey form (Hassell, 1978) is:

$$N_{t+1} = N_t \exp(r - aP_t) \qquad (1)$$

where N_t is the initial density of invading phytophagous insects; N_{t+1} is their density in the next generation; r is their intrinsic rate of increase; a is the attack rate of resident polyphagous parasitoids on the invading species and P_t is the density of the resident parasitoids in generation t.

N_t will inevitably be small, hence a "carrying capacity" term for the victim population is unnecessary. In the absence of parasitoids, the initial growth of the invading phytophage population is exponential.

For successful invasion, $N_{t+1} > N_t$, which is the same as the condition $r > a\ P_t$, i.e. the potential rate of increase of the victim population must be greater than the rate at which resident parasitoids find and kill them. Other things being equal, invasion will be more difficult:

1. When r is small
2. If invaders are particularly vulnerable to attack by resident parasitoids (high a), and
3. If the population of resident parasitoids (P_t) is large.

These simple theoretical considerations suggest that the more similar the potential invading species is to resident species of phytophages, the more vulnerable it will be to resident parasitoids (because a will be large); the problem is made worse if P_t is large, sustained by a large population of one or more species of resident phytophages. These are more or less identical to the conditions normally associated with conventional competitive exclusion of ecologically similar species. Only species with "sufficiently different" niches can invade.

The argument extends logically to attack from several species of polyphagous parasitoids, as well as to polyphagous predators. Note also that the ability to resist parasitoids by encapsulating them (see, for example, Herrebout, 1969; Carton and Kitano, 1981; Chapters 4 and 6) lowers the effective attack rate (Categories (4) and (d) in Section IV. A.), and would promote establishment in the community.

Characteristics favouring invasion (niche differences or parasitoid resistance) may have evolved as adaptations in response to selection from those parasitoids now encountered, or they may be fortuitous exaptations (*sensu* Gould and Vrba, 1982), whose evolution had nothing to do with the influence of contemporary enemies.

B. Species Invasion in Practice

Good data exist to show that phytophagous insects introduced as biological control agents against weeds are often killed by native parasitoids. As expected, the parasitoids are usually recruited from ecologically similar native insects, and levels of parasitism are often high. Goeden and Louda (1976) summarized examples from seven different weeds (see also Clausen, 1956; Wilson, 1965). The weevil *Rhinocyllus conicus* Froelich has been introduced into North America from Europe as a biological control agent for musk thistles, *Carduus nutans* L.; its larvae feed in the flower-heads. In both Virginia (Surles, 1974) and Missouri (Putter *et al.*, 1978) larvae are attacked by two native braconids, *Triaspis curculionis* (= *Aliolus curculionis* (Fitch)) and *Bracon mellitor* Say, normally associated with native weevils. In Virginia, it is also attacked by an ichneumon, *Campoplex polychrosidis* Viereck, a parasitoid of native plume-thistle moths.

However, there are *no* recorded cases in the biological control literature, that I can find, where parasitoids have unequivocally prevented a species from establishing (such data do exist for predators: see Goeden and Louda, 1976). But, consistent with Eq. 1, the net rate of increase ($r - aP_t$) of *Bactra venosana* (Zeller) (= *B. truculenta* Meyrick), an olethreutid moth introduced into Hawaii for the control of purple nutsedge, *Cyperus rotundus* L., was severely curtailed by the polyphagous egg parasitoid *Trichogramma minutum* Riley and two other unspecified parasitoids (Goeden and Louda, 1976). A lack of data showing that native parasitoids can drive introduced biological control agents to extinction and hence prevent their establishment is disappointing, but is probably to be expected given the extreme difficulty of finding out why introductions often fail.

One possible example, not involving biological control agents, is provided by Samways (1979). Adults of the sphingid *Erinnyis ello* (L.) oviposited on a small, isolated experimental plot of cassava, but the eggs and caterpillars were so heavily parasitized that no larvae reached the final instar. Eggs were particularly heavily attacked by *Trichogramma fasciatum* (Perkins), which used other hosts as well and was "probably present before *E. ello* arrived".

The obverse prediction, that invasion will be easiest for species immune to attack by resident parasitoids, is a more feeble test of theory. Knopper galls, the agamic generation of the cynipid *Andricus quercuscalicis* Burgsdorf, have no morphologically close counterpart in the British native oak-gall fauna. The wasp has spead rapidly in Britain after invasion in the early 1960s. No parasitoids have been recorded attacking the agamic generation at Silwood Park in southern England, where *A. quercuscalicis* is now abundant and well studied (Collins *et al.*, 1983), and there are few records of parasitism from elsewhere in Britain. These data are at least consistent with theoretical expectations.

This theory also receives support from an unexpected source, a comparison of the host plant ranges of lycaenid butterflies which do and do not associate with ants. The association of lycaenids with ants is generally regarded as a protection against larval and pupal enemies, particularly parasitoids (Atsatt, 1981). Some species associate with ants, and some do not. For example, ants protect *Glaucopsyche lygdamus* (Doubleday) against braconids and tachinids (Pierce and Mead, 1981). Pierce and Elgar's comparative study (1985) of 282 species of lycaenids confirms that in general, ant-attended species from Australia, South Africa and North America use a wider variety of host plants than do non-myrmecophilous species. Many groups of lycaenids also appear to have "host-shifted" in haphazard and unpredictable ways on to chemically and taxonomically unrelated food-plants, a feature again attributed by Atsatt (1981) to protection by ants. The inference must therefore be that enemies, particularly parasitoids, *prevent* the colonization of many potentially suitable host plants by Lepidoptera which do not enjoy the protection of ants.

Links with theory would be stronger if it could be shown that parasitoids are indeed important agents limiting host plant colonization by non-myrmecophilous Lepidoptera; and that the parasitoids involved are polyphagous and relatively "niche-specific" in their mode of attack.

C. Competition for Enemy-free Space in Equilibrium Communities—Theory

Community ecologists have generally been more interested in the sustained coexistence of species, rather than the simple ability to invade (which ignores what happens afterwards), though invasion and sustained coexistence are clearly closely linked. Holt (1977, 1984) used general Lotka-Volterra models of enemy-victim interactions to derive both the conditions for invasion by victim species (identical to those based on Eq. 1), and for stable, sustained coexistence in the community (these are more complex, and undoubtedly differ between generalized Lotka-Volterra models and Nicholson-Bailey type host-parasitoid models. As far as I know, nobody has carried out a detailed analysis, similar to Holt's, using host-parasitoid models.) Holt's qualitative and broad quantitative conclusions are, however, probably insensitive to model details.

Imagine populations of two victim species A and B, and a population of shared enemies. Suppose initially that only A and the enemy live in stable equilibrium. Now let B invade ($r_B > a_B P^*$: subscripts refer to species A or B; P^* is the equilibrium size of the enemy population coexisting with A alone). Once B invades, the total population of victims ($A + B$) increases, and the enemy also increases because it has more food. As Holt shows, the net effect on A is to *reduce* its equilibrium numbers, possibly to extinction. In his terminology there is "apparent competition" between A and B, via their shared enemy.

The outcome (which species coexist and which are eliminated) in this three- (or more) species community depends, among other things, on the relative values of the intrinsic rates of increase (r_j's) of victim species and on enemy attack rates upon victims (a_j's; Holt, 1977). It also depends on the degree of separation and rates of enemy movement between spatial niches supporting different victim species (Holt, 1984). That is, on the components of host niches already reviewed and known to influence vulnerability to shared parasitoids. Hence in principle, unless host-parasitoid models yield results startlingly different from those obtained by Holt (an unlikely event), equilibrium assemblages of phytophagous insects and polyphagous parasitoids, in which the parasitoids impose density dependent control on total host abundance, could be structured by apparent competition.

One other point is worth making about Holt's models. As enemies become more polyphagous (Holt, 1977) *or* less choosy in their spatial foraging niche (Holt, 1984), the permissible variance in the ratio r_j/a_j for stable coexistence of victim species *declines*. Under these conditions, coexistence is favoured by species *convergence* (r_j's and a_j's all similar), or by wide variation in a_j's (large niche differences between victims) and appropriate tuning of r_j's (with high a implying high r and vice versa). This makes qualitative testing of Holt's arguments very difficult.

D. Competition for Enemy-free Space—Field Data?

Data suitable for testing Holt's models in the field in part overlap with those discussed earlier on species invasion. The additional constraint is that now, parasitoids must be shown to impose density dependent control on host populations (see Chapter 7).

Then, the simplest prediction is that polyphagous parasitoids will have a greater impact on the abundance of host species A when host species B is also present, than when A occurs alone. Three possible examples will suffice. A polyphagous ichneumon in the genus *Pimpla* is reported to impose more effective control on the nun moth *Lymantria monacha* (L.) when it occurs in European pine woods with an understory suitable for alternative lepidopteran hosts (I. Trågardh in Watt, 1947). Second, failure of the tachinid *Bessa remota* (Aldrich) to control the moth *Levuana iridescens* Bethune-Baker on one Fijian island has been attributed to lack of alternative hosts (*Eublemma* sp. and *Plutella xylostella* (L.)) on the island (Andrewartha and Birch, 1954). Finally, *Anagrus epos* Girault, an egg parasitoid of the grape leafhopper *Erythroneura elegantula* Osborn, provided substantially better control of this pest species in vineyards close to hedges of blackberry, *Rubus* sp., the foodplant of an alternative species of leafhopper host, important for sustaining *Anagrus* populations over winter (Messenger, 1975). Unfortunately, critical data are often lacking and such examples may be consistent with more than one explanation. For example, parasitoids may be sustained by one host or set of hosts (without in any way regulating their populations) and impose sufficient density independent mortality on "target" species (*Lymantria monacha*, *Levuana iridescens* or *Erythroneura elegantula*) to greatly reduce their numbers. This is a very different situation to the one modelled by Holt (1977).

Murdoch *et al.* (1985) proposed this alternative scenario as one possible mechanism for the biological control of winter moth, *Operophtera brumata*

(L.), by the tachnid *Cyzenis albicans* Fallén in Nova Scotia, with a "reservoir" of *C. albicans* maintained in an alternative congeneric host, the Bruce spanworm, *O. bruceata* (Hulst). But even this reasonably well studied case is poorly understood and several quite different explanations are possible.

At the limit, and whatever the detailed mechanism, extreme depression of alternative host species by polyphagous parasitoids may cause extinction of the host, effectively constraining the number of coexisting species of phytophages in the community. *C. albicans* may well have exterminated the winter moth from hardwood forests in Nova Scotia (Murdoch *et al.*, 1985). Zwölfer (1979, and personal communication) provided a much clearer example. Experimental colonies of the thistle gallfly, *Urophora cardui* (L.), have been established on their normal host plant, *Cirsium arvense* Scopoli, beyond the limits of the natural European range of *U. cardui.* Some of these colonies have been exterminated by *Eurytoma robusta* Mayr, a widespread and important parasitoid of *Urophora* spp. with large populations sustained on alternative hosts. Natural populations of *U. cardui* also appear to suffer the same fate from time to time (Zwölfer, in Seitz and Komma, 1984). Finally, Gilbert (1984) described how polyphagous egg parasitoids, *Trichogramma* sp., can exterminate rare *Heliconius* species cultured with a common congener in a greenhouse.

Under some circumstances polyphagous parasitoids therefore appear able to drive established populations of one of their hosts to extinction, providing the parasitoid population is sustained by alternative hosts. This is clearly very similar to the theoretical situation (Section V. A.) where resident polyphagous enemies may prevent phytophagous insects from establishing themselves in a community. The mechanism of exclusion may or may not conform exactly to the situations modelled by Holt (1977, 1984), but the result is clearly very similar. Polyphagous parasitoids may greatly reduce the abundance of some victim species, and exclude others entirely from otherwise suitable habitats. The mechanism is competition for enemy-free space or "apparent competition", via polyphagous parasitoids shared with, and sustained by, alternative hosts.

The evidence gathered in the first part of this chapter suggests that such effects should be most severe when host insects have very similar niches, and should be progressively less likely as host niches diverge; the models presented in this section warn us that tests also require measurements of host rates of increase. As yet, such data are not available.

In this context, one of Askew's observations is intriguing. He notes (Askew, 1980) that many fewer species of *Phyllonorycter* leaf-miners generally coexist on any one tree than do species of cynipid gall wasp. He attributes this to the fact that "most *Phyllonorycter* mines are very similar and, as a rule, there is no discrimination by polyphagous parasites between those occurring on the same plant species" (see also (5) in Section IV. B.). The implication is that niche diversification by cynipids (in appearance, position on the host, etc.—see (2) in Section IV. B.) partitions parasitoids and permits coexistence of more species.

It would be enormously valuable, both to test existing theory more rigorously and to develop better theory, to have data on relative attack rates by polyphagous parasitoids on a range of coexisting host species (including those potential host insects on which effective attack rates are zero), together with information on niche differences between victim species. Herrebout's study (1969) of coexisting hosts and potential hosts of the tachinid *Eucarcelia rutilla* Villeneuve on *Pinus sylvestris* L. is unique in its detail and shows, as expected, a range of vulnerabilities depending upon host or potential host taxon, morphology, feeding site, feeding method, daytime hiding place, ability to remove or encapsulate parasitoid eggs and larvae, and so on. Coexisting caterpillars on these pine trees "partition" their vulnerability to *E. rutilla* in a way that is at least consistent with the hypothesis that they compete for enemy-free space.

Even more valuable would be comparable data for similar species which can feed on the host plant but do not do so normally; and if possible data on rates of increase of the insects. The task is daunting, but the potential rewards are very high.

E. Do Parasitoids Limit the Numbers of Coexisting Species in Phytophage Communities?

We have been working towards an answer to this final question. First we need one more piece of background information. Plants introduced into new geographic areas generally recruit phytophagous insects from the native fauna rather quickly, but the process eventually slows down. Long-established and native plants only occasionally acquire new species of phytophagous insects (Southwood and Kennedy, 1983); faunal recruitment tends to be asymptotic with time (Strong *et al.*, 1984). The question why is another matter. There are two possibilities: the "pool exhaustion hypothesis" and the "niche saturation hypothesis" (Lawton and Strong, 1981; Southwood and Kennedy, 1983; Strong *et al.*, 1984).

Under the pool exhaustion model, the supply of suitable colonists simply runs out, leaving many unfilled niches on plants. In marked contrast, the niche saturation model proposes that as successful colonists pre-empt available resources, invasion of the community by new species becomes progressively more difficult. Classical interspecific competition for food (or other resources) is one possible mechanism, but as we have seen (Section III) does not seem likely for most situations involving insects on plants.

Competition for enemy-free space, in contrast, is a very likely mechanism, at least in theory. As phytophages are recruited by plants, so are enemies, including parasitoids (Moran and Southwood, 1982). Progressively there builds up a web of enemy-mediated interactions that can, in theory at least, make it harder and harder for new species to invade; or if they do successfully invade they may displace existing species (Holt, 1977). The result could be asymptotic recruitment curves for insects on plants, and limits to the number of coexisting species.

VI. CONCLUSIONS

Data assembled in this paper suggest that many of the niche differences observed in coexisting assemblages of phytophagous insects may be a response, at least in part, to attack by parasitoids. Certainly, species with different spatial and temporal niches are differentially susceptible to attack by the local parasitoid complex, often conspicuously so. In marked contrast, good data unequivocally demonstrating that polyphagous parasitoids set limits to the *number* of coexisting species are much rarer and less satisfactory in many ways. Examples exist of species exclusion by polyphagous parasitoids, but they are few. Despite quite powerful theories that lead us to believe such effects should exist, and may be common, good data are frustratingly difficult to find. Either the models are misleading or we have not looked properly.

Parasitoids in any case are unlikely to have general effects constraining diversity across all members of the phytophage community. Rather, competition for parasitoid-free space is to be expected primarily within major guilds (leaf-miners, gall formers, external chewing folivores, and suckers) and even there, within taxonomically and more ecologically constrained subsets. From data on parasitoid specificity gathered in this chapter, we cannot really expect more. For this reason, it is my guess that polyphagous predators will generally be more important than parasitoids as agents of indirect competition within *entire* assemblages of phytophagous insects. But this is only a guess.

Paradoxically, polyphagous parasitoids attacking clearly defined, small guilds of phytophages may yet prove easier to work with in the field than polyphagous predators, and may ultimately teach us more about competition for enemy-free space.

Acknowledgements

I am grateful to Bob Holt for valuable discussions, and to Simon Fowler and Phil Heads for comments on the manuscript.

REFERENCES

Alghali, A. M. (1984). The selection of pupation sites by the stalk-eyed fly *Diopsis thoracica* (Diptera: Diopsidae) and pupal parasitism in some rice cultivars. *Annals of Applied Biology* **105,** 189–194.

Andrewartha, H. G., and Birch, L. C. (1954). "The Distribution and Abundance of Animals." University of Chicago Press, Chicago.

Arthur, A. P. (1981). Host acceptance by parasitoids. *In* "Semiochemicals: Their Role in Pest Control" (D. A. Norlund, R. L. Jones and W. J. Lewis, eds.), pp. 97–120. John Wiley, New York.

Askew, R. R. (1961). On the biology of the inhabitants of oak galls of Cynipidae (Hymenoptera) in Britain. *Transactions of the Society for British Entomology* **14,** 237–268.

Askew, R. R. (1971). "Parasitic Insects." Heinemann, London.

Askew, R. R. (1975). The organisation of chalcid-dominated parasitoid communities centred upon endophytic hosts. *In* "Evolutionary Strategies of Parasitic Insects and Mites" (P. W. Price, ed.), pp. 130–153. Plenum, New York.

Askew, R. R. (1980). The diversity of insect communities in leaf mines and plant galls. *Journal of Animal Ecology* **49,** 817–829.

Askew, R. R. and Shaw, M. R. (1974). An account of the Chalcidoidea (Hymenoptera) parasitising leaf-mining insects of deciduous trees in Britain. *Biological Journal of the Linnean Society* **6,** 289–335.

Atsatt, P. R. (1981). Lycaenid butterflies and ants: selection for enemy-free space. *American Naturalist* **118,** 638–654.

Booij, C. J. H. (1982). Biosystematics of the *Muellerianella* complex (Homoptera, Delphacidae): host-plants, habitats and phenology. *Ecological Entomology* **7,** 9–18.

Brower, L. P. (1958). Bird predation and food plant similarity in closely related procyptic insects. *American Naturalist* **92,** 183–187.

Bryan, G. (1983). Seasonal biological variation in some leaf-miner parasites in the genus *Achrysocharoides* (Hymenoptera, Eulophidae). *Ecological Entomology* **8,** 259–270.

Carton, Y., and Kitano, H. (1981). Evolutionary relationships to parasitism by seven species of the *Drosophila melanogaster* subgroup. *Biological Journal of the Linnean Society* **16,** 227–241.

Claridge, M. F., and Reynolds, W. J. (1972). Host plant specificity, oviposition behaviour and egg parasitism in some woodland leafhoppers of the genus *Oncopsis* (Hemiptera Homoptera: Cicadellidae). *Transactions of the Royal Entomological Society of London* **124,** 149–166.

Clausen, C. P. (1956). Biological control of insect pests in the continental United States. *US Department of Agriculture Technical Bulletin* **1139,** 151.

Collins, M., Crawley, M. J., and McGavin, G. C. (1983). Survivorship of the sexual and agamic generations of *Andricus quercuscalicis* on *Quercus cerris* and *Q. robur*. *Ecological Entomology* **8,** 133–138.

Connell, J. H. (1980). Diversity and the coevolution of competitors, or the ghost of competition past. *Oikos* **35,** 131–138.

Cornell, H. V. (1983). The secondary chemistry and complex morphology of galls formed by Cynipidae (Hymenoptera): why and how? *American Midland Nauralist* **110,** 225–234.

Cushman, R. A. (1926). Location of individual hosts versus systematic relation of host species as a determining factor in parasite attack. *Proceedings of the Entomological Society of Washington* **28,** 5–6.

DeBach, P. (1964). "Biological Control of Insect Pests and Weeds." Chapman and Hall, London.

Dempster, J. P. (1984). The natural enemies of butterflies. *In* "The Biology of Butterflies" (R. I. Vane-Wright and P. R. Ackery, eds.), pp. 97–104. Academic Press, London.

Doolan, J. M., and MacNally, R. C. (1981). Spatial dynamics and breeding ecology in the cicada *Cystosoma saundersii*: the interaction between distributions of resources and intraspecific behaviour. *Journal of Animal Ecology* **50,** 925–940.

Doutt, R. L. (1959). The biology of parasitic Hymenoptera. *Annual Review of Entomology* **4,** 161–182.

Elton, C. S. (1927). "Animal Ecology." Sidgwick and Jackson, London.

Freeland, W. J. (1983). Parasites and the coexistence of animal host species. *American Naturalist* **121,** 223–236.

Gardner, S. M., and Dixon, A. F. G. (1985). Plant structure and foraging success of *Aphidius rhopalosiphi* (Hymenoptera: Aphidiidae). *Ecological Entomology* **10,** 171–179.

Gibson, D. O., and Mani, G. S. (1984). An experimental investigation of the effects of selective predation by birds and parasitoid attack on the butterfly *Danaus chrysippus* (L.). *Proceedings of the Royal Society of London* **221,** 31–51.

Gilbert, L. E. (1984). The biology of butterfly communities. *In* "The Biology of Butterflies" (R. I. Vane-Wright and P. R. Ackery, eds.), pp. 41–54. Academic Press, London.

Gilbert, L. E., and Singer, M. C. (1975). Butterfly ecology. *Annual Review of Ecology and Systematics* **6,** 365–397.

Goeden, R. D., and Louda, S. M. (1976). Biotic interference with insects imported for weed control. *Annual Review of Entomology* **21,** 325–342.

Gould, S. J., and Vrba, E. S. (1982). Exaptation—a missing term in the science of form. *Paleobiology* **8,** 4–15.

Griffiths, G. C. D. (1964). The Alysiinae (Hym. Braconidae) parasites of the Agromyzidae (Diptera) I. General questions of taxonomy, biology and evolution. *Beitrage zur Entomologie* **14,** 823–914.

Griffiths, G. C. D. (1967). The Alysiinae (Hym. Braconidae) parasites of the Agromyzidae (Diptera) IV. The parasites of *Hexomya* Enderlein, *Melanagromyza* Hendel, *Ophiomyia* Braschnikov and *Napomyza* Westwood. *Beitrage zur Entomologie* **17,** 653–696.

Gross, S. W., and Fritz, R. S. (1982). Differential stratification, movement and parasitism of sexes of the bagworm, *Thyridopteryx ephemeraeformis* on redcedar. *Ecological Entomology* **7,** 149–154.

Hassell, M. P. (1978). "The Dynamics of Arthropod Predator-Prey Systems." Princeton University Press, Princeton, New Jersey.

Hawkins, B. A., and Goeden, R. D. (1984). Organization of a parasitoid community associated with a complex of galls on *Atriplex* spp. in southern California. *Ecological Entomology* **9,** 271–292.

Hebert, P. D. N., Ward, P. S., and Harmsen, R. (1974). Diffuse competition in Lepidoptera. *Nature* **252,** 389–391.

Herrebout, W. M. (1969). Some aspects of host selection in *Eucarcelia rutilla* Vill. (Diptera: Tachinidae). *Netherlands Journal of Zoology* **19,** 1–104.

Holliday, N. J. (1985). Maintenance of the phenology of the winter moth (Lepidoptera: Geometridae). *Biological Journal of the Linnean Society* **25,** 221–234.

Holt, R. D. (1977). Predation, apparent competition, and the structure of prey communities. *Theoretical Population Biology* **12,** 197–229.

Holt, R. D. (1984). Spatial heterogeneity, indirect interactions, and the coexistence of prey species. *American Naturalist* **124,** 377–406.

Howard, D. J., and Harrison, R. G. (1984). Habitat segregation in ground crickets: the role of interspecific competition and habitat selection. *Ecology* **65,** 69–76.

Hutchinson, G. E. (1978). "An Introduction to Population Ecology." Yale University Press, New Haven.

Jeffires, M. J., and Lawton, J. H. (1984). Enemy free space and the structure of ecological communities. *Biological Journal of the Linnean Society* **23,** 269–286.

Jervis, M. A. (1980). Ecological studies on the parasite complex associated with typhlocybine leafhoppers. (Homoptera, Cicadellidae). *Ecological Entomology* **5,** 123–136.

Klomp, H. (1961). The concepts "similar ecology" and "competition" in animal ecology. *Archives Neerlandaises de Zoologie* **14,** 90–102.

Lawton, J. H. (1978). Host-plant influences on insect diversity: the effects of space and time. *In* "Diversity of Insect Faunas" (L. A. Mound and N. Waloff, eds.), pp. 105–125. Blackwell Scientific Publications, Oxford.

Lawton, J. H. (1984a). Non-competitive populations, non-convergent communities and vacant niches: the herbivores of bracken. *In* "Ecological Communities: Conceptual Issues and the Evidence" (D. R. Strong, Jr., D. Simberloff, L. G. Abele and A. B. Thistle, eds.), pp. 67–100. Princeton University Press, Princeton, New Jersey.

Lawton, J. H. (1984b). Herbivore community organisation: general models and specific tests with phytophagous insects. *In* "A New Ecology. Novel Approaches to Interactive Systems" (P. W. Price, C. N. Slobodchikoff and W. S. Gaud, eds.), pp. 329–352. John Wiley, New York.

Lawton, J. H., and Hassell, M. P. (1984). Interspecific competition in insects. *In* "Ecological Entomology" (C. B. Huffaker and R. L. Rabb, eds.), pp. 451–495. John Wiley, New York.

Lawton, J. H., and Strong, D. R. Jr. (1981). Community patterns and competition in folivorous insects. *American Naturalist* **118,** 317–338.

Levin, S. A. (1983). Some approaches to the modelling of coevolutionary interactions. *In* "Coevolution" (M. H. Nitecki, ed.), pp. 21–65. University of Chicago Press, Chicago.

Levin, S. A., and Segel, L. A. (1982). Models of the influence of predation on aspect diversity in prey populations. *Journal of Mathematical Biology* **14,** 253–284.

Maguire, B. Jr. (1973). Niche response structure and the analytical potentials of its relationship to the habitat. *American Naturalist* **107,** 213–246.

Maynard Smith, J. (1976). A comment on the Red Queen. *American Naturalist* **110,** 325–330.

Messenger, P. S. (1975). Parasites, predators and population dynamics. *In* "Insects, Science and Society" (D. Pimentel, ed.), pp. 201–223. Academic Press, New York.

Moran, V. C., and Southwood, T. R. E. (1982). The guild composition of arthropod communities in trees. *Journal of Animal Ecology* **51,** 289–306.

Münster-Swendsen, M. (1980). The distribution in time and space of parasitism in *Epinotia tedella* (Cl.) (Lepidoptera: Tortricidae). *Ecological Entomology* **5,** 373–383.

Murdoch, W. W., Chesson, J., and Chesson, P. L. (1985). Biological control in theory and practice. *American Naturalist* **125,** 344–366.

Myers, J. H. (1981). Interactions between western tent caterpillars and wild rose: a test of some general plant herbivore hypotheses. *Journal of Animal Ecology* **50,** 11–25.

Orians, G. H. and Solbrig, O. T. (1977). "Convergent Evolution in Warm Deserts." Academic Press, London.

Otte, D., and Joern, A. (1977). On feeding patterns in desert grasshoppers and evolution of specialised diets. *Proceedings of the Academy of Natural Sciences of Philadelphia* **128,** 89–126.

Owen, J., Townes, H., and Townes, M. (1981). Species diversity of Ichneumonidae and Serphidae (Hymenoptera) in an English suburban garden. *Biological Journal of the Linnean Society* **16,** 315–336.

Pierce, N. E., and Elgar, M. A. (1985). The influence of ants on host plant selection by *Jalmenus evagoras*, a myrmecophilous lycaenid butterfly. *Behavioural Ecology and Sociobiology* **16,** 209–222.

Pierce, N. E., and Mead, P. S. (1981). Parasitoids as selective agents in the symbiosis between lycaenid butterfly larvae and ants. *Science* **211,** 1185–1187.

Porter, C. C. (1977). Ecology, zoogeography and taxonomy of the lower Rio Grande valley mesostenines (Hymenoptera, Ichneumonidae). *Psyche* **84,** 28–91.

Price, P. W. (1981). Semiochemicals in evolutionary time. *In* "Semiochemicals: Their Role in Pest Control" (D. A. Nordlund, R. L. Jones and W. J. Lewis, eds.), pp. 251–279. John Wiley, New York.

Price, P. W. (1984). "Insect Ecology" (2nd edn.). John Wiley, New York.

Price, P. W., Bouton, C. E., Gross, P., McPherson, B. A., Thompson, J. N., and Weiss, A. E. (1980). Interactions among three trophic levels: influence of plants on interactions between insect herbivores and natural enemies. *Annual Review of Ecology and Systematics* **11,** 41–65.

Pschorn-Walcher, H. (1977). Biological control of forest insects. *Annual Review of Entomology* **22,** 1–22.

Putter, B., Long, S. H., and Peters, E. J. (1978). Establishment in Missouri of *Rhinocyllus conicus* for the biological control of musk thistle (*Carduus nutans*). *Weed Science* **26,** 188–190.

Richards, O. W. (1940). The biology of the small white butterfly (*Pieris rapae*), with special reference to the factors controlling its abundance. *Journal of Animal Ecology* **9,** 243–288.

Ricklefs, R. E., and O'Rourke, K. (1975). Aspect diversity in moths: a temperate-tropical comparison. *Evolution* **29,** 313–324.

Rohde, K. (1979). A critical evaluation of intrinsic and extrinsic factors responsible for niche restriction in parasites. *American Naturalist* **114,** 648–671.

Root, R. B. (1973). Organization of a plant-arthropod association in simple and diverse habitats: the fauna of collards (*Brassica oleracea*). *Ecological Monographs* **43,** 95–124.

Samways, M. J. (1979). Immigration, population growth and mortality of insects and mites on cassava in Brazil. *Bulletin of Entomological Research* **69,** 491–505.

Schultz, J. C., Otte, D., and Enders, F. (1977). *Larrea* as a habitat component for desert arthropods. *In* "Creosote Bush: Biology and Chemistry of *Larrea* in New World Deserts" (T. J. Mabry, J. H. Hunzicker and D. R. Difeo, eds.), pp. 176–208. Dowden, Hutchinson and Ross, Stroudsburg, Pennsylvania.

Seitz, A., and Komma, M. (1984). Genetic polymorphism and its ecological back-ground in tephritid populations (Diptera: Tephritidae). *In* "Population Biology and Evolution" (K. Wohrmann and V. Loeschcke, eds.), pp. 143–158. Springer-Verlag, Berlin.

Shaw, M. R. (1977). On the distribution of some satyrid (Lep.) larvae at a coastal site in relation to their ichneumonid (Hym.) parasite. *Entomologists Gazette* **28,** 133–134.

Shaw, M. R., and Askew, R. R. (1976). Ichneumonoidea (Hymenoptera) parasitic upon leaf-mining insects of the orders Lepidoptera, Hymenoptera and Coleoptera. *Ecological Entomology* **1,** 127–133.

Sih, A. (1985). Evolution, predator avoidance, and unsuccessful predation. *American Naturalist* **125,** 153–157.

Southwood, T. R. E., and Kennedy, C. E. J. (1983). Trees as islands. *Oikos* **41,** 359–371.

Starý, P. (1970). "Biology of Aphid Parasites (Hymenoptera: Aphidiidae) with Respect to Integrated Control." Junk, The Hague.

Starý, P., and Rejmánek, M. (1981). Number of parasitoids per host in different systematic groups of aphids: the implications for introduction strategy in biological control (Homoptera: Aphidoidea; Hymenoptera: Aphidiidae). *Entomologica Scandinavica* **15** (suppl.), 341–351.

Strong, D. R., Jr. (1977). Insect species richness: hispine beetles of *Heliconia latispatha*. *Ecology* **58,** 573–582.

Strong, D. R., Lawton, J. H., and Southwood, T. R. E. (1984). "Insects on Plants. Community Patterns and Mechanisms." Blackwell Scientific Publications, Oxford.

Surles, W. W. (1974). Native hymenopteran parasitoids attacking an introduced weevil, *Rhinocyllus conicus* in Virginia. *Environmental Entomology* **3,** 1027–1028.

van Valen, L. (1973). A new evolutionary law. *Evolutionary Theory* **1,** 1–30.

Vinson, S. B. (1976). Host selection by insect parasitoids. *Annual Review of Entomology* **21,** 109–133.

Vinson, S. B. (1981). Habitat location. *In* "Semiochemicals: Their Role in Pest Control" (D. A. Nordlund, R. L. Jones and W. J. Lewis, eds.), pp. 51–77. John Wiley, New York.

Vinson, S. B. (1984a). How parasitoids locate their hosts: a case of insect espionage. *In* "Insect Communication" (T. Lewis, ed.), pp. 325–348. Academic Press, London.

Vinson, S. B. (1984b). Parasitoid-host relationship. *In* "Chemical Ecology of Insects" (W. J. Bell and R. T. Carde, eds.), pp. 205–233. Chapman and Hall, London.

Waage, J. K. (1979). The evolution of insect/vertebrate associations. *Biological Journal of the Linnean Society* **12,** 187–224.

Watt, A. S. (1947). Pattern and process in the plant community. *Journal of Ecology* **35,** 1–22.

Weis, A. E. (1982). Resource utilization patterns in a community of gall-attacking parasitoids. *Environmental Entomology* **11,** 809–815.

Weis, A. E., and Abrahamson, W. G. (1985). Potential selective pressures by parasites on the evolution of a plant-herbivore interaction. *Ecology* **66,** 1261–1269.

Weis, A. E., Abrahamson, W. G., and McCrea, K. D. (1985). Host gall size and oviposition success by the parasitoid *Eurytoma gigantea*. *Ecological Entomology* **10,** 341–348.

West, C. (1985). Factors underlying the late seasonal appearance of the lepidopterous leaf-mining guild on oak. *Ecological Entomology* **10,** 111–120.

Whittaker, R. H., Levin, S. A., and Root, R. B. (1973). Niche, habitat and ecotype. *American Naturalist* **107,** 321–338.

Wilson, F. (1965). Biological control and the genetics of colonizing species. *In* "The Genetics of Colonizing Species" (H. G. Baker and G. L. Stebbins, eds.), pp. 307–325. Academic Press, New York.

Zwölfer, H. (1971). The structure and effect of parasite complexes attacking phytophagous host insects. *In* "Proceedings of the Advanced Study Institute on 'Dynamics of Numbers in Populations' " (P. J. den Boer and G. R. Gradwell, eds.), pp. 405–416. Oosterbeek, The Netherlands.

Zwölfer, H. (1975). Artbildung und ökologische Differenzierung bei phytophagen Insekten. *Verhandlungsbericht der Deutschen Zoologischen Gesellschaft* **67,** 394–401.

Zwölfer, H. (1979). Strategies and counterstrategies in insect population systems competing for space and food in flower heads and plant galls. *Fortschritte der Zoologie* **25,** 331–353.

Zwölfer, H. (1983). Life systems and strategies of resource exploitation in tephritids. *In* CEC/IOBC Symposium, Athens, pp. 16–30.

Zwölfer, H., and Kraus, M. (1957). Biocoenotic studies on the parasites of two fir- and two oak-tortricids. *Entomophaga* **2,** 173–196.

10

Parasitoids in Classical Biological Control

D. J. GREATHEAD

I. INTRODUCTION

The introduction and permanent establishment of exotic species for the long-term suppression of pests is conveniently referred to as "classical biological control" to distinguish it from other applications of biotic agents in pest control which either involve periodic releases of native or exotic agents or are concerned with manipulations intended to enhance the impact of natural enemies already present in the crop environment.

The underlying assumption of classical biological control is that the populations of organisms are regulated by natural enemies and that pest species have escaped regulation because they have been introduced from a distance without their regulating natural enemies, or the environment has changed enabling them to escape regulation. In the first instance control is achieved by introducing the missing natural enemies where possible from the area of origin of the pest, and in the second by seeking natural enemies capable of effective action in the changed environment from elsewhere in the range of the pest or of related species. A corollary is that introduced natural enemies should be more effective in a new location if introduced free of their own natural enemies.

The possibility of introducing natural enemies to control pests had been considered, and introductions, some successful, were made before 1888 (DeBach, 1964; Hagen and Franz, 1973) but the very successful and well publicized introduction in that year of the predatory vedalia bettle, *Rodolia cardinalis* (Mulsant), from Australia into California for control of the cottony cushion scale, *Icerya purchasi* Maskell, (Doutt, 1958) inspired entomologists round the world to make introductions. Thus, the year 1888 is usually taken as the start of serious attempts at biological control. However, the first successful introduction of a parasitoid did not take place until 1906 when Italy imported *Encarsia berlesei* (Howard) from the USA (Berlese, 1915) for the control of the mulberry scale, *Pseudaulacaspis pentagona* Targioni-Tozzetti. The CAB International Institute of Biological Control (CIBC) has collected records of 860 successful establishments of 393 species of parasitoids against some 274 pest insects in 99 countries. Of these, on 216 occasions parasitoids alone or aided by introduced predators have been rated as achieving complete or satisfactory pest suppression and 52 others rated as having achieved a useful reduction in pest numbers. Failures are more difficult to document and analyse but the records suggest that overall at least 570 parasitoid species have been released in classical biological control attempts on 2110 occasions. Comparable figures for insect predators are 302 successful establishments achieving complete or satisfactory control alone or with parasitoids on 86 occasions, and a useful reduction on a further 12 occasions. Thus parasitoids have been established about three times as often as predators, and have achieved satisfactory control in the same proportion. This achievement represents an important contribution to pest control because the pest suppression obtained continues indefinitely with no further effort except where injudicious use of pesticides has upset the balance. However, a sound theoretical basis remains to be found to account for the results, and consequently biological control practitioners have developed a series of more or less controversial generalizations (listed by Munroe, 1971) as a basis for practice.

In this chapter I shall examine this achievement and some of the generalizations in more detail and attempt to draw some conclusions which may be of practical value to those contemplating using classical biological control.

II. THE DATA SET

In examining the records, it is important to appreciate that there is a great variation in the effort employed and in the degree to which pre-release studies were made to select the most promising candidates for introduction. At first a period of euphoria followed news of the success of the vedalia beetle in California. This led to the indiscriminate introduction of natural enemies, mostly predators, for control of aphids and scale insects, but with little success (Lounsbury, 1940). There followed a period when intensive pre-introduction studies were carried out, coupled with efforts to provide a theoretical framework for biological control (e.g. Howard and Fiske, 1911; and W. R. Thompson, see bibliography in Thorpe, 1973), which led to a number of important successes being achieved during the 1920s and 1930s. However, in spite of the much more scientific approach to biological control which was developed, many introductions are still made with inadequate preparation because money is short or natural enemy cultures happen to be available.

Equally, documentation of results varies from detailed book-length accounts of entire programmes (e.g. Tothill *et al.*, 1930; Taylor, 1937) to the barest mention in departmental reports—and in many instances no records are kept at all. Again, shortage of funding and expediency are reflected in these failings. Efforts have been made to remedy this lack of information by the publication of detailed regional reviews by the CIBC (Wilson, 1960; McLeod *et al.*, 1962; Canada, 1971; Greathead, 1971, 1976; Rao *et al.*, 1971; Kelleher and Hulme, 1984; Cock, 1985) and by the United States Department of Agriculture (USDA) in compiling a comprehensive world review (Clausen, 1978). However, CIBC coverage is still incomplete and the earlier volumes are out of date while the USDA volume includes records only to 1968 and for many countries the review is based on published documents only.

Assessment of results is often casual and tends to be over optimistic. These failings result from pressures on time, money and the desire to justify the effort which has been expended. A related problem is that success can be so spectacular that time spent on assessment appears to be unnecessary. More fundamental is the difficulty of applying objective techniques to the assessment of biological control. Controlled experiments with treated and untreated blocks are not possible because natural enemies spread rapidly. Practitioners have developed and employed other techniques but none is of universal value and none is free of criticism. Two direct methods have been used:

1. DeBach (1946) was the first to employ an insecticide check method, in which chemicals are used to eliminate natural enemies and the consequent pest resurgence is measured. This can be used only when the pest is resistant or protected from contact with the chemical and remains in place. Thus, in practice the method can be used realistically only for scale insects and other pests which are virtually sessile, and can be criticized because side-effects (e.g. phytotoxicity, tonic effects, elimination of non-target organisms) may distort the results
2. Exclusion of natural enemies using cages, first used by Smith and DeBach (1942), is also practised but again there may be side-effects (e.g. altered microclimate) and immigration and emigration are prevented. This can be overcome if natural enemies are systematically removed by hand (Fleschner *et al.*, 1955) rather than by caging.

More frequently, when these methods are not feasible or the resources for using them are not available the results may be estimated indirectly.

Changes in damage indices [e.g. sugarcane joints bored for measuring control of stemborers, *Diatraea saccharalis* (F.), in Barbados (Alam *et al.*, 1971)] or crop yield [e.g. to estimate effect of coconut scale, *Aspidiotus destructor* Signoret, on copra production in São Tome (Simmonds, 1960)] can be used. More convincingly, direct measurement of population changes [e.g. winter moth in Canada (Embree, 1966)] can be measured.

However, there is a danger that other contemporary changes may have taken place. For example, the decline of the sugarcane whitegrub in Mauritius was influenced by the results of plant breeding and changes in agronomic practice which cannot be isolated from the impact of introduced scoliid wasps (Greathead, 1971).

The difficulties are exemplified in the current efforts to measure the effect of *Epidinocarsis lopezi* (De Santis) on the cassava mealybug, *Phenacoccus manihoti* Matile-Ferrero, in Nigeria where a comparison of treated and untreated blocks 2 km apart was possible only for a year because of the rapid spread of the agent. Exclusion experiments by sleeving shoots are now being used to quantify the impact of *E. lopezi* (IITA, 1984, 1985).

Thus, attempts to analyse biological control programmes and their outcome are fraught with difficulty. Nonetheless, brave attempts have been made to assess the results and to generalize so as to extract guidelines for future effort. Important studies have been made by DeBach (1964) based on the first world survey and re-examined by Southwood (1977); by Greathead (1971) for Africa; by Turnbull and Chant (1961), Munroe (1971) and Beirne (1975) for Canada. More recently the USDA world review (Clausen, 1978) has been analysed by Hall, Ehler and co-workers in a series of papers (Hall and Ehler, 1979; Hall *et al.*, 1980; Ehler and Hall, 1982) and by Noyes (1985).

The CIBC databank includes the records in the publications listed above and subsequent published records. So far, only records of species that became established have been input, but checking of nomenclature and reconciliation of conflicting reports is not complete. However, some preliminary results are included in the following discussion to complement and extend published analyses.

There is an old controversy (see DeBach, 1964) as to the superiority of predators or parasitoids as biological control agents which must be examined first. Hall and Ehler (1979) found that there is no difference in the rate of establishment of parasitoids and predators or in the success rate, provided that the successful introductions of *Rodolia cardinalis* which bias results in favour of predators are excluded—overall 34% of attempts result in establishment; of these 16% achieved complete success and 58% a lesser benefit. However, in terms of numbers parasitoids have been established more than twice as often as predators and have been effective in the same ratio (Table I). Thus, in the absence of contrary evidence, the combined data for parasitoids and predators on which most assessments have been based can be used in subsequent discussion.

TABLE I. The contributions of predators and parasitoids to classical biological control

	Proportion		
	Established (Hall and Ehler, 1979)	Successful (Hall *et al.*, 1980)	
		Complete control	All results
Predators	0.34	0.26*	0.55*
Parasitoids	0.34	0.14	0.60

*0.06 and 0.43 excluding repeated use of *Rodolia cardinalis* (Mulsant), then all comparisons are not significant.

Note that successful natural enemies in both categories are effectively host-specific or have a circumscribed host range and that truly polyphagous species are seldom selected or effective.

III. SOME INFERENCES AND THEIR BEARING ON THE GENERALIZATIONS EMPLOYED BY PRACTITIONERS OF BIOLOGICAL CONTROL

A. Pest Systematics

Hall and Ehler (1979) showed that there is a substantially higher establishment rate for Homoptera and other Exopterygota over the main orders of Endopterygota, and Hall *et al.* (1980) that the success rates are also higher even when *Rodolia cardinalis* is excluded (Table II).

The number of Exopterygota, other than Homoptera, is too small to draw conclusions. The successes among Homoptera are dominated by attempts against Aleyrodidae and above all Coccoidea. DeBach (1964) and Munroe (1971) note that successes against Coccoidea comprise some 40% of the total and that the disproportion is increased when the relative species richness of the main orders of insects is considered (Table II): however, a more valid comparison would be between the ranking in terms of numbers of pests in each order. The reasons put forward for the disproportion include:

TABLE II. Results of classical biological control introductions in relation to the orders of insects

Order	Proportion			Rank**
	Established (Hall and Ehler, 1979)	Successful (Hall *et al.*, 1980)		
		Complete	All	
Misc Endopterygota	0.48	0.05	0.37	5
Homoptera	0.43	0.30*	0.80	5
Diptera	0.37	0.00	0.31	3
Hymenoptera	0.34	0.00	0.56	2
Lepidoptera	0.27	0.06	0.48	4
Coleoptera	0.23	0.04	0.36	1

*0.23 excluding repeated use of *Rodolia cardinalis* (Mulsant); **rank in order of species-richness, adapted from Munroe (1971).

1. Ease of transport on planting material and produce, therefore Coccoidea are among the most frequently accidentally introduced pests
2. Frequency as pests on valuable cash crops, e.g. citrus
3. Difficulty of applying chemical control
4. The historical impetus of early successes, notably against *Icerya purchasi*
5. Inherent biological attributes of the pests, e.g. sessile feeding stages.

Doubtless all these factors have been influential; certainly Sternorhyncha, and more particularly Coccoidea, comprise a disproportionately large number of introduced major pests of important cash crops (Simmonds and Greathead, 1977).

B. Islands *vs* Continents

Longstanding controversy, based on the preponderance of early successes on islands (e.g. Fiji, Hawaii, Mauritius), concerns the merits of islands, or by extension, ecological islands (e.g. isolated valleys, mountains) as sites for successful biological control (e.g. Taylor, 1955). Hall and Ehler (1979) found that the rate of establishment is higher on islands, but Hall *et al.* (1980) showed that there is no significant difference in the success rates (Table III). This confirms the conclusion of Greathead (1971) for Africa and its islands, based on the island biogeographic theory of Mac Arthur and Wilson (1967) that islands have a lower species/area ratio, i.e. species are less densely packed and have broader niches, therefore there is less resistance to the establishment of additional species. However, the equilibrium species density is less on islands so that the establishment of additional species is more likely to lead to the

TABLE III. Comparison of results of classical biological control on islands and continents

	Proportion			
		Successful		
	Established (Hall and Ehler, 1979)	World (Hall *et al.*, 1980)		Africa (Greathead, 1971)
		Complete	All	All
Islands	0.40	0.14	0.60	0.31
Continents	0.30	0.17	0.56	0.19
	($P<0.0005$)	($P=0.40$)	($P=0.40$)	

extinction of less competitive ones. On the other hand, he showed that the species density on ecological islands increases little with area, presumably because surrounding land is less hostile than surrounding sea, allowing a greater influx of species into ecological islands, enabling them to maintain a higher species density than true islands. Therefore, there should be less difference in ease of establishment of species between ecological islands and large uniform continental areas than between the islands and continents. Besides, since pests also have a greater chance of colonizing islands, introduced species form a higher proportion of the pest fauna on islands (Sailer, 1978) and therefore islands present more opportunities for classical biological control.

C. Temperate *vs* Tropical

It has often been asserted (e.g. Taylor, 1955) that biological control is more suited to the tropics than to temperate regions, but Munroe (1971) found no compelling evidence in support of this hypothesis. The notion seems to be a peculiarly European view since there are abundant examples of successful biological control in temperate North America and Australasia. An explanation is that Europe has colonized other temperate regions and has been a source of pests which are obvious targets for classical biological control and has received few exotic major pests in return. Indeed, many of those it has received have been targets for classical biological control (Greathead, 1976).

D. Direct *vs* Indirect Pests

Lloyd (1960) contended that indirect pests (those that do not damage the product, e.g. defoliators on fruit trees) are more amenable to biological control than direct pests (those that damage the product, e.g. fruit) because the former can be tolerated at a higher density than the latter. This conclusion is pertinent to analyses of results in that a level of suppression by natural enemies that is rated complete control for an indirect pest might well be rated a failure or partial success for a direct pest. Thus, in analyses, success cannot be equated with the degree of regulation or the mean population reduction achieved by a particular agent. Rather it must mean that the control agents brought the level of damage to below the economic threshold.

E. New *vs* Old Associations

Pimentel (1963) proposed that homeostasis develops between natural enemies and their hosts leading to loss of "virulence" so that natural enemies of related species may be expected to be more effective control agents than those with a long association with their hosts. This argument was developed by Hokkanen and Pimentel (1984) who attempted to demonstrate that new associations are 75% more likely to be successful than old associations. However, the case is poorly presented. Their list of "new associations" embraces phytophages, birds, diseases and snails which have diverse relations with their hosts, and many examples are dubious since the rating of results and the presumed origins of both hosts and natural enemies can be disputed. The data set is further confused as there is no distinction between examples where several natural enemies combined to achieve the result, sometimes involving both new and old associations. Thus, on my assessment, their list includes only seven undisputed complete successes (three of which are against native pests) against insects by parasitoids and three by predators as a result of establishing "new associations". New associations have undoubtedly provided successful controls, but is there any evidence that new associations are inherently better? On the present evidence the case can only be considered "not proven".

F. Native *vs* Exotic Pests

Most effort has been directed against introduced pests, which are the obvious target for the classical approach, but some notable successes have been achieved against native pests. Hall and Ehler (1979) and Hall *et al.* (1980) have shown that both establishment and success rates are higher for introduced pests (Table IV). These results run counter to the hypothesis discussed above, since introductions against introduced pests have been

TABLE IV. Comparison of the results of classical biological control against native and exotic pests

	Proportion		
	Established (Hall and Ehler, 1979)	Successful (Hall *et al.*, 1980)	
		Complete	All
Native	0.25	0.06	0.29
Exotic	0.34	0.17	0.60
	$(0.05 < P < 0.025)$	$(P = 0.09)$	$(P < 0.01)$

chiefly of natural enemies from the area of origin of the pest, while introductions against native pests will usually be species that have evolved on related hosts (Carl, 1982).

G. Size and Number of Releases

Beirne (1975) analysed the effect of the numbers of a natural enemy released on the rate of establishment and found a striking increase with numbers, a trend confirmed by Ehler and Hall (1982) who also found that the success rate rises in the same way (Table V). These comparisons relate to total numbers of each species released and do not distinguish between the various combinations of repeated releases in time and space which will affect the chance of establishment. Nor is the degree of effort employed in making the release considered. Intuitively it seems likely that usually more effort will be put into choice of site and ensuring the best possible conditions for establishment for a planned release campaign involving several sites and several releases over a season or successive seasons, than for single small releases often of, for example, material which happened to be readily available—the survivors of an unsatisfactory shipment or species which were just not available in large numbers and were not considered to have much chance of success.

There are exceptions to the "the more the better" rule as noted by

TABLE V. Number of natural enemies released in relation to establishment and result in classical biological control

(a) Ehler and Hall (1982)

	Rate of establishment			
No. released	1–999	10^3–9999	10^4–10^5	$>10^5$
Result				
Failure to control	0.18	0.17	0.27	0.31
Successful control	0.33	0.39	0.56	0.62
All	0.23	0.26	0.38	0.41

(b) Beirne (1975)

No. released	No. of occasions	Rate of establishment
<5000	98	0.10
5000–31,200*	33	0.40
>31,200*	22	0.78

*31,200 is the median for established species.

Greathead (1971) in questioning the response to failure of simply releasing more of the same species rather than turning to alternatives. For example, in the unsuccessful campaign to control the Karoo caterpillar, *Loxostege frustralis* Zeller, in South Africa parasitoids of the beet web-worm, *L. sticticalis* (L.), were imported from the USA. *Chelonus texanus* Cresson bred readily in the laboratory on a factitious host and between 1942 and 1954 almost six million were released and only four authentic recoveries were made (Greathead, 1971). Similarly, in Réunion millions of parasitoids were released from 1965 to 1969 against *Chilo sacchariphagus* (Bojer) in sugarcane, to no avail (Etienne, 1973). These examples illustrate the point made by Beirne (1980) that the breeding and release of readily bred species tends to become an end in itself which may be driven by the desire to justify employment of staff and produce impressive annual reports. On the other hand, there are examples of species established from single releases of less than 50 individuals, for example, in Mauritius alone—*Xanthopimpla stemmator* (Thunberg) (13), *Tiphia parallela* Smith (10), *Campsomeris coelebs* Sichel (40), *C. lachesis* Saussure (34) (Greathead, 1971).

H. Competitive Exclusion

Ehler and Hall (1982) also compared rate of establishment with the number of species released simultaneously and sequentially and in each instance found a declining rate of establishment as the number increased. They also detected a higher rate of establishment in the absence of incumbent natural enemies. The implication that competitive exclusion might have taken place was criticized by Keller (1984) who pertinently pointed out that the choice of species and the sequence in which they are released is not random. Usually species are released in sequence partly because of the logistics of handling more than one at a time, and are released in what is considered to be decreasing order of likelihood of success. Further, simultaneous releases of several species result from the receipt of mixed shipments, often of small numbers of each species, and are unlikely to be as well planned.

Another possibility, discussed by Tallamy (1983), is the interesting proposition that the equilibrium model developed by Mac Arthur and Wilson (1967) may be applied to introduced pests as "islands". Thus he shows that the colonization rate for parasitoids introduced into North America against the gypsy moth, *Lymantria dispar* (L.), falls as the extinction rate rises, suggesting that it can support only 11 species of parasitoids. If this can be substantiated it would provide a more fundamental explanation for the declining establishment rate underlying the biases introduced by biological control practitioners.

I. Decreasing Success with Time

Hall and Ehler (1979) detected a decrease in the rate of establishment of natural enemies with time since 1888 and Hall *et al.* (1980) found indications that the success rate has also declined. This would seem to support Taylor's (1955) contention that the "cream has been skimmed" and that good targets are getting fewer as time goes on. However, new pest introductions of major importance continue to be reported and provide promising new targets. A more credible explanation is that the number of introductions has increased as pest control activity has increased, and transfer of natural enemies has become easier with rapid air travel. Formerly species had to survive lengthy sea journeys, often being bred en route which effectively eliminated species difficult to rear in cages, with the result that greater care was taken in the selection of species and only the most resilient (?good colonizers) survived to be released. Now it is possible to obtain numbers of species of natural enemies cheaply and make releases just to see what happens, as was happening in the campaign to control *Diatraea saccharalis* in Barbados when the successful parasite *Apanteles flavipes* Cameron was introduced (Alam *et al.*, 1971).

J. Egg Parasitoids *vs* Parasitoids of Other Stages

It has been suggested that egg parasitoids alone will be ineffective (see DeBach, 1964). However, the control of the eucalyptus weevil, *Gonipterus scutellatus* Gyllenhal, by a mymarid egg parasitoid, *Patasson nitens* (Girault), has been very successful in several countries in Africa and on nearby islands (Tooke, 1955; Greathead, 1971). It is probably significant that host plant, pest and parasitoid are all native to Australia and that eucalyptus attracts few insects in Africa so that plantations are ecological islands with an extremely depauperate fauna. Further, the weevil is an indirect pest and forestry pests can be tolerated at higher densities than can pests of high value crops.

This is an unusually straightforward example where other biotic mortality factors seem to have been absent, but there are other examples where introduction of an egg parasitoid alone has been effective. Thus *Trissolcus basalis* (Wollaston) (Scelionidae) has controlled the immigrant vegetable bug, *Nezara viridula* (L.), in Australia (Wilson, 1960) and elsewhere in the Pacific in the absence of competition from other parasitoids. However the absence of competing parasitoids is not a condition for success: *Spodoptera* spp. have been suppressed by the introduced scelionid, *Telenomus remus*

Nixon, in Barbados, where larval parasitoids were already present (Cock, 1985).

There are also examples of satisfactory control being achieved by egg parasitoids introduced in combination with parasitoids of other stages, e.g. banana leaf-roller, *Erionota thrax* (L.) by *Ooencyrtus erionotae* Ferrière (Encyrtidae) and *Apanteles erionotae* Wilkinson (Braconidae) in Mauritius and Hawaii (CAB, 1980).

On the other hand, van Hamburg and Hassell (1984), considering stem-borers, *Chilo partellus* (Swinhoe), on maize have shown that at least in theory raising egg-parasitism could actually increase the number of borers surviving to the feeding stage.

The most successful families of egg parasitoids are Scelionidae and Mymaridae. Trichogrammatidae, which have been the main preoccupation of attempts at inundative control, have been less successful (although exotic species have been released for classical biological control and others used for inundative control have had the opportunity to become established). Possible reasons include polyphagy and the distribution of progeny which results in incomplete parasitism of pest egg-masses and a poor functional response.

IV. THE APPLICATION OF ECOLOGICAL INSIGHTS TO THE SELECTION OF TARGETS

A. Choice of Pest

Biological principles and human foibles are so confounded in the results of biological control programmes that the analyses of results which have been attempted are of limited value as guidelines for future action or in testing theory. Further examples of the effect of administrative and political factors on the outcome of biological control have been discussed by Beirne (1980) who believes that they are a major obstacle to success. DeBach (1964) concluded his analysis by suggesting that success is achieved in proportion to the effort applied. I believe that the success rate can also be improved by a better and more scientific choice of targets aided by ecological theory.

Southwood (1977) applied the *r-K* selection theory to biological control and showed that the success rate seems to change with the stability of the environment as exemplified by crop type, and compared this result with his "synoptic model" which suggests that the populations of extreme *r* and *K* type animals are not limited by natural enemies but rather by their ability to over-run the food supply on the one hand and to stabilize numbers within the carrying capacity on the other. In both instances (species or crop type) those that are intermediate favour biological control. The analyses of Hall and

Ehler (1979) for rates of establishment and of Hall *et al.* (1980) for success rates support this view, as does the analysis of results in Canada by Beirne (1975) (Table VI).

B. Choice of Crop

Greathead and Waage (1983) and Greathead (1984) attempted to extend the argument to the crop environment, combining climate, duration of crop, area and level of cultivation into an index of stability. The results, which need refining, do seem to indicate that these factors do indeed influence the outcome of biological control programmes and again suggest that the success rate falls at the upper extreme as well as at the lower extreme but to a lesser extent. These conclusions support the views of Way (1977) and van Emden and Williams (1974) on managed diversity to optimize natural enemy impacts on farms.

C. Effects of Crop Improvement

Another point relevant in rating prospects is the effect of crop improvement on the pest and its natural enemies. Latière (1917) suggested that the failure of parasitoids to control olive fly, *Dacus oleae* (Gmelin), was a result of the

TABLE VI. Influence of stability of the environment in terms of crop type on the results of classical biological control introductions

	Proportion			
	Established		Successful (Hall *et al.*, 1980)	
Category	Canada (Beirne, 1975)	World (Hall and Ehler, 1979)	Complete	All
Unstable (annual and short cycle)	0.16	0.28	0.03	0.43
Intermediate (orchard and other perennial)	0.43	0.32	0.30*	0.72
Stable (forests and rangeland)	0.23	0.36	0.08	0.47
			($P<0.01$)	($P<0.01$)

*0.22 excluding repeated use of *Rodolia cardinalis* (Mulsant) ($P<0.01$).

tough skins of cultivated olives compared with their wild counterparts. Surely, breeding of unnaturally large fruits has also made parasitoids less effective by increasing the distance between the host and the plant surface and so taking the host out of reach of the ovipositor of the parasitoid. This may account for the rarity and ineffectiveness of parasitoids of such pests as the codling moth, pink bollworm and cocoa pod borer. However, this hypothesis has not been tested. Other changes such as altered growth habit, increased palatability and nutritional value of crop plants may also be important in raising the multiplication rate of pests beyond the ability of parasitoids to respond, as has been suggested in attempting to explain the recent devastating outbreaks of leafhoppers and planthoppers on rice (Kenmore, 1980).

D. Choice of Natural Enemy

So far most discussion on the rationale of biological control has centred on the pest and its environment rather than the natural enemy. Parasitoids have proved outstanding control agents, as have insect predators. In only a few instances have micro-organisms been effective "classical" agents. Anderson (1982) has shown that the dynamics of disease tend to give rise to cyclical epidemics. There are other problems—notably the need for an efficient dispersal mechanism (often a vector) and the tendency to lose virulence with time. At the other end of the evolutionary scale vertebrates have seldom been effective because they are not sufficiently host-specific. As a result of switching prey to exploit the most "profitable" source, they seldom regulate pests and often become pests themselves. Insects include many highly polyphagous species with similarly undesirable characteristics but I believe monophagous and oligophagous species, especially parasitoids, exhibit ideal characters:

1. Power of flight which enables good dispersal
2. Limited central nervous system which ensures a rigidity of behaviour that results in a long-term constancy in host selection which in turn
   a. prevents changes of host and so minimizes risk to non-target species
   b. promotes a desirable coupling of population fluctuations of host and parasitoid that stabilizes pest density
3. High fecundity and short generation time which enable a rapid temporal response
4. Comparable rates of genetic change to their hosts, enabling them to co-evolve with their hosts and avoid loss of efficiency with time.

V. PARASITOIDS USED IN CLASSICAL BIOLOGICAL CONTROL

Of course, the species which have been used in classical biological control programmes are not chosen at random from the families of insects which include parasitoids. An indication of the overall number of species by families, including those that have failed to become established, is given by Noyes (1985). Table VII lists the families of parasitoids which have contributed species that have been released and indicates the number of successful establishments. Three taxa stand out as major sources of agents—Ichneumonoidea, Chalcidoidea, Tachinidae. Within superfamilies and families the choice of species is also uneven. Some of these differences are readily explained:

1. They may not attack pest species, e.g. Conopidae, Rhipiphoridae, Gasteruptiidae, Eucharitidae and Leucospidae attack Aculeata; Heloridae attack Chrysopidae
2. They may be predominantly or exclusively hyperparasitic, e.g. Ceraphronidae, Perilampidae, Signiphoridae, *Marietta* (Aphelinidae)
3. They may contain few parasitoids, e.g. Phoridae, Muscidae (only *Acridomyia*), Staphylinidae (only *Aleochara* s.l.), Eurytomidae
4. They may contain few species which specialize on a restricted range of hosts, e.g. Pyrgotidae on Scarabaeidae, Evaniidae on cockroach eggs, Diapriidae on muscoid puparia, Dryinidae on Auchenorhyncha, Cryptochetidae on Margarodidae.

Difficulty in collecting, handling, shipping and breeding in captivity explains other gaps, e.g. Nemestrinidae (on grasshoppers) and other Diptera, and above all Strepsiptera which seem to have been introduced (unsuccessfully) on only one occasion.

Conversely some taxa of pests are poor targets. Also difficulty in collecting and rearing parasitoids has contributed. Thus, for example, there is only one case of successful biological control of a timber borer, viz. *Sirex noctilio* F. in Tasmania (Taylor, 1978).

The contrast between Aphidoidea and Coccoidea as targets for classical biological control is pertinent and is reflected in the low numbers of Aphidiinae and *Aphelinus* spp. compared with the large numbers of Encyrtidae, *Aphytis* spp. and *Coccophagus* spp. which have been used.

Allowing for these biases, some interesting observations can be made which are helpful in deciding what makes a good classical biological control agent.

TABLE VII. Families of parasitoids established in classical biological control

	Genera	Species	Pests	Occasions	Effective control
Hymenoptera					
Evanoidea					
Stephanidae	1	1	1	1	—
Ichneumonoidea					
Braconidae	23	66	59	158	53
Ichneumonidae	30	45	28	72	22
Procotrupoidea					
Proctotrupidae	—	—	—	—	—
Scelionidae	4	12	11	23	6
Platygasteridae	4	7	6	12	5
Diapriidae	—	—	—	—	—
Cynipoidea	3	4	4	7	1
Chalcidoidea					
Trichogrammatidae	2	12	12	24	—
Eulophidae	21	36	47	72	23
Mymaridae	4	7	9	15	9
Chalcididae	2	3	2	4	—
Eurytomidae	—	—	—	—	—
Torymidae	1	1	1	2	—
Pteromalidae	15	26	22+	49	17
Encyrtidae	34	61	40+	132	53
Aphelinidae	13	59	32	185	90
Eupelmidae	2	4	3	4	—
Bethyloidea					
Bethylidae	2	2	2	3	—
Dryinidae	2	2	1	2	—
Scolioidea	3	13	10	21	3
Diptera					
Pyrgotidae	—	—	—	—	—
Cryptochetidae	1	2	2	5	5
Tachinidae	27	30	27	69	35
Muscidae (*Acridomyia*)	—	—	—	—	—
Sarcophagidae	—	—	—	—	—
Strepsiptera	—	—	—	—	—
Totals	194	393	274+*	860	216*

*Totals are reduced because more than one parasitoid has often been established on the same host in a single country.

A. Ichneumonoidea

The Braconidae and Ichneumonidae are large and diverse families with a wide range of hosts and life cycles. However, the successful genera and species are restricted. Among Braconidae, it is striking that the successes have been predominantly achieved against Lepidoptera, whereas Ichneumonidae have been proportionately more successful against Symphyta (Table VIII).

Another contrast is that in the Braconidae, successful establishments are predominantly in the genera *Apanteles* s.l. (16 spp.), *Bracon* s.l. (11 spp.) and *Opius* s.l. (10 spp.), whereas the generic distribution of Ichneumonidae shows no such dominance. Further, although *Apanteles* spp. and *Bracon* spp. have similar life histories, except that *Bracon* spp. feed externally, *Apanteles* spp. have been successful against at least 14 pests but *Bracon* spp. have been successful against only some three pests (Table IX). No explanation can be offered but a detailed comparison would be rewarding.

Opius spp. are parasitic on Diptera and have been applied chiefly against tephritid fruitflies with little success except in Hawaii (Clausen *et al.*, 1965). The poor results can be explained by the direct damage caused by the pests and the effects of plant breeding on the fruits rather than by inherent characteristics of the parasitoids. In fact *Opius* spp. parasitic on bean fly, *Ophiomyia phaseoli* (Tryon), an indirect pest, have been very successful in Hawaii (Greathead, 1975).

The Aphidiinae are exclusively aphid parasitoids and in spite of the

TABLE VIII. Contrasting performance of Braconidae and Ichneumonidae in contributing to effective classical biological control of Lepidoptera and Symphyta

Pest/No.	Parasitoid family	Species established	Successful	% Successful
Lepidoptera 39	Braconidae	37	20	54
	Ichneumonidae	22	8	36
Symphyta 11	Braconidae	1	1	—
	Ichneumonidae	13	6	46

TABLE IX. Contrasting performance of *Apanteles* s.l. and *Bracon* s.l. established and contributing to effective classical biological control

	Total species	Successful species	Hosts	Occasions
Apanteles	16	11	21	25
Bracon	11	3	12	11

supposed unsuitability of their hosts as targets, they have a good record in relation to the number of attempts to utilize them. Host lists (Mackauer and Starý, 1967) have suggested that they are mostly polyphagous, but Pungerl (1983) has shown that this may be an artefact of inadequate taxonomy.

B. Proctotrupoidea

The importance of Scelionidae, egg parasitoids, is noted in Section III.J. Platygasteridae have contributed to successes against whiteflies (*Amitus hesperidum* Silvestri) and *Pseudococcus comstocki* (Kuwana) (*Allotropa* spp.) in combination with chalcidoids, but otherwise have not been successful (Table X). Notable failures are the absence of any Proctotrupidae or Diapriidae from the table of establishments. Are Proctotrupoidea inherently unsuitable compared with Chalcidoidea?

C. Cynipoidea

Only one species, *Eucolia impatiens* (Say), has been credited with contributing to control of pests—blowflies in Hawaii. However, cynipoids have been little used, probably because they are not frequently encountered parasitizing important pests and even then are seldom responsible for significant mortality.

D. Chalcidoidea

The Chalcidoidea are the most frequently used and successful superfamily (Noyes, 1985) but again the importance of the families is very variable. The most important contributions have been made by the Eulophidae, Pteromalidae, Encyrtidae and Aphelinidae. The first three are very diverse in their host relations but the hosts of the Aphelinidae are confined to Sternorhyncha. Within families certain genera are again predominant.

Among the Eulophidae, species of *Tetrastichus* and *Pediobius* predominate (Table XI). Many species in both genera are obligate or facultative hyperparasitoids and are avoided. However, *T. sokolowskii* Kurdyumov which attacks both *Plutella xylostella* (L.) and its parasitoid *Apanteles plutellae* Kurdyumov but does not preferentially attack the latter has been deliberately introduced on the grounds that the combined effect of the two parasitoids will be greater than either acting alone. It is of interest that *Pediobius parvulus* (Ferrière), first used successfully against coconut leaf-miner, *Promecotheca*

TABLE X. Summary of parasitoids contributing to the successful control of Aleurodidae

	Species of		
	Parasitoids	Pests	Occasions
Peromalidae			
Amitus	2	2	3
Aphelinidae			
Cales	1	1	5
Encarsia	6	6	13
Eretmocerus	2	2	12
Totals	11	7	26

TABLE XI. Contribution of the eulophid genera *Pediobius* and *Tetrastichus* to classical biological control

	Established (successful)		
	Species	Pests	Occasions
Tetrastichus	8 (5)	10 (7)	15 (10)
Pediobus	5 (3)	12 (6)	8 (5)

coerulipennis Blanchard, in Fiji (Taylor, 1937) has been equally successful against the same or other *Promecotheca* spp., except in Sri Lanka where *Dimmockia javana* Ferrière which had failed in Fiji was the successful agent (Dhamadhikari *et al.*, 1977).

The principal contribution of Pteromalidae is as parasitoids of house and stable flies, chiefly *Musca domestica* L. and *Stomoxys calcitrans* (L.). *Spalangia* spp., *Muscidifurax raptor* Girault and Saunders superspecies and *Pachycrepoideus vindemiae* (Rondani) have been widely used and various combinations have been effective in the USA (mostly requiring inoculative release), Hawaii and Mauritius (Patterson *et al.*, 1981; Greathead and Monty, 1982). Among Lepidoptera the chief success has been *Pteromalus puparum* L. against the cabbage white, *Artogeia rapae* (L.) [= *Pieris rapae* (L.)] in New Zealand, elsewhere it was less successful. Introductions for bark beetle control have been made recently and *Rhopalicus tutele* (Walker) established against *Hylastes ater* (Paykull) shows promise.

The Encyrtidae are chiefly known as control agents for Pseudococcidae and Coccidae but one of the most useful parasitoids of the potato tuber moth, *Phthorimaea operculella* (Zeller), has been the polyembryonic larval parasitoid *Copidosoma koehleri* (Blanchard) established in Cyprus, Africa, Mauritius and India and contributing to control in combination with other species. Curiously, *C. desantisi* Annecke and Mynhardt, also widely released, has succeeded only in Australia (Sankaran and Girling, 1980). Also successful has been *Ooencyrtus erionotae* in Mauritius and Hawaii along with *Apanteles erionotae* in controlling banana skipper, *Erionota thrax*. *Tachinaephagus zealandicus* Ashmead has contributed to the control of flies breeding in dung in California but like the pteromalid parasitoids of flies in this habitat it is not effective in other situations. Thus, *Stomoxys nigra* (Macquart) which breeds in rotting vegetation is parasitized by *T. stomoxicida* Subba Rao in Uganda and this species has proved partially successful when introduced into Mauritius (Greathead and Monty, 1982).

As well as being the chief control agents for Diaspididae (Table XII), the Aphelinidae have provided the principal control agents for Aleyrodidae (Table X) and *Aphelinus* spp. are useful parasitoids of Aphidoidea. One of them, *A. mali* (Haldeman), which attacks the apple woolly aphid, *Eriosoma lanigerum* (Hausman), has been established in at least 35 countries. Where infestation is principally on twigs it has usually given satisfactory control, but it fails (e.g. in Zimbabwe) where root infestation is important since it does not attack hosts underground.

The Encyrtidae and *Coccophagus* spp. (Aphelinidae) have controlled Pseudococcidae and Coccidae whereas Aphelinidae and *Comperiella bifasciata* Howard (Encyrtidae) have controlled Diaspididae. In each instance there is a dominant genus, five *Anagyrus* spp. for Pseudococcidae, five *Metaphycus* spp. for Coccidae, and nine *Aphytis* spp. for Diaspididae (Table XII).

No particular reason can be deduced for the success of these genera except for *Aphytis*. This genus, because of its exceptional performance, has been the most intensively studied (Rosen and DeBach, 1979). The species are difficult to define from morphological characters and cluster into superspecies, the best known being the *Aphytis lignanensis* group parasitizing *Aonidiella aurantii* (Maskell). Successive introductions into California, and later elsewhere, demonstrated that each has different optimal environmental requirements which resulted in the classic example of competitive displacement. Rosen and DeBach (1979) present a number of reasons for their success, most of which are common to other successful chalcidoids—good searching ability, shorter generation time than the host, development of several offspring on a single host etc. However, alone among the successful parasitoids of Coccoidea, they develop as external parasitoids, thus avoiding hyperparasitism.

The internal parasitoids of Coccoidea are frequently subject to a high rate

TABLE XII. Summary of parasitoids involved in effective classical biological control of Coccoidea

	No. of species	Host	No. of host spp.	Occasions
Encyrtidae				
Achrysopophagus	1	Pseudococcid	1	1
Anagyrus	6	Pseudococcid	6	6
Aneristus	1	Coccid	1	1
Arhopoideus	2	Pseudococcid	2	4
Clausenia	1	Pseudococcid	1	1
Comperiella	1	Diaspidid	3	5
Hambletonia	1	Pseudococcid	1	1
Leptomastidea	1	Pseudococcid	2	2
Leptomastix	2	Pseudococcid	3	4
Metaphycus	5	Coccid	5	6
Neodusmetia	1	Pseudococcid	1	4
Pseudaphycus	2	Pseudococcid	2	2
Trichomasthus	1	Coccid	1	1
Aphelinidae				
Aphytis	9	Diaspidid	10	14
Coccobius	2	Diaspidid	2	2
Coccophagus	2	Pseudococcid	2	4
Encarsia	3	Diaspidid	3	9
Pteropterix	1	Diaspidid	1	1
Totals				
Encyrtidae	24	Pseudococcid	12	24
		Coccid	8	10
Aphelinidae	17	Diaspidid	12	37

of mortality from hyperparasitoids and parasitism by males of those Aphelinidae which develop as hyperparasitoids on females of their own and other species within the same host, sometimes to the tertiary level (e.g. Williams and Greathead, 1973). As Williams (1977) has argued, the interactions are complex but although male hyperparasitism may have a damping effect on the impact of other hyperparasitoids and thus overall host mortality may not be severely impaired, obligate hyperparasitoids must reduce the efficiency of primary parasitoids.

E. Aculeata

Aculeates here considered to be parasitoids are those that develop on a single host which is not removed by the parent to a special site for oviposition.

The Bethyliidae develop as gregarious external parasitoids of stem-boring Lepidoptera and wood-boring Coleoptera. Species of *Goniozus* and *Parasierola* have been employed against cereal and sugarcane borers but so far without success. *Prorops nasuta* Waterston attacking the coffee berry borer, *Hypothenemus hampei* (Ferrari), was successfully established in Brazil and provided partial control until the advent of synthetic pesticides. This and a second species, *Cephalodromia stephanoderis* Bertrem, are now being reinvestigated.

The Dryinidae, parasitic on Auchenorhyncha, are potentially good control agents for leaf- and planthoppers since the females are predatory on the larval hosts, but so far they have been employed only once, against *Perkinsiella saccharicida* Kirkaldy in Hawaii.

Scolioidea which attack Scarabaeoidea have been widely used in attempts to control white grubs in sugarcane and pastures. Species have been established in Hawaii and the USA, against Japanese beetle, *Popillia japonica* Newman, and in Mauritius against *Clemora smithi* (Arrow), but although they are considered to have contributed to the decline of their hosts, other factors are believed to have been more important. The only clearcut success has been control of a native white grub, *Oryctes tarandus* (Olivier), in Mauritius by *Scolia oryctophaga* Coquerel from related hosts in Madagascar.

F. Diptera

Cryptochetidae and Tachinidae have provided successful control agents. Cryptochetidae are a small highly specialized family parasitizing Margarodidae. The contribution of *Cryptochetum iceryae* (Williston) to the control of *Icerya purchasi* in California, Bermuda, Chile and Egypt is often overlooked. Another species, *C. monophlebi* Skuse, has been successful against *I. seychellarum* (Westwood) in Mauritius. The ease with which *Rodolia cardinalis* can be handled and established and its usual excellent performance compared with the difficulties experienced in handling *Cryptochetum* spp. are undoubtedly responsible for this neglect.

Tachinidae rank with the more important families of Hymenoptera as biological control agents. The key to their successful use is the ability to find techniques for mating in captivity. The development of the Scaramuzza-Box technique initiated their successful use against *Diatraea saccharalis* (Bennett, 1969). Although they were the only successful control agents for this pest until the importation of *Apanteles flavipes* into Barbados, attempts to use them and Old World tachinids against sugarcane borers in the Old World have failed (except possibly in Taiwan) although the principal species have been reared and released in large numbers (Cock, 1985). Tachinidae have also achieved outstanding results against a coconut defoliator, *Levuana iridescens* Bethune-Baker, in Fiji; a sugarcane weevil, *Rhabdoscelis obscurus* (Boisduval), and the green vegetable bug, *Nezara viridula*, in Hawaii; and winter moth, *Operophtera brumata* (L.), in Canada.

A diversity of oviposition or larviposition strategies is a feature of Tachinidae and all the main types (macrotype eggs or larvae on the host, larvae or ovoviviparous eggs close to the host, microtype eggs on the food plant) are represented by the successful species. Thus the strategy employed does not seem to be a constraint on use of Tachinidae. However, many species considered as control agents have had to be rejected because of failure to achieve mating in captivity, e.g. *Jaynesleskia jaynesei* (Aldrich) and *Miobiopsis diadema* (Wiedemann), two potentially useful parasitoids of *Diatraea* spp. (Cock, 1985).

G. Conclusion

It is difficult to generalize from this survey of the families and genera of parasitoids used in biological control in terms of why certain taxonomic groupings have predominated over others as the whole range of size, oviposition strategy, fecundity, stage attacked etc. is exhibited. Perhaps the most important factor in choice of natural enemy has been ease of handling and, if necessary, a technique for laboratory breeding. Successful establishment depends on colonizing ability which will be influenced by dispersal behaviour and mating strategy in bisexual species since in the early stages of colonization the numbers of males and females will be very small and location of a mate will be critical.

VI. SOME LIKELY FUTURE TRENDS

This survey of classical biological control has attempted to evaluate the dogmas developed over the years by practitioners of biological control and to seek principles which can be applied in decision making at the various stages of a control programme that will enhance the chances of success. However, the ultimate test remains outcome of the release, the results of which still cannot be anticipated with confidence because of the complex of physical and biological factors involved. However, over a period the number of successes is likely to be proportional to the amount of research and importation work carried out (DeBach, 1964).

In the past it was considered sufficient that if biological control agents caused no damage to economically important species and no concern was expressed about their likely impact on other non-target species they should be released. In fact it was often noted as an advantage that agents should have alternative hosts on which to carry over during temporary scarcity of the target pest (DeBach, 1964). However, now increasing concern for the environment has led to criticism relating to the impact of non-specific introduced biological control agents on native species (Howarth, 1983). While it can be disputed that natural enemies, rather than loss of habitat, are the cause of the loss of rare or endangered species, the criticism is valid even in strictly agricultural terms. For example, Howarth (1983) notes that introduced parasitoids are preventing introduced Lepidoptera from controlling weeds in Hawaii. For these reasons I believe that in future it will be necessary to screen parasitoids, as is now done for weed control agents, before introductions are made and that there will be a demand for environmental impact statements before programmes involving release of natural enemies are approved. Thus, practitioners will need to be more scrupulous in selection of natural enemies and the research input required in a biological control programme will be increased.

At a time when public expenditure is increasingly being scrutinized, there will be an increasing need for economic justification before programmes are funded. Here again biological weed control has led the way. Harris (1979) has discussed ways of costing biological control research in terms of scientist years per species in relation to the number of species likely to be screened and released before control is achieved. These projected costs are measured against the losses caused by the weed and the cost of alternative methods of control (Harris and Cranston, 1979).

These requirements will tend to reduce the number of programmes but, hopefully, focus resources on the more worthwhile ones. They will reduce the haphazard "shotgun" approach to introductions which has been responsible for many introductions made without adequate prior research and subsequent assessment which were the target of the criticisms made by Howarth (1983).

Although it is usually difficult to obtain funds for thorough assessment of the results sufficient to enable cost/benefit calculations, such assessments are desperately needed not only to show that biological control pays but also to provide facts essential to understanding the conditions required for success, both to improve the performance of biological control and also to gain insights which will advance biological science. For the ecologist, biological control introductions are experiments on a grand scale which would not be permitted for any other purpose and so are potentially very valuable as tests of hypotheses.

The importance of precision in the taxonomy of both pests and natural enemies to the efficient conduct of biological control requires much greater collaboration of taxonomists or, more precisely, biosystematists with biological control workers (Sabrosky, 1955; Greathead, in press) to meet demands for the more professional conduct of biological control programmes.

More careful selection of control agents will be not only for host specificity and environmental matching between source areas and target areas but also for tolerance to pesticides and other factors which will enable classical biological control to be incorporated in integrated pest management schemes. So far, efforts to improve strains by selective breeding in captivity have not been successful, but eventually genetic engineering may achieve the desired results.

Far from there being fewer opportunities for classical biological control in the future, improved understanding will open up possibilities for control of native pests and there will continue to be new introductions of pests. Thus, one of the largest programmes ever undertaken is currently in progress for the control of cassava mealybug, *Phenacoccus manihoti* Matile-Ferrero, and cassava green mites, *Mononychellus* spp., which reached Africa in the early 1970s. *Liriomyza trifolii* (Burgess), which continues to spread through the Old World since its discovery in Kenya about 10 years ago is an obvious target for a well-funded programme but has received only piecemeal attention so far. *Prostephanus truncatus* (Horn), imported in grain from Central America, which has become established in East and West Africa since 1979, has had a devastating impact.

In conclusion, classical biological control has achieved some important successes but the record can be improved in the future by increased understanding of the underlying mechanisms as well as better funded and more carefully researched programmes, which will together lead to a higher success rate of introductions. The effort is well justified because of the very important advantage of achieving permanent control with minimal environmental disturbance, a result which cannot be achieved by any other means except total eradication.

Acknowledgements

I am most grateful to Annette Greathead for preparing the CIBC databank and for editorial assistance, without which this review could not have been completed. Its value has been enhanced by the constructive criticism of Jeff Waage and Matthew Cock, who have also drawn my attention to pertinent examples and literature.

REFERENCES

Alam, M. M., Bennett, F. D., and Carl, K. P. (1971). Biological control of *Diatraea saccharalis* (F.) in Barbados by *Apanteles flavipes* Cam. and *Lixophaga diatraeae* T.T. *Entomophaga* **16,** 151–158.

Anderson, R. M. (1982). Theoretical basis for the use of pathogens as biological control agents of pest species. *Parasitology* **84,** 3–33.

Beirne, B. P. (1975). Biological control attempts by introductions against pest insects in the field in Canada. *Canadian Entomologist* **107,** 225–236.

Beirne, B. P. (1980). Biological control: benefits and opportunities. *In* "Perspectives in World Agriculture" pp. 307–321. Commonwealth Agricultural Bureaux, Farnham Royal, England.

Bennett, F. D. (1969). Tachinid flies as biological control agents for sugar cane moth borers. *In* "Pests of Sugar Cane" (J. R. Williams, J. R. Metcalfe, R. W. Mungomery and R. Mathes, eds.), pp. 117–148. Elsevier, Amsterdam.

Berlese, A. (1915). La distruzione della *Diaspis pentagona* a mezzo della *Prospaltella berlesei. Redia* **10,** 151–218.

CAB (1980). "Biological Control Service. 25 Years of Achievement." Commonwealth Agricultural Bureaux, Farnham Royal, England.

Canada (1971). Biological control programmes against insects and weeds in Canada 1959–1968. *Technical Communication of the Commonwealth Institute of Biological Control* **4**.

Carl, K. P. (1982). Biological control of native pests by introduced natural enemies. *Biocontrol News and Information* **3,** 191–200.

Clausen, C. P. (1978). Introduced parasites and predators of arthropod pests and weeds: a world review. *Agriculture Handbook of the United States Department of Agriculture* **480**.

Clausen, C. P., Clancy, D. W., and Chock, Q. C. (1965). Biological control of the oriental fruit fly (*Dacus dorsalis* Hendel) and other fruit flies in Hawaii. *Technical Bulletin of the United States Department of Agriculture* **1322**.

Cock, M. J. W. (1985). A review of biological control of pests in the Commonwealth Caribbean and Bermuda up to 1982. *Technical Communication of the Commonwealth Institute of Biological Control* **9**.

DeBach, P. (1946). An insecticidal check method for measuring the efficacy of entomophagous insects. *Journal of Economic Entomology* **39,** 695–697.

DeBach, P. (1964). "Biological Control of Insect Pests and Weeds." Chapman and Hall, London.

Dharmadhikari, P. R., Perera, P. A. C. R., and Hassen, T. M. F. (1977). A short account of the biological control of *Promecotheca cumingi* (Col.: Hispidae) the coconut leaf-miner, in Sri Lanka. *Entomophaga* **22,** 3–18.

Doutt, R. L. (1958). Vice, virtue, and the vedalia. *Entomological Society of America Bulletin* **4,** 119–123.

Ehler, L. E., and Hall, R. W. (1982). Evidence for competitive exclusion of introduced natural enemies in biological control. *Environmental Entomology* **11,** 1–4.

Embree, D. G. (1966). The role of introduced parasites in the control of the winter moth in Nova Scotia. *Canadian Entomologist* **98,** 1159–1167.

van Emden, H. F., and Williams, G. F. (1974). Insect stability and diversity in agro-ecosystems. *Annual Review of Entomology* **19,** 455–475.

Etienne, J. (1973). Lutte biologique et aperçu sur les études entomologiques diverses effectuées ces dernières années à la Réunion. *Agronomie Tropicale* **28,** 683–687.

Fleschner, C. A., Hall, J. C., and Ricker, D. W. (1955). Natural balance of mite pests in an avocado grove. *California Avocado Society Yearbook* **39,** 155–162.

Greathead, D. J. (1971). A review of biological control in the Ethiopian Region. *Technical Communication of the Commonwealth Institute of Biological Control* **5**.
Greathead, D. J. (1975). Biological control of the beanfly, *Ophiomyia phaseoli* (Dipt.: Agromyzidae), by *Opius* spp. (Hym.: Braconidae) in the Hawaiian islands. *Entomophaga* **20,** 313–316.
Greathead, D. J. (1976). A review of biological control in western and southern Europe. *Technical Communication of the Commonwealth Institute of Biological Control* **7**.
Greathead, D. J. (1984). Biological control constraints to agricultural production. *In* "Advancing Agricultural Production in Africa." (D. L. Hawksworth, ed.), pp. 200–206. Commonwealth Agricultural Bureaux, Farnham Royal, England.
Greathead, D. J. (in press). Biosystematics and biological control: the need for close collaboration. USDA Agricultural Service Publication.
Greathead, D. J., and Monty, J. (1982). Biological control of stableflies (*Stomoxys* spp.): results from Mauritius in relation to fly control in dispersed breeding sites. *Biocontrol News and Information* **3,** 105–109.
Greathead, D. J., and Waage, J. K. (1983). Opportunities for biological control of agricultural pests in developing countries. *World Bank Technical Paper* **11**.
Hagen, K. S., and Franz, J. M. (1973). A history of biological control. *In* "History of Entomology" (R. F. Smith, T. E. Mittler and C. N. Smith, eds.), pp. 433–476. Annual Reviews, Palo Alto, California.
Hall, R. W., and Ehler, L. E. (1979). Rate of establishment of natural enemies in classical biological control. *Entomological Society of America Bulletin* **25,** 280–282.
Hall, R. W., Ehler, L. E., and Bisabri-Ershadi, B. (1980). Rate of success in classical biological control of arthropods. *Entomological Society of America Bulletin* **26,** 111–114.
van Hamburg, H., and Hassell, M. P. (1984). Density dependence and the augmentative release of egg parasitoids against graminaceous stalkborers. *Ecological Entomology* **9,** 101–108.
Harris, P. (1979). Cost of biological control of weeds by insects in Canada. *Weed Science* **27,** 242–250.
Harris, P., and Cranston, R. (1979). An economic evaluation of control methods for diffuse and spotted knapweeds in western Canada. *Canadian Journal of Plant Science* **59,** 175–382.
Hokkanen, H., and Pimentel, D. (1984). New approach for selecting biological control agents. *Canadian Entomologist* **116,** 1109–1121.
Howard, L. O., and Fiske, W. F. (1911). The importation into the United States of the parasites of the gispy moth and the brown-tail moth. *Bulletin of the Bureau of Entomology United States Department of Agriculture* **91,** 344.
Howarth, F. G. (1983). Classical biocontrol: panacea or Pandora's box. *Proceedings of the Hawaiian Entomological Society* **24,** 239–244.
IITA (1984). "Annual Report for 1983." International Institute of Tropical Agriculture, Ibadan, Nigeria.
IITA (1985). "Annual Report for 1984." International Institute of Tropical Agriculture, Ibadan, Nigeria.
Kelleher, J. S., and Hulme, M. A. (1984). "Biological Control Programmes against Insects and Weeds in Canada 1969–1980." Commonwealth Agricultural Bureaux, Farnham Royal, England.
Keller, M. A. (1984). Reassessing evidence for competitive exclusion of introduced natural enemies. *Environmental Entomology* **13,** 192–195.
Kenmore, P. E. (1980). "Ecology and Outbreaks of a Tropical Insect Pest of the Green Revolution, the Rice Brown Planthopper, *Nilaparvata lugens* (Stal)." PhD Thesis, University of California, Berkeley.
Latière, H. (1917). La lutte contre les maladies des plantes en Italie. *Annales Epiphyties* **4,** 76–114.

Lloyd, D. C. (1960). The significance of the type of host plant crop in successful biological control of insect pests. *Nature, London* **187,** 430–431.

Lounsbury, C. P. (1940). The pioneer period of economic entomology in South Africa. *Journal of the Entomological Society of South Africa* **3,** 9–29.

Mac Arthur, R. H., and Wilson, E. O. (1967). "The Theory of Island Biogeography." Princeton University Press, Princeton, New Jersey.

Mackauer, M., and Starý, P. (1967). Hym. Ichneumonoidea world Aphidiidae. *In* "Index of Entomophagous Insects" (V. Delucchi and G. Remaudière, eds.), Vol. 2. Le François, Paris.

McLeod, J. H., McGugan, B. M., and Coppel, H. C. (1962). A review of the biological control attempts against insects and weeds in Canada. *Technical Communication of the Commonwealth Institute of Biological Control* **2**.

Munroe, E. G. (1971). Status and potential of biological control in Canada. *Technical Communication of the Commonwealth Institute of Biological Control* **4,** 213–255.

Noyes, J. S. (1985). Chalcidoids and biological control. *Chalcid Forum* **5,** 5–13.

Patterson, R. S., Koehler, P. G., Morgan, P. B., and Harris, R. L. (1981). "Status of Biological Control of Filth Flies." US Department of Agriculture, New Orleans.

Pimentel, D. (1963). Introducing parasites and predators to control native pests. *Canadian Entomologist* **95,** 785–792.

Pungerl, N. B. (1983). Variability in characters commonly used to distinguish *Aphidius* species (Hymenoptera: Aphidiidae). *Systematic Entomology* **8,** 425–430.

Rao, V. P., Ghani, M. A., Sankaran, T., and Mathur, K. C. (1971). A review of the biological control of insects and other pests in south-east Asia and the Pacific region. *Technical Communication of the Commonwealth Institute of Biological Control* **6**.

Rosen, D., and DeBach, P. (1979). "Species of *Aphytis* of the World (Hymenoptera: Aphelinidae)." Junk, The Hague.

Sabrosky, C. W. (1955). The interrelations of biological control and taxonomy. *Journal of Economic Entomology* **48,** 710–714.

Sailer, R. I. (1978). Our immigrant insect fauna. *Entomological Society of America Bulletin* **24,** 3–11.

Sankaran, T., and Girling, D. J. (1980). The current status of biological control of the potato tuber moth. *Biocontrol News and Information* **1,** 207–211.

Simmonds, F. J. (1960). Biological control of the coconut scale, *Aspidiotus destructor* Sign., in Principe, Portuguese West Africa. *Bulletin of Entomological Research* **51,** 223–237.

Simmonds, F. J., and Greathead, D. J. (1977). Introductions and pest and weed problems. *In* "Origins of Pest, Parasite, Disease and Weed Problems" (J. M. Cherrett and G. R. Sagar, eds.), pp. 109–124. Blackwell Scientific Publications, Oxford.

Smith, H. S., and DeBach, P. (1942). The measurement of the effect of entomophagous insects on population densities of their hosts. *Journal of Economic Entomology* **35,** 845–849.

Southwood, T. R. E. (1977). The relevance of population dynamic theory to pest status. *In* "Origins of Pest, Parasite, Disease and Weed Problems" (J. M. Cherrett and G. R. Sagar, eds.), pp. 35–54. Blackwell Scientific Publications, Oxford.

Tallamy, D. W. (1983). Equilibrium biogeography and its application to insect host-parasite systems. *American Naturalist* **121,** 244–254.

Taylor, K. L. (1978). Evaluation of the insect parasitoids of *Sirex noctilio* (Hymenoptera: Siricidae) in Tasmania. *Oecologia* **32,** 1–10.

Taylor, T. H. C. (1937). "The Biological Control of an Insect in Fiji." Imperial Institute of Entomology, London.

Taylor, T. H. C. (1955). Biological control of insect pests. *Annals of Applied Biology* **42,** 190–196.

Thorpe, W. H. (1973). William Robin Thompson 1887–1972. *Biographical Memoirs of Fellows of the Royal Entomological Society* **19,** 655–678.

Tooke, F. G. C. (1955). The eucalyptus snout-beetle, *Gonipterus scutellatus* Gyll. *Entomology Memoirs of the South Africa Department of Agriculture* **3**.

Tothill, J. D., Taylor, T. H. C., and Paine, R. W. (1930). "The Coconut Moth in Fiji." Imperial Bureau of Entomology, London.

Turnbull, A. L., and Chant, D. A. (1961). The practice and theory of biological control of insects in Canada. *Canada Journal of Zoology* **39,** 697–753.

Way, M. J. (1977). Pest and disease status in mixed stands *vs* monocultures; the relevance of ecosystem stability. *In* "Origins of Pest, Parasite, Disease and Weed Problems" (J. M. Cherrett and G. R. Sagar, eds.), pp. 127–138. Blackwell Scientific Publications, Oxford.

Williams, J. R. (1977). Some features of sex-linked hyperparasitism in Aphelinidae (Hymenoptera). *Entomophaga* **22,** 345–350.

Williams, J. R., and Greathead, D. J. (1973). The sugarcane scale insect *Aulacaspis tegalensis* (Zhnt.) and its biological control in Mauritius and East Africa. *Pest Articles and News Summaries* **19,** 353–367.

Wilson, F. (1960). A review of the biological control of insects and weeds in Australia and Australian New Guinea. *Technical Communication of the Commonwealth Institute of Biological Control* **1**.

11

Enhancing Parasitoid Activity in Crops

W. POWELL

I. INTRODUCTION

In the development of biological control techniques over the last 30 years, considerable attention has been paid to what has become known as the "classical" approach (see Chapter 10). This usually involves the introduction of exotic parasitoids or predators into areas in attempts to control pests which themselves are often introductions, generally unintentional ones. However, there is a growing realization that much can be gained from the exploitation of naturally occurring enemies of indigenous or long-established pests. This normally entails efforts to conserve and enhance the activity of these natural enemies by manipulating their environment.

Aspects of this approach are discussed in this review and include the provision of supplementary resources such as alternative hosts and adult food sources, modification of cultural practices to reduce parasitoid mortality and improve their effectiveness and the manipulation of parasitoids using behaviour-controlling chemicals (semiochemicals). Although these techniques are often aimed at "wild" parasitoid populations, they may also be of value used in conjunction with inundative releases or to manipulate recently introduced parasitoid populations. Many of the topics discussed could individually form the subject of a lengthy review and this paper does not claim to be exhaustive.

II. HABITAT MANIPULATION

Annual crop monocultures are amongst the most difficult of environments in which to induce the efficient operation of biological control agents. This is because they usually lack adequate resources for the effective performance of natural enemies (Rabb *et al.*, 1976) and many of the cultural practices used in annual cropping are damaging to natural enemy populations. Additional resources necessary for improving parasitoid effectiveness can often be provided by increasing habitat diversity. These include:

1. Provision of alternative hosts at times when the pest host is scarce
2. Provision of food (pollen and nectar) for adult parasitoids
3. Provision of refugia (e.g. for overwintering)
4. Maintenance of small populations of the pest host over extended periods to ensure the continued survival of the parasitoid population.

Increased diversity within the agroecosystem may be produced by multiple cropping, intercropping, strip harvesting, selective retention of weeds within the crop or conservation of wild plants at field margins. However, it is not the establishment of diversity *per se* which is important but rather the addition of specific resources to the ecosystem via diversity. It is therefore important that changes in habitat diversity are purposely designed to obtain specific effects within the relevant socio-economic constraints of the crop.

The advantages and disadvantages to pest control of increasing diversity in agroecosystems have been discussed at some length (Dempster and Coaker, 1972; van Emden and Williams, 1974; Murdoch, 1975; Way, 1979; Speight, 1983) and specific implications for biological control by habitat manipulation have been reviewed by van den Bosch and Telford (1964), Rabb *et al.* (1976), Ables and Ridgway (1981) and Altieri and Letourneau (1982).

A. Alternative Hosts

Many insect pests are not continuously present in crops, particularly in annual crops, and their parasitoids must survive elsewhere during their absence. The availability of alternative hosts in or around the crop can help to ensure the survival of viable parasitoid populations. Efficient control requires a parasitoid to reduce the numbers of its pest host sufficiently early during the pest infestation to prevent economic damage, which usually means operating effectively at low host densities. The provision of alternative hosts can help to improve synchrony between parasitoids and their pest hosts, improve parasitoid distribution and reduce intraspecific competition in the parasitoid population (van den Bosch and Telford, 1964).

Alternative hosts may be innocuous species feeding on wild plants or they may be other pests on a different crop. Examples of situations in which increased parasitism has been attributed to the presence of alternative hosts are given in Table I. The control of the oriental fruit moth, *Cydia molesta* (Busck), in peach orchards in the USA was greatly improved by the presence of several weed species which supported alternative hosts for the parasitoid *Macrocentrus ancylivorus* Rohwer. These allowed large numbers of the parasitoid to pass the winter in the absence of its fruit moth host (Peppers and Driggers, 1934). Parasitism of the grape leafhopper, *Erythroneura elegantula* Osborn by *Anagrus epos* Girault and of the apple maggot, *Rhagoletis pomonella* (Walsh), by braconids was increased in the presence of certain weeds which harboured alternative hosts (Doutt and Nakata, 1973; Maier, 1981). Starý (1974) examined the parasitoid spectrum associated with aphids on *Galium* spp. in Czechoslovakia. Aphids feeding on *Galium* were mostly innocuous species but several of their parasitoids also attacked aphids of economic importance and the author concluded that these weeds can provide a useful reservoir of aphid parasitoids in cultivated areas.

Pests on different crops may act as alternative hosts for the same parasitoid. Planting flax with cotton in Peru allowed tachinid parasitoids to survive on flax cutworms at times when their main host, the bollworm *Heliothis virescens* (F.), was absent from cotton (Hambleton, 1944). In parts of the USA the sunflower aphid, *Aphis helianthi* Monell, acts as a useful alternative host for several parasitoids of *Schizaphis graminum* (Rondani) (Eikenbary and Rogers, 1974). Parasitism of *S. graminum* on sorghum, particularly by *Lysiphlebus testaceipes* (Cresson), increased when sunflowers infested with *A. helianthi* were growing nearby. The parasitoids were able to maintain high populations on *A. helianthi* during two critical periods in the year when *S. graminum* was scarce or absent.

The value of alternative hosts for enhancing parasitoid activity in the field depends upon the facility with which the parasitoid switches between the alternative host and the target host. Reluctance to switch hosts and impaired performance after switching can be problems. This is exemplified by the aphid parasitoid *Aphidius ervi* Haliday which attacks hosts on both crop plants and on weeds. In Europe its principal hosts are *Microlophium*

TABLE I. Examples of increased parasitism due to the presence of alternative hosts

Parasitoids	Pests	Crops	Alternative hosts	References
Macrocentrus spp. (Braconidae)	*Cydia molesta* (Busck)	Peaches	Lepidoptera on weeds	Peppers and Driggers (1934)
Archytas spp. (Tachinidae)	*Heliothis virescens* (F.)	Cotton	Cutworms on flax	Hambleton (1944)
Scelionids	*Eurygaster integriceps* Puton	Cereals	Pentatomidae in nearby natural habitats	Shapiro (1959)
Lydella grisescens Robineau-Desvoidy (Tachinidae)	*Ostrinia nubilalis* (Hübner)	Maize	*Papaipema nebris* (Guenée) on giant ragweed	Hsiao and Holdaway (1966)
Anagrus epos Girault (Mymaridae)	*Erythroneura elegantula* Osborn	Vines	*Dikrella cruentata* (Gillette) on blackberry	Doutt and Nakata (1973)
Lysiphlebus testaceipes (Cresson) (Aphidiidae)	*Schizaphis graminum* (Rondani)	Sorghum	*Aphis helianthi* Monell on sunflowers	Eikenbary and Rogers (1974)
Emersonella niveipes Girault (Eulophidae)	*Chelymorpha cassidea* (F.)	Sweet potato	*Stolas* sp. on morning glory	Carroll (1978)
Braconids	*Rhagoletis pomonella* (Walsh)	Apples	Tephritidae on weeds	Maier (1981)

carnosum (Buckton) on perennial stinging nettle, *Urtica dioica, Acyrthosiphon pisum* (Harris) on leguminous crops and *Sitobion avenae* (F.) on cereals (Starý, 1978, 1983a). Stinging nettle is an abundant weed on farmland and it has been suggested that *M. carnosum* could act as a reservoir host for *Aphidius ervi*, helping to build up parasitoid populations when pest hosts are scarce, particularly early in the year (Perrin, 1975; Starý, 1983a). Perrin (1975) suggested that patches of nettles should be cut in mid-June to prevent parasitoids and other natural enemies from continuing to concentrate on the alternative hosts in preference to pest hosts on neighbouring crops. Also the observations of Starý (1978) suggest that *Aphidius ervi* could be induced to switch between *Acyrthosiphon pisum* and *S. avenae* if lucerne and wheat are grown in close proximity.

However, in laboratory trials *Aphidius ervi* has shown a reluctance to switch between hosts, often manifested by reduced mummy production on the new host (Cameron *et al.*, 1984; Pungerl, 1984; Starý *et al.*, 1985). Data from olfactometric tests show that female *Aphidius ervi* reared on *Acyrthosiphon pisum* respond to odours from this host but not to those from *Microlophium carnosum* or *S. avenae*, although some response was obtained to another cereal aphid, *Metopolophium dirhodum* (Walker) (Powell and Zhang, 1983). In the same tests both sexes responded to odours from bean, *Vicia faba* and wheat leaves but not to those from nettle leaves. Conversely, female *Aphidius ervi* reared on *Microlophium carnosum* responded more strongly to this host on nettle leaves than to *Acyrthosiphon pisum* on bean leaves in similar olfactometer trials (Starý *et al.*, 1985). These observations, together with data from the study of colour patterns (Starý, 1983b) and from electrophoretic enzyme analyses (Němec and Starý, 1983a, 1983b) suggest that two or more biotypes or races of *Aphidius ervi*, each preferring different hosts, exist in the field.

The parasitic Hymenoptera appears to be a group which is rapidly speciating (Askew, 1968; Gordh, 1977). This has led to the existence of many sibling species and biotypes which are morphologically identical but are partially or completely isolated reproductively (Gordh, 1977). It is, therefore, easy to assume that a single species is utilizing several different hosts within one area or ecosystem when, in reality, there may be several distinct populations or sibling species which intermix to a very limited extent and utilize different hosts.

Recently, host-switching trials have been done at Rothamsted using the parasitoid *Aphidius rhopalosiphi* De Stef which attacks several aphid species on cereal crops. Mummy production was reduced considerably when parasitoids reared on *Metopolophium dirhodum* through several generations were switched to *Sitobion avenae* (Powell, unpublished data). Switches done in the opposite direction, from *S. avenae* to *M. dirhodum*, resulted in an increase in mummy production. Paradoxically, Ankersmit (1983) recorded reduced mummy production when *Aphidius rhopalosiphi* was switched from *S. avenae* to *M. dirhodum*. Laboratory host-switching trials using *Aphidius ervi* have also given contrasting results. Pungerl (1984) and Starý *et al.* (1985) both found that parasitoids reared on *Microlophium carnosum* produced very few mummies when switched to *Acyrthosiphon pisum*. However, in trials at Rothamsted the same switch had no significant effect on mummy production (Powell, unpublished data).

Reluctance to switch hosts and concomitant, deleterious effects on reproduction obviously would limit the usefulness of alternative hosts in pest management if they are widespread phenomena amongst polyphagous parasitoids. However, most of these switching data come from laboratory studies and laboratory cultures of parasitoids are known to suffer rapid reductions in genetic diversity as a result of genetic drift (Unruh *et al.*, 1983). This could easily confound the results of host-switching and host preference trials and may explain the inconsistent results mentioned above. Differences in laboratory strains of the hosts may also play a part. More work is urgently needed, especially studies of host-switching in the field since the ease with which host-switching occurs is not only important with respect to the role of alternative hosts but also with respect to the mass rearing and release of parasitoids for biological control.

B. Adult Food Sources

Most adult hymenopteran parasitoids require food, usually in the form of nectar or pollen, to ensure effective reproduction (van Emden, 1962; van den Bosch and Telford, 1964). Adult parasitoids provided with sources of nectar have shown increases in longevity, fecundity and attack rates compared with unfed controls (Syme, 1975; Foster and Ruesink, 1984). Unfortunately, many crop monocultures, especially of annual crops, do not provide an adequate supply of these food resources and adult parasitoids may be unwilling or unable to commute over long distances between nectar sources and host habitats. Topham and Beardsley (1975) estimated that the effective range of *Lixophaga sphenophori* (Villeneuve), a dipteran parasitoid of sugarcane weevils, was about 45–60 m from nectar sources at field margins.

Retention or encouragement of flowering weeds in and around crops has often been associated with increased levels of parasitism in pest populations (Allen and Smith, 1958; Leius, 1967; Topham and Beardsley, 1975; Foster and Ruesink, 1984; Table II). Considerably more eggs, larvae and pupae of lepidopteran pests were parasitized in apple orchards with rich undergrowths of wild flowers than in clean cultivated orchards (Leius, 1967). Also, greater

TABLE II. Examples of increased parasitism due to the presence of adult food sources

Parasitoids	Pests	Crops	Food sources	References
Tiphia popilliavora Rohwer (Tiphiidae)	*Phyllophaga* spp. *Lachnosterna* spp.	Various	Nectar from weeds; honeydew from scale-insects	Wolcott (1942)
Aphelinus mali (Haldeman) (Aphelinidae)	Aphids	Apples	Nectar from the honey-plants *Phacelia* and *Eryngium*	Telenga (1958)
Apanteles medicaginis (Muesebeck) (Braconidae)	*Colias philodice* Godart	Alfalfa	Nectar from weeds; honeydew from aphids	Allen and Smith (1958)
Aphytis proclia (Walker) (Aphelinidae)	*Quadraspidiotus perniciosus* (Comstock)	Orchards	Nectar from the honey-plant *Phacelia tanacetifolia* Bentham	Chumakova (1960)
Various	*Malacosoma americanum* (F.) *Cydia pomonella* (Busck)	Apples	Nectar from weeds	Leius (1967)
Lixophaga sphenophori (Villeneuve) (Tachinidae)	*Rhabdoscelus obscurus* (Boisduval)	Sugar-cane	Nectar from *Euphorbia* spp. weeds	Topham and Beardsley (1975)

levels of parasitism of alfalfa caterpillars, *Colias philodice* Godart, by *Apanteles medicaginis* Muesebeck were evident in small fields with abundant marginal vegetation than in large fields with few weeds (Allen and Smith, 1958). These increases in parasitism have been attributed to a more plentiful supply of nectar and honeydew on the wild plants. Random retention of weeds in and around crops could cause problems, such as the creation of reservoirs of pest species and damage to crop yields, which would outweigh any advantages of increasing food resources for adult parasitoids. However, weed floras can be manipulated both quantitatively and qualitatively and the potential benefits of weeds in pest management, including their role in providing nectar sources and alternative hosts for parasitoids, have recently been reviewed by Altieri *et al.* (1977), Zandstra and Motooka (1978), Altieri and Whitcomb (1979) and Altieri and Letourneau (1982).

Plants which are rich sources of nectar, such as the honey plants *Phacelia* spp. and *Eryngium* spp. have been planted between orchard trees as a food resource for adult parasitoids. Resulting enhancement of parasitoid activity has been reported in several instances, including those of *Aphelinus mali* (Haldeman) attacking apple aphids, *Aphytis proclia* (Walker) attacking San José scale, *Quadraspitiotus perniciosus* (Comstock), and *Trichogramma* spp. attacking lepidopteran eggs (Telenga, 1958; Chumakova, 1960).

Some crop plants are also rich sources of nectar. As early as the 1930s cotton was suggested as a valuable crop to use in intercropping for this reason (Marcovitch, 1935). More recently, Gallego *et al.* (1983), recommended the planting of leguminous crops such as *Cajanus cajan* (L.) Millspaugh in and around coconut plantations to encourage immigration and survival of nectar-feeding parasitoids of the coconut leaf-miner, *Promecotheca cumingii* Baly.

C. Cropping Techniques

The variety and abundance of parasitoids in agroecosystems can be strongly influenced by the degree of diversity of the crops grown. Intercropping or mixed cropping of two or more crops often results in increased natural enemy activity and higher levels of parasitism in pest populations compared with the situation in monocultures of the same crops (Altieri *et al.*, 1978; Risch, 1979; Altieri and Todd, 1981; Table III). Mixed cropping of corn and sweet potato in Costa Rica increased the species of parasitic Hymenoptera present in the crops by 75% and the number of individuals by 100% compared with those in monocultures, contributing to a much higher natural enemy to pest ratio in the mixed cultures (Risch, 1979). Parasitism of leafhoppers, *Empoasca kraemeri* Ross and Moore, by egg parasitoids,

TABLE III. Examples of increased parasitism due to the adoption of certain cropping practices

Parasitoids	Pests	Crops	Cropping practices	References
Macrocentrus ancylivorus Rohwer (Braconidae)	*Ancylis comptana* (Froelich) & *Cydia molesta* (Busck)	Strawberries Peaches	Intercropping	Marcovitch (1935)
Bathyplectes curculionis (Thomson) (Ichneumonidae)	*Hypera postica* (Gyllenhal)	Alfalfa	Timing of first cutting of crop	Hamlin *et al.* (1949)
Aphidiids	Aphids	Alfalfa	Strip harvesting	van den Bosch *et al.* (1967) Schlinger and Dietrick (1960)
Lysiphlebus testaceipes (Cresson (Aphidiidae)	*Schizaphis graminum* (Rondani)	Barley; sorghum	Resistant cultivars	Starks *et al.* (1972)
Anagrus sp. (Mymaridae)	*Empoasca kraemeri* Ross & Moore	Beans	Mixed cropping with maize	Altieri *et al.* (1978)
Various	Various	Maize; sweet potato	Mixed cropping	Risch (1979)
Trichogramma spp. (Trichogrammatidae)	*Heliothis zea* (Boddie)	Soybean	Mixed cropping with maize	Altieri and Todd (1981)

Anagrus sp., was greater on beans which were grown in polycultures with maize than on beans grown alone (Altieri *et al.*, 1978). Similarly, artificially placed *Heliothis zea* (Boddie) eggs were more heavily parasitized by *Trichogramma* spp. in corn-soybean polycultures than in soybean monocultures (Altieri and Todd, 1981).

The harvesting of lucerne several times during the season seriously damages natural enemy populations which do not recover as quickly as those of their pest hosts. This can be alleviated by a system of strip cutting so that the uncut strips act as a reservoir of natural enemies which can rapidly recolonize harvested strips. Several of the aphidiid parasitoid species which attack the spotted alfalfa aphid, *Therioaphis trifolii* (Monell) *forma maculata* (Buckton), were only found in lucerne fields which had been strip-harvested (Schlinger and Dietrick, 1960) whilst van den Bosch *et al.* (1967) recorded better host-parasitoid synchrony and improved parasitoid efficacy for *Aphidius smithi* Sharma & Subba Rao attacking *Acyrthosiphon pisum* as a result of strip-harvesting.

There are other examples of the adoption of specific management techniques which have enhanced parasitoid activity in crops. Some Australian apple growers stored wood prunings infested with woolly aphid, *Eriosoma lanigerum* Hausmann, over winter to conserve diapausing *Aphelinus mali*, and then returned them to the orchards in spring (Wilson, 1966). The planting of windbreaks in citrus orchards in Texas helped to increase parasitism of soft scales, *Coccus hesperidum* L., by *Metaphycus stanleyi* (Compere) because the parasitoid's efficiency was apparently hindered in high winds (Reed *et al.*, 1970).

The impact of parasitoids on pest populations can also be enhanced by the use of cultural practices which hinder the establishment and growth of pest infestations. Partial resistance to pest attack by crop cultivars, either due to non-preference during colonization or to antibiosis, may interact with parasitoids or other natural enemies to allow effective control. In greenhouse trials the aphidiid *Lysiphlebus testaceipes* failed to control *Schizaphis graminum* on susceptible cultivars of sorghum and barley but succeeded on resistant cultivars (Starks *et al.*, 1972). Similar observations were made by Wyatt (1970) for *Aphidius matricariae* Haliday attacking *Myzus persicae* (Sulzer) on chrysanthemum cultivars.

Klassen (1981) outlined five ways in which resistant cultivars may complement the action of biological control agents:

1. By reducing the rate of development of immature stages of the pest so that they are at risk for longer
2. Antibiosis may help to synchronize parasitoids and their hosts
3. Tolerance increases the economic threshold

4. By providing protection when environmental conditions prevent natural enemies from attacking the pest
5. Pests may be more exposed to their enemies on resistant cultivars.

Intercropping and mixed cropping can also hinder the development of pest infestations by interfering with initial colonization (Perrin and Phillips, 1978). However, host habitat location by some parasitoids could be impaired by the presence of other plants alongside the food-plants of their hosts, especially if they are using the same olfactory and visual cues as their prey, and in such cases intercropping or the presence of weeds would hinder parasitoid efficiency. Monteith (1960) found that the host-finding ability of *Bessa harveyi* (Townsend), a tachinid parasitoid of the larch sawfly, *Pristiphora erichsonii* (Hartig) was seriously impaired by the presence of a variety of plants in the forest understory which masked the odours of the host larvae and their food-plant.

Some cultural practices and the timing of certain cultural operations can be damaging to parasitoid populations. An obvious example is the use of broad-spectrum insecticides at times of optimum parasitoid activity. The timing of chemical control measures against pests other than the parasitoid's host and against the host itself should be adapted to minimize damage to parasitoid populations as far as possible (van den Bosch and Stern, 1962). Damage can be minimized if it is feasible for pesticide applications to coincide with the least susceptible stage in the parasitoid's life cycle, which is often the pupal stage. The pupae of aphid parasitoids, for example, are protected from chemicals to some extent by the mummified remains of the host and thus survive some pesticide treatments very well (Süss, 1983). The use of insect growth regulators against gypsy moth, *Lymantria dispar* (L.), can be integrated with biological control by adjusting application times to avoid damage to larvae of its parasitoid, *Apanteles melanoscelus* (Ratzeburg), as they emerge from their hosts (Granett *et al.*, 1976). Increased use of minimal cultivation techniques and direct drilling of some annual crops undoubtedly aid the survival of parasitoids during cultivation but damage to parasitoid populations is sometimes unavoidable and many are killed during the harvesting of annual crops. The density of mummified cereal aphids, containing diapausing parasitoids, was reduced by over 90% during the harvesting of a winter wheat field (Powell, unpublished data).

III. BEHAVIOUR-CONTROLLING CHEMICALS

The potential of behaviour-controlling chemicals [termed semiochemicals by Law and Regnier (1971)] for use in pest management has long been recognized, but most work has been done on chemicals, notably sex pheromones, which affect the behaviour of the pest species themselves. More recently, it has been recognized that considerable opportunities exist for the manipulation of natural enemies using semiochemicals and the topic has been discussed by Vinson (1977, 1981), Lewis (1981), Gross (1981), Lewis *et al.* (1982a) and Nordlund *et al.* (1981).

A. Improving Parasitoid Efficiency

The main objective of a female parasitoid is to locate and oviposit in or on to its host. Unless emergence occurs alongside a suitable host, the parasitoid responds to a series of environmental cues which lead it to the appropriate habitat, host locality (food-plant) and host and stimulate it to oviposit (Vinson, 1977). Many of these cues are chemical in nature and many originate from the host (kairomones) or from the host's food-plant (synomones). If identified and synthesized, these chemicals could be useful in pest management programmes for attracting parasitoids into crops at risk from pest hosts and for stimulating and prolonging searching behaviour.

The existence of chemicals which stimulate host-searching behaviour has been established for a variety of parasitoids. Vinson and Lewis (1965) elicited a host-seeking response from *Cardiochiles nigriceps* Viereck using a substance extracted from the body and salivary secretions of the tobacco budworm, *Heliothis virescens*. Kairomones from ovipositing female stinkbugs (Pentatomidae) attracted egg parasitoids of the genera *Trissolus* and *Telenomus* (Shumakov, 1977). Searching activity by the aphid parasitoid *Aphidius nigripes* Ashmead was increased by the presence of aphid honeydew on plants (Bouchard and Cloutier, 1984).

It is important for parasitoids to begin to have a significant impact on their host population towards the start of a pest infestation if they are to prevent the pest from exceeding economic thresholds. However, if host density is too low at this time there is a danger that the parasitoids will disperse due to a lack of sufficient host encounters to maintain searching behaviour. Treatment of the crop with appropriate chemical cues could theoretically prolong searching activity and so increase the chances of parasitoid/host encounters.

Lewis *et al.* (1972) made one of the first attempts to influence parasitoid behaviour in the field using semiochemicals. In laboratory trials they demonstrated that moth scales from *Heliothis zea* females induced host-searching behaviour in *Trichogramma evanescens* Westwood, a parasitoid of *H. zea* eggs. Following a field application to cotton of a hexane extract of moth scales, host-finding and percentage parasitism of host eggs by released populations of *T. evanescens* were improved. The kairomones involved are believed to stimulate and continuously reinforce the searching behaviour of the parasitoid rather than directly attract them to their hosts (Lewis *et al.*, 1975). Later, Lewis *et al.* (1979) used artificial moth scales, in the form of diatomaceous earth particles which they impregnated with moth scale extract, to investigate the effects of treatment patterns on parasitism of *H. zea* eggs by *Trichogramma* spp. By regulating application density in this way they were able to arrive at an optimum rate which would retain parasitoids in the target area but not inhibit movement between oviposition sites.

This highlights one of the likely problems with field applications of search-inducing kairomones. Over-stimulation of parasitoids could cause them to search intensively in restricted, kairomone-treated areas over long periods, reducing the probability of encounters with patchily-distributed hosts. Chiri and Legner (1983) treated cotton with foliar sprays of hexane and methanol extracts of host-moth body-scales in an attempt to increase parasitism of the pink bollworm by *Chelonus* sp. The treatment was unsuccessful even though laboratory bioassays had given promising results. The authors suggest that the blanket coverage of the crop with the spray retained parasitoids in areas where host eggs were absent and led to habituation due to prolonged exposure to the kairomones. There is an obvious danger of damaging wild parasitoid populations in the long term if kairomone sprays are used extensively. By diverting their searching activities into crop situations where hosts are relatively scarce the overall reproductive rate of the population will be reduced. Also, such diversion of foraging activities may reduce the impact of the parasitoids on pest reservoirs outside the target crop. Parasitoids exhibit success-motivated searching (Vinson, 1977) so that in the absence of hosts a parasitoid will eventually disperse from an area, even in the presence of a search-stimulating kairomone (Vinson, 1981). The kairomone, however, will usually prolong the period before the parasitoid leaves, termed the "giving-up time" (Waage, 1979; see Chapter 2).

A possible way of preventing parasitoid dispersal would be to release real or factitious hosts together with a kairomone. In field trials using hexane extracts of *H. zea* moth scales the distribution of supplemental host eggs on to the cotton crop improved the effects of the kairomone on the parasitization rate by released *Trichogramma pretiosum* Riley (Gross *et al.*, 1984). The efficacy of the kairomone spray also appeared to be dependent on the absence of high levels of naturally occurring kairomone. Release of real hosts would also alleviate problems of reduced reproductive success in the parasitoid population and so may be necessary to ensure the long-term success of kairomone use. The potential of kairomone use in conjunction with inundative parasitoid release has been emphasized by Nordlund *et al.* (1981).

Parasitoids respond to both short-range or contact cues and to long-range, volatile, chemical cues during host location (Vinson, 1976; Weseloh, 1981).

The kairomones from moth scales to which *Trichogramma* spp. respond are short-range cues. Interestingly, some of the long-range kairomones affecting parasitoid behaviour have been identified as the sex pheromones of their hosts. Sex pheromones of the citrus red scale, *Aonidiella aurantii* (Maskell), attract *Aphytis* spp. (Sternlicht, 1973) and a sex pheromone from male stinkbugs, *Nezara viridula* (L.), attracts a tachinid parasitoid, *Trichopoda pennipes* (F.) (Mitchell and Mau, 1971). *Trichogramma pretiosum* responds to a blend of chemicals from the ovipositor gland of *H. zea* (Lewis *et al.*, 1982b), which has been identified as the moth's sex pheromone (Klun *et al.*, 1980). Synthetic sex pheromone of *H. zea* applied in slow-release hollow fibres to field plots of cotton significantly increased parasitism by wild *Trichogramma* spp. populations (Lewis *et al.*, 1982b).

Since these long-range cues do not stimulate the same intensive searching of small areas as do the short-range cues, they are likely to present fewer problems in terms of application pattern. It has also been suggested that in the case of sex pheromones acting as parasitoid kairomones there could be possibilities for combining parasitoid manipulation with mating disruption of the host in pest management strategies (Lewis *et al.*, 1982b).

Plants also provide chemical cues to which parasitoids respond (synomones), helping them to locate the appropriate host habitat (Vinson, 1976, 1984). Aphid parasitoids, for example, respond to odours from the food-plants of their hosts in olfactometry tests (Read *et al.*, 1970; Powell and Zhang, 1983). Two different parasitoids of *Heliothis* spp., *Eucelatoria* sp. (Tachinidae) and *Cardiochiles nigriceps* (Braconidae), are both attracted to volatiles from host food-plants but the volatiles involved appear to be different for the two (Nettles, 1979, 1980; Vinson, 1975). *Eucelatoria* sp. responds to volatiles from cotton, corn and okra whereas *C. nigriceps* attacks *Heliothis* spp. on cotton and tobacco but not on corn or peanuts. Parasitism of *H. zea* eggs by *Trichogramma* spp. was enhanced when water extracts of corn and of a weed, *Amaranthus* sp., were applied to a range of crops as sprays or on wicks suspended from crop plants (Altieri *et al.*, 1981).

The use of semiochemicals for manipulating parasitoids seems to hold considerable potential as a technique for incorporation in pest management strategies. However, it is obvious from the few field trials carried out so far that there are problems to be overcome. Vinson (1977) has discussed some potential problems, stressing the importance of the influence of host density at the time of application of search-stimulating kairomones. Timing and mode of application are also probably critical factors for success. It is apparent that a thorough knowledge of the parasitoid-pest-crop system concerned is necessary in order to develop the technique efficiently. Also it may be rewarding to investigate the use of kairomones in combination with other enhancement techniques. For example, could they be used to induce parasitoids to switch from alternative hosts to the target host at the appropriate time? Is it necessary to provide additional resources such as adult food or factitious hosts alongside kairomone application? We are only scratching the surface in this topic and much more work, especially field trials involving naturally-occurring parasitoid populations, is needed before an adequate evaluation of potential practical applications can be made.

B. Monitoring Parasitoid Populations

Parasitoid enhancement in crops should be regarded as part of a broader pest management approach rather than as a complete pest control measure in itself. Efficient pest management depends upon adequate monitoring of the relevant components of the agroecosystem concerned. Monitoring is important in order to determine parasitoid:host ratios and to aid decisions concerning the necessity for, and optimal timing of, enhancement measures. The maximum impact of parasitoids on the host population does not necessarily coincide with maximum parasitoid abundance but rather depends upon the parasitoid:host ratio. In Britain for example, parasitoids appear to have their most useful impact upon cereal aphid populations early in the season, at a time when the parasitoid:aphid ratio is usually high and mortality factors which act upon the parasitoid population, such as hyperparasitoids, are at a low level (Powell, 1983; Powell *et al.*, 1983).

Semiochemicals, such as sex pheromones, are potentially useful for monitoring parasitoid populations in the same way as they are used to monitor pests. Recent work at Rothamsted has demonstrated the existence of sex pheromones in cereal aphid parasitoids (Powell and Zhang, 1983; Decker, unpublished data). Pheromone traps baited with live virgin females attracted large numbers of male parasitoids when placed in cereal fields. Two species were tested, *Aphidius rhopalosiphi* and *Praon volucre* Haliday, and individual traps caught only males of the species with which they were baited in numbers significantly higher than those caught in unbaited control traps. Powell and King (1984) attracted male *Microplitis croceipes* (Cresson) to traps baited with virgin females and placed in cotton fields. However, although trapping male parasitoids can indicate the presence of that species in the target crop or the start of parasitoid activity, unless stable sex ratios are known to exist in the field it does not give reliable information on female parasitoid abundance. The use of traps baited with host-derived kairomones or plant synomones may provide better short-term monitoring data and merits investigation.

IV. PARASITOIDS AS COMPONENTS OF AGROECOSYSTEMS

It is important to remember when considering ways of enhancing parasitoid activity that the parasitoids are operating in complex agroecosystems and that they interact with many other components in those ecosystems. Some of these interactions can be exploited to improve parasitoid effectiveness in ways already discussed. It is necessary to be aware of the role of a particular parasitoid in the ecosystem as a whole if an adequate assessment is to be made of the expediency of attempting enhancement and the optimum time for implementing enhancement measures. The ways in which the parasitoid interacts with other parasitoid species and with other natural enemies of the target host could be important.

For example, in Britain cereal aphids are attacked by a range of natural enemies including parasitoids, pathogenic fungi and a variety of predators. In the laboratory, aphids infected with a fungal pathogen less than four days after being parasitized died before the parasitoids could complete their development (Powell *et al.*, in press). Therefore, it would be of little use trying to enhance cereal aphid parasitoids during periods of high fungal activity and the best time for enhancement appears to be early in the season when other natural enemies are scarce. Levels of parasitism in field populations of cereal aphids were highest at the beginning of the season, gradually declining thereafter, whereas fungal infection began at a low level and increased through the season (Powell *et al.*, in press). Interference between parasitoids and fungal pathogens has been recorded with respect to a number of other crop pests (King and Bell, 1978; Los and Allen, 1983; Velasco, 1983). Similarly, parasitized insects may be more vulnerable to predator attack which would partly nullify parasitoid impact on host populations (Tostowaryk, 1971; Rathcke and Price, 1976).

A lot of data have accumulated concerning the relationships between parasitoid and host populations but too few of these data have been collected from field situations. Parasitoid population field data are usually recorded during studies of the population dynamics of their hosts in which the parasitoids feature as mortality factors in host species life tables. There is an urgent need for more life table studies centred on the parasitoids themselves if we are to gain a better perception of manipulation possibilities. Perhaps investigations of parasitoid-host relationships involving non-pests in natural habitats would be instructive.

Since parasitoids and their pest hosts function as constituent parts of complex agroecosystems, the success of enhancement measures could be influenced by the scale upon which they are implemented. Obviously, the importance of scale will depend upon the particular measures used and the agricultural environment in which they are attempted. It would be helpful to have some assessment of the relative chances of success of different approaches when implemented on an individual field scale, over an individual farm unit or over a wider geographical area, but little information on this exists.

Different approaches are appropriate to different agricultural situations; each situation must be assessed on its own merits and decisions made on the basis of sound background data. For example, mixed cropping is standard practice in many tropical areas and information about the parasitoid faunas in these areas, their relationships with crop pests and their basic ecology is necessary so that the most appropriate cropping systems can be designed. Other approaches are more applicable to the intensive annual crop monocultures which occur over large areas of the temperate region. There is much scope for the manipulation and conservation of parasitoids but it would be naive to assume that this would be profitable in all parasitoid-host-crop situations. Ultimately, the value of attempting to enhance parasitoid activity to aid pest control will be governed by the socio-economic restrictions operating on the cropping situation in question.

REFERENCES

Ables, J. R., and Ridgway, R. L. (1981). Augmentation of entomophagous arthropods to control pest insects and mites. *In* "Biological Control in Crop Production" (G. C. Papavizas, ed.), pp. 273–303. Allanheld, Osmun, USA.

Allen, W. W., and Smith, R. F. (1958). Some factors influencing the efficiency of *Apanteles medicaginis* Muesebeck (Hymenoptera: Braconidae) as a parasite of the alfalfa caterpillar, *Colias philodice eurytheme* Boisduval. *Hilgardia* **28,** 1–42.

Altieri, M. A., and Letourneau, D. K. (1982). Vegetation management and biological control in agroecosystems. *Crop Protection* **1,** 405–430.

Altieri, M. A., and Todd, J. W. (1981). Some influences of vegetational diversity on insect communities of Georgia soybean fields. *Protection Ecology* **3,** 333–338.

Altieri, M. A., and Whitcomb, W. H. (1979). The potential use of weeds in the manipulation of beneficial insects. *HortScience* **14,** 12–18.

Altieri, M. A., van Schoonhoven, A., and Doll, J. (1977). The ecological role of weeds in insect pest management systems: a review illustrated by bean (*Phaseolus vulgaris*) cropping systems. *Pest Abstracts and News Summaries* **23,** 195–205.

Altieri, M. A., Francis, C. A., van Schoonhoven, A., and Doll, J. (1978). A review of insect prevalence in maize (*Zea mays* L.) and bean (*Phaseolus vulgaris* L.) polycultural systems. *Field Crops Research* **1,** 33–49.

Altieri, M. A., Lewis, W. J., Nordlund, D. A., Gueldner, R. C., and Todd, J. W. (1981). Chemical interactions between plants and *Trichogramma* wasps in Georgia soybean fields. *Protection Ecology* **3,** 259–263.

Ankersmit, G. W. (1983). Aphidiids as parasites of the cereal aphids *Sitobion avenae* and *Metopolophium dirhodum*. *In* "Aphid Antagonists" (R. Cavalloro, ed.), pp. 42–49. Balkema, Rotterdam.

Askew, R. R. (1968). Considerations on speciation in Chalcidoidea (Hymenoptera). *Evolution* **22,** 642–645.

van den Bosch, R., and Stern, V. M. (1962). The integration of chemical and biological control of arthropod pests. *Annual Review of Entomology* **7,** 367–381.

van den Bosch, R., and Telford, A. D. (1964). Environmental modification and biological control. *In* "Biological Control of Insect Pests and Weeds" (P. DeBach, ed.), pp. 459–488. Chapman and Hall, London.

van den Bosch, R., Lagace, C. F., and Stern, V. M. (1967). The interrelationship of the aphid, *Acyrthosiphon pisum*, and its parasite, *Aphidius smithi*, in a stable environment. *Ecology* **48,** 993–1000.

Bouchard, Y., and Cloutier, C. (1984). Honeydew as a source of host-searching kairomones for the aphid parasitoid *Aphidius nigripes* (Hymenoptera: Aphidiidae). *Canadian Journal of Zoology* **62,** 1513–1520.

Cameron, P. J., Powell, W., and Loxdale, H. D. (1984). Reservoirs for *Aphidius ervi* Haliday (Hymenoptera: Aphidiidae), a polyphagous parasitoid of cereal aphids (Hemiptera: Aphididae). *Bulletin of Entomological Research* **74,** 647–656.

Carroll, C. R. (1978). Beetles, parasitoids and tropical morning glories: a study in host discrimination. *Ecological Entomology* **3,** 79–85.

Chiri, A. A., and Legner, E. F. (1983). Field applications of host-searching kairomones to enhance parasitization of the pink bollworm (Lepidoptera: Gelechiidae). *Journal of Economic Entomology* **76,** 254–255.

Chumakova, B. M. (1960). Supplementary feeding as a factor increasing the activity of parasites of harmful insects. *Trudy Vsesoyuznogo Nauchno-issledovatelscogo Instituta Zashchity Rastenii* **15,** 57–70.

Dempster, J. P., and Coaker, T. H. (1972). Diversification of crop ecosystems as a means of controlling pests. *In* "Biology in Pest and Disease Control" (D. Price Jones and M. E. Solomon, eds.), pp. 106–114. Blackwell, Oxford.

Doutt, R. L., and Nakata, J. (1973). The *Rubus* leafhopper and its egg parasitoid: an endemic biotic system useful in grape-pest management. *Environmental Entomology* **2,** 381–386.

Eikenbary, R. D., and Rogers, C. E. (1974). Importance of alternate hosts in establishment of introduced parasites. *In* "Proceedings of the Tall Timbers Conference on Ecological Animal Control by Habitat Management", pp. 119–133. Talahassee, Florida.

van Emden, H. F. (1962). A preliminary study of insect numbers in field and hedgerow. *Entomologist's Monthly Magazine* **98,** 255–259.

van Emden, H. F., and Williams, G. F. (1974). Insect stability and diversity in agro-ecosystems. *Annual Review of Entomology* **19,** 455–475.

Foster, M. A., and Ruesink, W. G. (1984). Influence of flowering weeds associated with reduced tillage in corn on a black cutworm (Lepidoptera: Noctuidae) parasitoid, *Meteorus rubens* (Nees von Esenbeck). *Environmental Entomology* **13,** 664–668.

Gallego, V. C., Baltazar, C. R., Cadapan, E. P., and Abad, R. G. (1983). Some ecological studies on the coconut leafminer, *Promecotheca cumingii* Baly (Coleoptera: Hispidae) and its hymenopterous parasitoids in the Philippines. *Philippine Entomologist* **6,** 471–493.

Gordh, G. (1977). Biosystematics of natural enemies. *In* "Biological Control by Augmentation of Natural Enemies" (R. L. Ridgway and S. B. Vinson, eds.), pp. 125–148. Plenum, New York.

Granett, J., Dunbar, D. M., and Weseloh, R. M. (1976). Gypsy moth control with Dimilin sprays timed to minimize effects on the parasite *Apanteles melanoscelus*. *Journal of Economic Entomology* **69,** 403–404.

Gross, H. R. (1981). Employment of kairomones in the management of parasitoids. *In*

"Semiochemicals: Their Role in Pest Control" (D. A. Nordlund, R. L. Jones and W. J. Lewis, eds.), pp. 137–150. John Wiley, New York.

Gross, H. R., Lewis, W. J., Beevers, M., and Nordlund, D. A. (1984). *Trichogramma pretiosum* (Hymenoptera: Trichogrammatidae): effects of augmented densities and distributions of *Heliothis zea* (Lepidoptera: Noctuidae) host eggs and kairomones on field performance. *Environmental Entomology* **13,** 981–985.

Hambleton, E. J. (1944). *Heliothis virescens* as a pest of cotton, with notes on host plants in Peru. *Journal of Economic Entomology* **37,** 660–666.

Hamlin, J. C., Lieberman, F. V., Bunn, R. W., McDuffie, W. C., Newton, R. C., and Jones, L. J. (1949). Field studies of the alfalfa weevil and its environment. *USDA Technical Bulletin* **975**.

Hsiao, T. H., and Holdaway, F. G. (1966). Seasonal history and host synchronization of *Lydella grisescens* (Diptera: Tachinidae) in Minnesota. *Annals of the Entomological Society of America* **59,** 125–133.

King, E. G., and Bell, J. V. (1978). Interactions between a braconid, *Microplitis croceipes*, and a fungus, *Nomuraea rileyi* in laboratory-reared bollworm larvae. *Journal of Invertebrate Pathology* **31,** 337–340.

Klassen, W. (1981). The role of biological control in integrated pest management systems. *In* "Biological Control in Crop Production" (G. C. Papavizas, ed.), pp. 433–445. Allanheld, Osmun, USA.

Klun, J. A., Plimmer, J. R., Bierl-Leonhardt, B. A., Sparks, A. N., Primiani, M., Chapman, O. L., Lee, G. H., and Lepone, G. (1980). Sex pheromone chemistry of female corn earworm moth, *Heliothis zea. Journal of Chemical Ecology* **6,** 165–175.

Law, J. H., and Regnier, F. E. (1971). Pheromones. *Annual Review of Biochemistry* **40,** 533–548.

Leius, K. (1967). Influence of wild flowers on parasitism of tent caterpillar and codling moth. *Canadian Entomologist* **99,** 444–446.

Lewis, W. J. (1981). Semiochemicals: their role with changing approaches to pest control. *In* "Semiochemicals: Their Role in Pest Control" (D. A. Nordlund, R. L. Jones and W. J. Lewis, eds.), pp. 3–12. John Wiley, New York.

Lewis, W. J., Jones, R. L., and Sparks, A. N. (1972). A host-seeking stimulant for the egg parasite *Trichogramma evanscens*: its source and a demonstration of its laboratory and field activity. *Annals of the Entomological Society of America* **65,** 1087–1089.

Lewis, W. J., Jones, R. L., Nordlund, D. A., and Gross, H. R. (1975). Kairomones and their use for management of entomophagous insects: II. Mechanisms causing increase in rate of parasitization by *Trichogramma* spp. *Journal of Chemical Ecology* **1,** 349–360.

Lewis, W. J., Beevers, M., Nordlund, D. A., Gross, H. R., and Hagen, K. S. (1979). Kairomones and their use for management of entomophagous insects. IX. Investigations of various kairomone-treatment patterns for *Trichogramma* spp. *Journal of Chemical Ecology* **5,** 673–680.

Lewis, W. J., Nordlund, D. A., and Gueldner, R. C. (1982a). Semiochemicals influencing behaviour of entomophages: roles and strategies for their employment in pest control. *In* "Les Médiateurs Chimiques" pp. 225–242. INRA.

Lewis, W. J., Nordlund, D. A., Gueldner, R. C., Teel, P. D., and Tumlinson, J. H. (1982b). Kairomones and their use for management of entomophagous insects. XIII. Kairomonal activity for *Trichogramma* spp. of abdominal tips, feces, and a synthetic sex pheromone blend of *Heliothis zea* (Boddie) moths. *Journal of Chemical Ecology* **8,** 1323–1332.

Los, L. M., and Allen, W. A. (1983). Incidence of *Zoophthora phytonomi* (Zygomycetes: Entomophthorales) in *Hypera postica* (Coleoptera: Curculionidae) larvae in Virginia. *Environmental Entomology* **12,** 1318–1321.

Maier, C. T. (1981). Parasitoids emerging from puparia of *Rhagoletis pomonella* (Diptera:

Tephritidae) infesting hawthorn and apple in Connecticut. *Canadian Entomologist* **113,** 867–870.

Marcovitch, S. (1935). Experimental evidence on the value of strip farming as a method for the natural control of injurious insects with special reference to plant lice. *Journal of Economic Entomology* **28,** 62–70.

Mitchell, W. C., and Mau, R. F. L. (1971). Response of the female southern stink bug and its parasite *Trichopoda pennipes* to male stink bug pheromones. *Journal of Economic Entomology* **64,** 856–859.

Monteith, L. G. (1960). Influence of plants other than the food plants of their host on host-finding by tachinid parasites. *Canadian Entomologist* **92,** 641–652.

Murdoch, W. W. (1975). Diversity, complexity, stability and pest control. *Journal of Applied Ecology* **12,** 795–807.

Němec, V., and Starý, P. (1983a). Elpho-morph differentiation in *Aphidius ervi* Hal. biotype on *Microlophium carnosum* (Bckt.) related to parasitization on *Acyrthosiphon pisum* (Harr.) (Hym., Aphidiidae). *Zeitschrift für angewandte Entomologie* **95,** 524–530.

Němec, V., and Starý, P. (1983b). Genetic polymorphism in *Aphidius ervi* Hal. (Hym., Aphidiidae), an aphid parasitoid on *Microlophium carnosum* (Bckt.). *Zeitschrift für angewandte Entomologie* **95,** 345–350.

Nettles, W. C. (1979). *Eucelatoria* sp. females: factors influencing response to cotton and okra plants. *Environmental Entomology* **8,** 619–623.

Nettles, W. C. (1980). Adult *Eucelatoria* sp.: response to volatiles from cotton and okra plants and from larvae of *Heliothis virescens*, *Spodoptera eridania*, and *Estigmene acrea*. *Environmental Entomology* **9,** 759–763.

Nordlund, D. A., Lewis, W. J., and Gross, H. R. (1981). Elucidation and employment of semiochemicals in the manipulation of entomophagous insects. *In* "Management of Insect Pests with Semiochemicals—Concepts and Practice" (E. R. Mitchell, ed.), pp. 463–475. Plenum, New York.

Peppers, B. B., and Driggers, B. F. (1934). Non economic insects as intermediate hosts of parasites of the oriental fruit moth. *Annals of the Entomological Society of America* **27,** 593–598.

Perrin, R. M. (1975). The role of the perennial stinging nettle, *Urtica dioica*, as a reservoir of beneficial natural enemies. *Annals of Applied Biology* **81,** 289–297.

Perrin, R. M., and Phillips, M. L. (1978). Some effects of mixed cropping on the population dynamics of insect pests. *Entomologia Experimentalis et Applicata* **24,** 385–393.

Powell, J. E., and King, E. G. (1984). Behaviour of adult *Microplitis croceipes* (Hymenoptera: Braconidae) and parasitism of *Heliothis* spp. (Lepidoptera: Noctuidae) host larvae in cotton. *Environmental Entomology* **13,** 272–277.

Powell, W. (1983). The role of parasitoids in limiting cereal aphid populations. *In* "Aphid Antagonists" (R. Cavalloro, ed.), pp. 50–56. Balkema, Rotterdam.

Powell, W., and Zhang, Z. L. (1983). The reactions of two cereal aphid parasitoids, *Aphidius uzbekistanicus* and *A. ervi* to host aphids and their food plants. *Physiological Entomology* **8,** 439–443.

Powell, W., Dewar, A. M., Wilding, N. and Dean, G. J. (1983). Manipulation of cereal aphid natural enemies. *Proceedings of the 10th International Congress of Plant Protection* **1,** 780.

Powell, W., Wilding, N., Brobyn, P. J., and Clark, S. J. (in press). Interference between parasitoids (Hym., Aphidiidae) and fungi (Entomophthorales) attacking cereal aphids. *Entomophaga.*

Pungerl, N. B. (1984). Host preferences of *Aphidius* (Hymenoptera: Aphidiidae) populations parasitising pea and cereal aphids (Hemiptera: Aphididae). *Bulletin of Entomological Research* **74,** 153–161.

Rabb, R. L., Stinner, R. E., and van den Bosch, R. (1976). Conservation and augmentation of natural enemies. *In* "Theory and Practice of Biological Control" (C. B. Huffaker and P. S. Messenger, eds.), pp. 233–254. Academic Press, New York.

Rathcke, B. J., and Price, P. W. (1976). Anomalous diversity of tropical ichneumonid parasitoids: a predation hypothesis. *American Naturalist* **110,** 889–893.

Read, D. P., Feeny, P. P., and Root, R. B. (1970). Habitat selection by the aphid parasite *Diaeretiella rapae* (Hymenoptera: Braconidae) and hyperparasite *Charips brassicae* (Hymenoptera: Cynipidae). *Canadian Entomologist* **102,** 1567–1578.

Reed, D. K., Hart, W. G., and Ingle, S. J. (1970). Influence of windbreaks on distribution and abundance of brown scale in citrus groves. *Annals of the Entomological Society of America* **63,** 792–794.

Risch, S. J. (1979). A comparison, by sweep sampling, of the insect fauna from corn and sweet potato monocultures and dicultures in Costa Rica. *Oecologia* **42,** 195–211.

Schlinger, E. I., and Dietrick, E. J. (1960). Biological control of insect pests aided by strip-farming alfalfa in experimental program. *California Agriculture* **14,** 8–9; 15.

Shapiro, V. A. (1959). Effect of agrotechnical measures on the effectiveness of oophages of the harmful stinkbug. *In* "Biological Methods of Controlling Pests of Plants", pp. 182–191. Kiev.

Shumakov, E. M. (1977). Ecological principles associated with augmentation of natural enemies. *In* "Biological Control by Augmentation of Natural Enemies" (R. L. Ridgway and S. B. Vinson, eds.), pp. 39–78. Plenum, New York.

Speight, M. R. (1983). The potential of ecosystem management for pest control. *Agriculture, Ecosystems and Environment* **10,** 183–199.

Starks, K. J., Muniappan, R., and Eikenbary, R. D. (1972). Interaction between plant resistance and parasitism against the greenbug on barley and sorghum. *Annals of the Entomological Society of America* **65,** 650–655.

Starý, P. (1974). Parasite spectrum (Hym., Aphidiidae) of aphids associated with *Galium*. *Entomologica Scandinavica* **5,** 73–80.

Starý, P. (1978). Seasonal relations between lucerne, red clover, wheat and barley agro-ecosystems through the aphids and parasitoids (Homoptera, Aphididae; Hymenoptera, Aphidiidae). *Acta Entomologica Bohemoslovaca* **75,** 296–311.

Starý, P. (1983a). The perennial stinging nettle (*Urtica dioica*) as a reservoir of aphid parasitoids (Hymenoptera, Aphidiidae). *Acta Entomologica Bohemoslovaca* **80,** 81–86.

Starý, P. (1983b). Colour patterns of adults as evidence on *Aphidius ervi* biotypes in field environments (Hymenoptera, Aphidiidae). *Acta Entomologica Bohemoslovaca* **80,** 377–384.

Starý, P., Pospisil, J., and Němec, V. (1985). Integration of olfactometry and electrophoroesis in the analysis of aphid parasitoid biotypes (Hym., Aphidiidae). *Zeitschrift für angewandte Entomologie* **99,** 476–482.

Sternlicht, M. (1973). Parasitic wasps attracted by the sex pheromone of their coccid host. *Entomophaga* **18,** 339–342.

Süss, L. (1983). Survival of pupal stage of *Aphidius ervi* Hal. in mummified *Sitobion avenae* F. to pesticide treatment. *In* "Aphid Antagonists" (R. Cavalloro, ed.), pp. 129–134. Balkema, Rotterdam.

Syme, P. D. (1975). The effects of flowers on the longevity and fecundity of two native parasites of the European pine shoot moth in Ontario. *Environmental Entomology* **4,** 337–346.

Telenga, N. A. (1958). Biological method of pest control in crops and forest plants in the USSR. *In* "Report of the Soviet Delegation, Ninth International Conference on Quarantine and Plant Protection." pp. 1–15. Moscow.

Topham, M., and Beardsley, J. W. (1975). An influence of nectar source plants on the New

Guinea sugarcane weevil parasite, *Lixophaga sphenophori* (Villeneuve). *Proceedings of the Hawaiian Entomological Society* **22,** 145–155.

Tostowaryk, W. (1971). Relationship between parasitism and predation of diprionid sawflies. *Annals of the Entomological Society of America* **64,** 1424–1427.

Unruh, T. R., White, W., Gonzalez, D., Gordh, G., and Luck, R. F. (1983). Heterozygosity and effective size in laboratory populations of *Aphidius ervi* (Hym., Aphidiidae). *Entomophaga* **28,** 245–258.

Velasco, L. R. I. (1983). Field parasitism of *Apanteles plutellae* Kurdj. (Braconidae, Hymenoptera) on the diamond-back moth of cabbage. *Philippines Entomologist* **6,** 539–553.

Vinson, S. B. (1975). Biochemical coevolution between parasitoids and their hosts. *In* "Evolutionary Strategies of Parasitic Insects and Mites" (P. W. Price, ed.), pp. 14–48. Plenum, New York.

Vinson, S. B. (1976). Host selection by insect parasitoids. *Annual Review of Entomology* **21,** 109–134.

Vinson, S. B. (1977). Behavioural chemicals in the augmentation of natural enemies. *In* "Biological Control by Augmentation of Natural Enemies" (R. L. Ridgway and S. B. Vinson, eds.), pp. 237–279. Plenum, New York.

Vinson, S. B. (1981). Habitat location. *In* "Semiochemicals: Their Role in Pest Control" (D. A. Nordlund, R. L. Jones and W. J. Lewis, eds.), pp. 51–77. John Wiley, New York.

Vinson, S. B. (1984). How parasitoids locate their hosts: a case of insect espionage. *In* "Insect Communication" (T. Lewis, ed.), pp. 325–348. Academic Press, London.

Vinson, S. B., and Lewis, W. J. (1965). A method of host selection by *Cardiochiles nigriceps*. *Journal of Economic Entomology* **58,** 869–871.

Waage, J. K. (1979). Foraging for patchily distributed hosts by the parasitoid *Nemeritis canescens* (Grav.). *Journal of Animal Ecology* **48,** 353–371.

Way, M. J. (1979). Significance of diversity in agroecosystems. *9th International Congress of Plant Protection*, Washington DC. Proccedings Opening Session and Plenary Session, 9–12.

Weseloh, R. M. (1981). Host location by parasitoids. *In* "Semiochemicals: Their Role in Pest Control" (D. A. Nordlund, R. L. Jones and W. J. Lewis, eds.), pp. 79–95. John Wiley, New York.

Wilson, F. (1966). The conservation and augmentation of natural enemies. *Proceedings FAO Symposium on Integrated Pest Control* **3,** 21–26.

Wolcott, G. N. (1942). The requirements of parasites for more than hosts. *Science* **96,** 317–318.

Wyatt, I. J. (1970). The distribution of *Myzus persicae* (Sulz.) on year-round chrysanthemums. II. Winter season: the effect of parasitism by *Aphidius matricariae* Hal. *Annals of Applied Biology* **65,** 31–41.

Zandstra, B. H., and Motooka, P. S. (1978). Beneficial effects of weeds in pest management—a review. *Pest Abstracts and News Summaries* **24,** 333–338.

12

Parasitoids in the Greenhouse: Successes with Seasonal Inoculative Release Systems

J. C. VAN LENTEREN

I. INTRODUCTION

Biological control of pests has been defined in at least 30 different ways. The broadest definition embraces all non-conventional chemical and non-chemical methods of pest control: the narrowest definition refers only to the use of predators, parasitoids and pathogens as natural enemies against pest species. I prefer the earliest definition as given by Smith in 1919: "the use of natural enemies (whether introduced or otherwise manipulated) to control insect pests". Nowadays the term also includes the control of other animal pests, diseases and weeds. This definition implies activity by man. In nature many potential pests are kept at low densities by their natural enemies, in which situation we speak of natural control. Disturbance, for example through the application of pesticides, may lead to pest outbreaks when natural enemies are killed to a greater extent than the pest organism; examples can be found in papers by DeBach (1964) and Huffaker and Messenger (1976).

Estimates of the total number of animal species vary between two and eight million and it is usually agreed that 75% are insects. In 1978 1,067,000 animal species had been described: 768,240 of these were insects (Huffaker and Rabb, 1984). The number of important pest species for agriculture at world level is estimated to be 350 (Simmonds *et al.*, 1976). I present these figures to show the relatively low number of pest species. To combat them about 200 active chemical ingredients are used in an array of formulations (Besemer, 1985).

A. A Change from Chemical to Biological Control?

Only a few pests are controlled through deliberate introduction of natural enemies but if one compares statistics related to the development and application of chemical and biological control, one wonders why biological control is not applied on a much larger scale (Table I).

Both the success ratio and the economic data show that biological control can be very profitable. Further, side-effects, on the environment and natural enemies, can be neglected but they are of major concern in chemical control. Application of biological control can, in my opinion, be greatly increased if we adopt the public relation methods used by chemical companies. Biological control with parasitoids seems to be unattractive to large industries because:

1. It is impossible to patent natural enemies. Every competitor can start producing once an effective natural enemy has been found
2. Mass production is complicated. Methods for mass-rearing natural

TABLE I. Comparison of aspects related to development and application of chemical and biological control

	Chemical control	Biological control
Number of "ingredients" tested	$>1.10^6$	5000
Success ratio	1:10,000	1:100
Developmental costs	55.10^6 Dfl.	2.10^6 Dfl.
Development time	10 years	10 years
Benefit per unit of money invested	2.5–5	30
Risk of resistance	Large	Small
Specificity	Small	Large
Harmful side-effects	Many	Few

enemies differ in many ways from the production of chemical pesticides. The know-how and rearing facilities are usually not available

3. Parasitoids have a short shelf-life. Most natural enemies have to be used very soon after production (often within a week) which demands a different distribution network from that employed by chemical control companies
4. Specificity to a single pest. Each pest organism has its own natural enemy complex. A natural enemy can usually be applied against only one pest species but to control a pest complex a number of different natural enemies have to be reared
5. Different guidance to growers. More entomological knowledge in the field of taxonomy and ecology is required by crop protection specialists.

Classical biological control may never interest industry but programmes requiring mass production and frequent releases offer better possibilities.

Chemical companies will not start producing other than broad-spectrum pesticides, unless their use is prohibited or pest organisms develop resistance. Since the Second World War, many agricultural entomologists have dealt mainly with technical problems of application and the development and testing of new chemical pesticides. Much information on the biology of pest organisms remained unused because development of ideas on how pests are caused and how they may be prevented is not necessary when cheap and powerful chemical pesticides are available. Application of, for example, cultural pest control methods and host-plant resistance has declined.

The availability of cheap and effective chemicals resulted in a lowering of economic damage thresholds, which made and will make the application of other controls even more difficult. During the first euphoria, insect pests were expected to be eradicated but negative aspects soon became clear: risks for producer, applier, consumer and environment. However, the main problem for chemical industry is the development of resistance. In 1914 the first case of arthropod resistance occurred. By 1980, 428 arthropod species had developed resistance against one or more pesticides (Anonymous, 1981). Diseases like malaria are on the increase, and the production of some agricultural crops is falling, all due to the development of resistance (Metcalf, 1980; Dover and Croft, 1984).

Furthermore, development of new pesticides becomes increasingly difficult. Because of the more complicated structure of new potential pesticides their production is more expensive. More research is also necessary before they can be registered. These factors combined will lead to increasing costs for chemical control.

I think the time has come to change from opportunism in pest control—whereby problems are awaited, solutions are hoped for and danger signals are neglected—to biologically based pest control. Many alternatives are at hand and one of these is biological control.

B. Biological Control in Greenhouses

In the Netherlands, as in many western European countries, research was on an *ad hoc* basis until about the end of the 1950s and few successes were obtained. Since then a number of national organizations have been created which are active in the development of integrated pest management, and a global organization, the International Organization for Biological Control (IOBC), has been established. In the Netherlands all research, development and application of alternative methods is organized within the Working Party on Integrated Control of Pests (Minks and Gruys, 1980).

Since 1960 this Dutch working party has developed successful biological control programmes for four pests in greenhouses and two in orchards. On an international level application of biological control in greenhouses has also developed rapidly, mainly due to good cooperation within the IOBC working group on "Integrated Pest Control in Protected Crops" (Hussey and Scopes, 1985).

Biological control programmes for greenhouses differ in several ways from those applied in field crops. Therefore, it is useful to describe first the different ways in which natural enemies can be utilized. The *inoculative release method* or one-time release method, also known as classical biological control, is most often applied in forest and orchard ecosystems where long-term survival of natural enemies can be guaranteed (see Chapter 10). The *inundative release method* or repeated releases, with short-term effect, can be applied where long-term control cannot be obtained (e.g. when parasitoid occurrence is not synchronized with host, parasitoid does not survive winter, or parasitoid population does not develop sufficiently fast). Then parasitoids have to be introduced regularly and in large numbers. This method is, among others, applied against univoltine pests in annual crops. The *seasonal inoculative release method*, or repeated releases with a long-term effect, can be applied when the culture method of a crop prevents control extending over many years. For example, in greenhouses, the crop, together with the pests and natural enemies, is removed at the end of the growing season.

The sequence of procedures in any biological control project is: (1) search, evaluation, selection and collection of natural enemies in one area; (2) shipment and screening and sometimes mass propagation, and in the second area (3) release and colonization and evaluation of the consequences (Beirne, 1980). More details on the planning of a programme can be found in analyses by Zwölfer *et al.*, 1976; Pschorn-Walcher, 1977; van Lenteren, 1980, 1986. Experience in developing biological control projects has taught me that most problems are encountered in the *selection* of parasitoids prior to introduction, and in the *mass production* of these organisms.

II. SELECTING PARASITOIDS

A. Value of Pre-introduction Studies: General Considerations

I estimate that some 5000 species have been tested for use in biological control programmes. Out of these, 270 species have led to partial (100), substantial (100) or complete (70) control. More than 80% of the successful natural enemies are parasitoids; the remainder are predators (17%) and pathogens (1%) (DeBach, 1964; Laing and Hamai, 1976; Clausen, 1978).

Most natural enemies have been found through trial and error. Many researchers have thought about ways of optimizing the pre-introductory studies so as to increase the predictability of success before introductions are made. We should keep in mind, however, that the success ratio and economic evaluation are both strongly in favour of biological control when compared with chemical control, even with the trial and error method.

I can best summarize the present-day extremes in point of view by quoting Ehler (1982):

> "(1) conduct pre-introduction studies to determine the 'best' species or complex of species to introduce; or (2) conduct minimal pre-introduction studies and simply release all suitable natural enemies and hope that the 'best' species or combination of species will be sorted out in the field."

B. General Criteria for Selection in Pre-introduction Studies

The criteria for selecting parasitoids are listed in Table II, collected from the following references: Varley, 1951; Flanders, 1957; Sweetman, 1958; Andrewartha, 1961; DeBach, 1964, 1971, 1974; Askew, 1971; Huffaker *et al.*, 1971, 1976; Krebs, 1972; Hassell and Rogers, 1972; Hassell and May, 1973; Varley *et al.*, 1973; van Emden, 1974; Huffaker, 1976; Coppel and Mertins, 1977; Ridgway and Vinson, 1977; Waage and Hassell, 1982.

TABLE II. Criteria for pre-introductory evaluation of natural enemies

	Release programme		
Criterion	Inoculative	Seasonal inoculative	Inundative
1. Seasonal synchronization with host	+	−	−
2. Internal synchronization with host	+	+	−
3. Climatic adaptation	+	+	+
4. No negative effects	+	+	+
5. Good culture method	−	+	+
6. Host specificity	+	−	−
7. Great reproductive potential	+	+	−
8. Good density responsiveness	+	+	±

+ = Important; − = not important.

Density responsiveness seems to be the most difficult to determine. Firstly, it is not absolute but obtains its value only when compared with values for other enemies. Secondly, many methods for determining a parasitoid's response to host density have been proposed but most of them are difficult to apply and do not lead to conclusive answers.

A new controversy is developing on the importance of density responsiveness. The central idea behind theoretical explanations for successful biological control has been that efficient natural enemies operate by creating a stable pest-enemy equilibrium at low densities.

Waage and Hassell (1982) concluded that "models have identified searching efficiency, fecundity, larval survival, sex ratio, interference and spatial heterogeneity as key contributors to the depression of host equilibria and/or to the stability of the interaction". Three of these factors—searching efficiency, interference and spatial heterogeneity—relate particularly to density responsiveness. To this list the developmental time of the natural enemy should be added, but it is not mentioned by Waage and Hassell (1982) because in the models they use, developmental time plays no role. The main conclusion of this approach is that aggregation of natural enemies at patches with high host densities is the critical feature in natural enemy behaviour that results both in stability and in successful biological control.

Recently, Murdoch *et al.* (1984, 1985) challenged the central role of a low stable pest equilibrium and suggested different, stochastic models. They examined the role of parasitoid aggregation independently of pest distribution and looked at field examples of successful biological control with the following conclusions and suggestions:

1. Successful control does not require a control agent to aggregate at areas of high host density, nor does it require aggregation independently of host density, or other mechanisms that create "partial refuges". Therefore, random movement by parasitoids is capable of producing excellent control
2. Control appears possible in the absence of conditions leading to local stable equilibria
3. In model systems control is compatible with local extinction of the pest and polyphagy in the natural enemy.

To which they add: "It is still possible, of course, that the aggregate result of unstable local dynamics, when combined with movement, may produce stable dynamics over large areas."

The quality criteria for natural enemies recommended by Waage and Hassell's deterministic approach (1982) are different from those suggested by a stochastic approach, especially with regard to density responsiveness. Murdoch *et al.* (1985) proposed stochastic models which suggest that parasitoids drive the pest as low as possible in any area, even to extinction, and present field data which do not support the view of many biological control practitioners that control is achieved by the establishment of a low stable pest-enemy equilibrium. The field evidence also fails to confirm that all criteria listed in Table II are essential in successful control agents.

Instead, if one believes biological control to be a stochastic or non-equilibrium process, other criteria become important. Murdoch *et al.* (1985) reasoned: "If local extinction of the pest is both a possible way to control pests and perhaps a desirable goal . . . two strategies are apparent, given that most of the time during successful biological control the predator or parasitoid is preventing outbreaks rather than reducing outbreak populations. The first strategy we might term 'lying-in-wait', the second 'search and destroy'."

As an example of lying-in-wait they propose that continuously present polyphagous natural enemies are able to survive brief periods of starvation and show an adequate attack when the pest re-invades or begins to increase. For search and destroy it is assumed that natural enemies are monophagous (or oligophagous) on the pest, and highly adapted to finding and attacking it. Spatial patchiness, heterogeneity and migration allow the pest (and hence the natural enemy) to survive. This requires some of the characteristics of Table II: host specificity, a high search rate and a great reproductive capacity.

What does this mean for the biological control practitioner whose aim is to obtain a pest density below an economic threshold all or most of the time following introduction of a natural enemy? At the moment we cannot conclude more than that different theories produce different, or even conflicting, advice on selection criteria. The practitioner is not only confused but may also delay application of biological control measures while he is determining the wrong criteria.

Thus Way's opinion (1973) is still applicable:

> "At present too few research workers are able to bridge the gap between pest-control practitioners who are blinkered by short-term needs, and theoreticians or pure ecologists who have inadequate appreciation of practical realities".

C. Selecting Parasitoids for Greenhouse Systems

I will now discuss the importance of the criteria listed in Table II in relation to the seasonal inoculative release programmes, which are predominantly used in greenhouse biological control:

1. Seasonal synchronization of the parasitoid with its host is not important in greenhouses since synchronization can be obtained by the grower through releasing parasitoids when most hosts are in the developmental stage for parasitization. Adjustments can be made throughout the growing season
2. The ability of the parasitoid to develop in the host to the adult stage is essential for ongoing control. If it kills the host but does not develop, parasitoids will have to be re-introduced in each subsequent host generation, which requires an inundative programme. These have proved to be unattractive in greenhouses because they are too expensive. Further, parasitoid development should be synchronous with that of the host so that adult parasitoids are available when suitable host stages are present. This is especially important at the start of the growing season when host generations are still discrete, although poor

synchronization can be corrected in part through repeated introductions. Later in the growing season this problem ceases to be important

3. At an early stage of pre-introductory research, tests should be performed to determine whether the parasitoids are able to develop, reproduce and migrate given the weather conditions under which they are to be used
4. The parasitoids should not attack other beneficial organisms in the same environment or non-pest hosts of importance in the area where they are to be introduced.

   Some natural enemies introduced into western European greenhouses, for example, *Encarsia formosa* (Gahan), are so specific that only a few related host species are attacked. Others (e.g. leaf-miner parasitoids) are native to western Europe, so their use involves no extra risks as long as we know that they are not facultatively hyperparasitic as are some leaf-miner parasitoids (Takada and Kamijo, 1979). The parasitoids we use in the greenhouse system are all known to attack potential pest species only. Furthermore they do not attack parasitoids or predators used against other pests.

   The only negative interference known in greenhouses is between the parasitoid *Encarsia formosa* and the entomopathogenic fungi *Verticillium lecanii* (Zimmermann) Viégas and *Aschersonia aleyrodis* Webber. Combined application of the fungi and parasitoids leads to the death of some of the developing parasitoids by fungus infection or to elimination of the fungi by the parasitoid (Fransen, personal communication). However, this does not result in lower overall percentage death of the greenhouse whitefly.
5. Good culture methods for natural enemies are the basis for a successful greenhouse biological control programme. Culture methods largely determine the eventual cost of the natural enemy and so the probability of application. Simple differences in biology of host or parasitoid may lead to a 100-fold difference in rearing costs, as was found when comparing the cost of mass rearing of *Encarsia formosa* with that of parasitoids of leaf-miners: rearing up to 1000 *Encarsia formosa* per leaf is no problem, whereas only about five to 10 leaf-miner parasitoids can be "harvested" from the same leaf area
6. In crops where different host species (both non-pest and pest species) may occur it is of importance to introduce natural enemies that preferentially attack the pest species in order to obtain a sufficient pest reduction. However, this problem does not occur in greenhouses where parasitoids are released into a closed environment and usually only one host species is parasitized by each parasitoid
7. It is frequently stated that an efficient parasitoid should have a potential

maximum rate of population increase (r_m) equal to or larger than that of its host. If the parasitoid does not only oviposit in the host but also causes other substantial mortality (e.g. through host feeding or host mutilation) we should reformulate the previous sentence to: "an efficient parasitoid should cause an overall host death rate larger than the rate of population increase of the host in the absence of the parasitoid."

However, an r_m larger than the r_m of the host is not by itself sufficient for parasitoid efficiency, because at low host densities the full potential may not be realized. Then searching efficiency is also of great importance

8. Good density responsiveness is also a necessity and the difficulty of measuring it has already been discussed.

D. The Problem of Density Effects: An Example from Greenhouse Systems

A completely successful greenhouse biological control programme of the seasonal inoculative type is characterized by a reduction in pest numbers after introduction of the parasitoids and persistence of the pest below the economic threshold density. Until recently density dependent processes were supposed to be responsible for the persistence of low parasitoid and host densities, but how they actually work is usually unknown.

Our research group has tried to analyse the stabilizing mechanism in the *Encarsia formosa–Trialeurodes vaporariorum* (Westwood) system, a successful and most frequently applied biological control system in greenhouses (Eggenkamp-Rotteveel Mansveld *et al.*, 1982). We hoped to find an explanation for the observation that good control in small greenhouses was difficult, whereas control in large greenhouses succeeded in most cases. We compared the system with a predatory mite-phytophagous mite (*Typhlodromus occidentalis* Nesbitt–*Eotetranychus sexmaculatus* (Riley)) system (Huffaker, 1958). Using this model we supposed that the heterogeneity in a large greenhouse (1 ha) would lead to overall stability, although in a single subpopulation (the size of a small greenhouse; 1000 m^2) extermination of the whitefly and eventually also of the parasitoid may occur. Subsequent immigration of whiteflies, which always occurs, may then lead to uncontrolled whitefly development in small greenhouses. However, the asynchrony of the changes in size of subpopulations in a large greenhouse may give the idea that population fluctuations are not strong.

There are, however, reasons to believe that less dramatic changes take place since *Encarsia formosa* shows a sigmoid functional response when host density increases (Veerkamp-Van Baarle, unpublished). Such a functional response would prevent complete extermination of subpopulations, because parasitoids leave when the host density becomes very low.

Absolute counting of a population in a large greenhouse over 16 weeks showed that extinction of subpopulations did not occur. (A subpopulation is defined as a group of hosts that occurs in areas with a maximal spatial discontinuity of 3 m and a maximal temporal discontinuity of three weeks from other groups.) However, whiteflies were exterminated frequently on individual plants while on other plants new whitefly infestations developed. As a result the percentage parasitism was 0% on some plants, whereas it reached 100% on other plants. The average percentage parasitism over the entire greenhouse fluctuated around 55% and was rather constant, although the average was computed from very different degrees of parasitism in different subpopulations, which is in agreement with Huffaker's data. However, our data are insufficient to support his hypothesis that coexistence of predator and prey (or parasitoid and host) is possible only if special requirements of space and dispersal capacity of both prey and predator are fulfilled.

If we try to relate our data to the ideas of Murdoch *et al.* (1985) we must conclude that their description of population events shows great similarity with that of Eggenkamp-Rotteveel Mansveld *et al.* (1982):

> "If we define our universe to be small enough—an individual in the extreme—we expect to see extinction in the short run, with probability close to one. As we increase the size of the universe this probability of extinction will decrease. This decrease results from the fact that population fluctuations and environmental events may show a degree of asynchrony or statistical independence in space, and as the area is increased the local fluctuations are added together and tend to cancel each other out to yield a much stabler situation for the population on a large scale."

However, since we are not yet able to differentiate between the conflicting theories with our data, we propose a two-step analysis for the density responsiveness.

E. A Procedure for Selection

As pre-introductory evaluation criteria for natural enemies to be used in seasonal inoculative releases, I suggest studies on points 2–5 and 7 of Table II.

In Fig. 1 a flow diagram is presented outlining an evaluation programme for natural enemies. We hope to extend the end part of the flow diagram—the determination of searching efficiency—soon. Our ideas are that it will be a stepwise procedure also based on questions like:

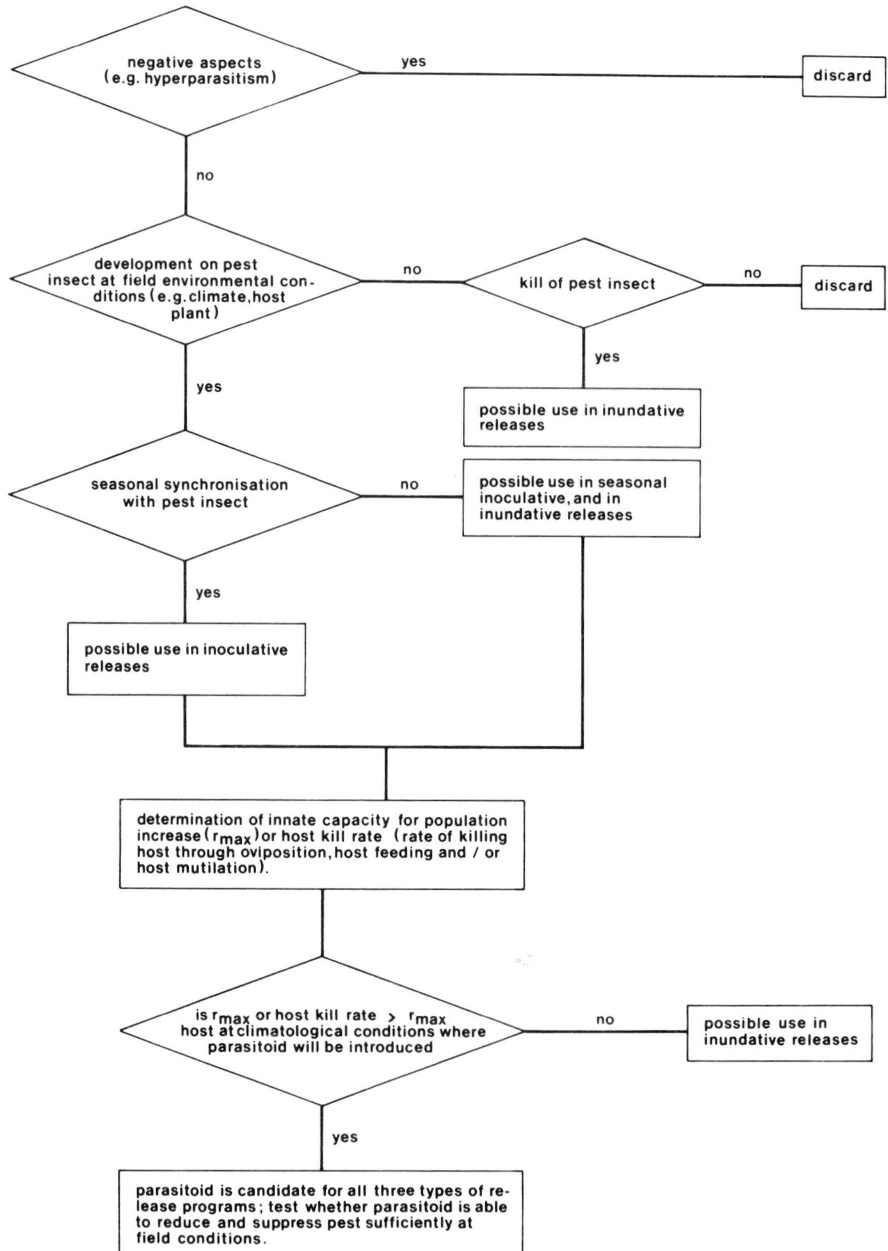

Fig. 1. Flow diagram depicting an evaluation programme for natural enemies to be used in inoculative, seasonal inoculative and inundative release.

1. Is the parasitoid capable of locating host-infested patches by means of host-related stimuli?
2. Is the parasitoid's search more efficient than random?
3. What is the efficiency of the parasitoid in allocating searching time over patches with different host densities?

Before answering such questions, which would certainly require quite intensive behavioural research, we may find easier solutions. We are not interested in long-term stability *per se* but merely aim at suppression of host numbers below the economic threshold. Therefore, we may be able to estimate the power of a parasitoid to suppress its host, starting with parasitoids searching at random by developing system-specific simulation models. In these parasitoids a range of parameter values could be given in relation to fecundity, life span, walking speed, handling time of host and patch, and which are fairly easy to measure. For the hosts, parameter values can be varied for host and patch density and fertility. If the results suggest insufficient killing power such models would allow us to study the effect of more efficient search strategies both between and within patch. Concurrent experiments would inform us whether parasitoids possess any characteristics which make them better than random searchers. Thus simulation models will tell us whether random search is sufficient for pest suppression over the growing season. If so, searching efficiency does not have to be measured in more detail. In this case, when different natural enemies are available, selection on the basis of the reproductive capacity (or killing rate) will suffice.

If random search is not sufficient, the selection criteria will need to be more rigorous and include searching efficiency within and between patches. Intensive behavioural studies would then be needed to determine which species searches most efficiently and makes best use of direct host-derived cues (see Chapter 2).

The importance of the last item of Table II, a good density responsiveness, has been challenged very seriously by Murdoch *et al.* (1984, 1985) and is so difficult to estimate that I would discourage detailed study of it in present-day evaluation programmes. Instead I propose tests in the greenhouse at realistic pest densities to determine whether the natural enemy is able to:

1. Discover pest-infested patches throughout the greenhouse before the economic threshold has been crossed
2. Reduce pest number sufficiently after having found an infested patch
3. Keep the overall pest density under the economic threshold value during the growing season.

III. EXAMPLES OF SPECIFIC SELECTION PROGRAMMES

If we study the history of greenhouse biological control programmes we must conclude that until very recently, successful parasitoids were merely lucky hits rather than selected in the way described (van Lenteren, 1983). For the greenhouse pre-introduction evaluations are conducted at our department as described in the following paragraphs.

A. Selection of Parasitoids for Low Temperature Conditions

Work on parasitoids of the greenhouse whitefly to develop a reliable control method for production of tomato at low greenhouse temperatures is in its final stage. The control potential of several *Encarsia* spp. or strains of the same species and other species of whitefly parasitoids have been compared. Published data on the developmental characteristics and reproductive capacities of *E. formosa* and *Trialeurodes vaporariorum* indicated that *E. formosa* would not be able to control *T. vaporariorum* because the values for the intrinsic rate of increase [calculated using Lewontin's (1965) age-specific fecundity equation] for *E. formosa* are lower than those for *T. vaporariorum* at temperatures below 20°C. A search for other parasitoids was performed in the native area of *T. vaporariorum* (southwestern USA; Vet *et al.*, 1980). Eight primary parasitoids and one obligate hyperparasitoid of *E. formosa* were collected. Four species were cultured successfully and the fecundity, paraistization rate, developmental period and longevity were determined at 17 ± 1°C (Vet, 1980; Vet and van Lenteren, 1981). The temperature thresholds for oviposition and egg maturation were measured (Kajita and van Lenteren, 1982; van Lenteren and van der Schaal, 1981), host selection and sex allocation behaviour were studied (Vet and van Lenteren, 1981; Buys *et al.*, 1981). The same data were obtained for *E. tricolor* Foerster collected in Europe (Christochowitz *et al.*, 1981). It was concluded that *E. formosa* was the best candidate for control, even at low temperatures. Our work further showed that previous data on *E. formosa* were not correct. A recalculation of the intrinsic rate of increase (r_m) with the new data in the range of 12–20°C showed that r_m values for *E. formosa* are always higher than those for *T. vaporariorum* (Christochowitz *et al.*, 1981; van der Laan *et al.*, 1982; van Lenteren and Hulspas-Jordaan, 1983). The methodology of the parasitoid fecundity tests excluded measuring the searching capacity which certainly is an important factor for successful whitefly control (Eggenkamp-Rotteveel Mansveld *et al.*, 1982). According to Madueke (1979), *E. formosa* does not fly and shows hardly any activity when temperatures are lower than 21°C, which would mean that in spite of the data on the intrinsic rate of increase of

E. formosa, biological control of *T. vaporariorum* at a temperature regime of day 18°C/night 7°C would be virtually impossible because the parasitoids would then be unable to migrate from low host density patches to high density ones. New experiments (Christochowitz *et al.*, 1981; van der Laan *et al.*, 1982) proved that *E. formosa* females are capable of flying at temperatures as low as 13°C and that migration is common at temperatures of 17–18°C. Finally, a large-scale test in a greenhouse under practical tomato growing conditions and with a low temperature regime made it clear that *E. formosa* can be used in such conditions (Hulspas-Jordaan *et al.*, in press).

B. Selection of Parasitoids for Crops with Different Leaf Structures

Parasitoids of the greenhouse whitefly were sought for improvement of whitefly control on cucumber. All those we have studied show basically the same host-searching behaviour as *E. formosa.* As their developmental and reproductive characteristics are unfavourable compared with those of *E. formosa,* we do not expect them to be of use. The factors limiting a good control on cucumber seem to be twofold. The intrinsic rate of increase of *E. formosa* is similar to that of *T. vaporariorum* (on tomato it is much higher) and the hairiness of cucumber leaves impairs searching (Hulspas-Jordaan and van Lenteren, 1978). An attempt to improve the reproductive capacity of *E. formosa* by selecting for females with a high number of ovarioles did not succeed (van Vianen and van Lenteren, in press). In cooperation with the Institute for Horticultural Plant Breeding, we are presently breeding for plants which are partly resistant to whiteflies, and for plants with fewer hairs to improve the searching efficiency of *E. formosa* (de Ponti and van Lenteren, 1981).

C. Selection of Leaf-miner Parasitoids

Leaf-miners *Liriomyza bryoniae* Kaltenbach and *L. trifolii* Burgess were the first greenhouse pests for which we had to start a parasitoid evaluation programme from the beginning, and for which we followed the schedule presented in Fig. 1. The work on *L. bryoniae* started in 1977 and resulted in the discovery of efficient parasitoids and development of a mass-rearing programme within five years (Hendrikse, 1980; Hendrikse *et al.*, 1980). During the development of this programme the accidentally-introduced *L. trifolii* achieved pest status. Since the most effective *L. bryoniae* parasitoid, *Opius pallipes* Wesmael, could not be used against *L. trifolii* because it is encapsulated by this host, we had to start searching for new parasitoids. At present we are comparing three species and hope to have an effective parasitoid next year (Minkenberg, personal communication).

IV. MASS PRODUCTION OF PARASITOIDS

Since the beginning of this century mass production of natural enemies has been considered as a means of improving biological control programmes, especially those based on inundative and seasonal inoculative releases. For examples of detailed mass-production techniques of natural enemies see studies by DeBach (1964), Finney and Fisher (1964), Smith (1966), Coppel and Mertins (1977), Morrison and King (1977) and King and Morrison (1984).

A. Reducing the Cost of Rearing

The first step in a mass-rearing programme is usually an attempt to rear the natural enemy on a natural host in an economical way and most are reared in this way. For several, mass-rearing on their natural host is too expensive or impossible because of the risk of contamination with the pest organism or with other pests or diseases.

In these cases the possibilities of rearing on unnatural hosts are explored. This usually means that the host plant on which a natural enemy is reared is also unnatural. The next step is usually a change to an artificial medium for rearing the host. Rearing on artificial diets economizes on space and air conditioning costs.

The historical development and future prospects of insect diets have recently been summarized by Singh (1984), from which we conclude that:

1. Many (some 750 species, mainly phytophagous) insects can be reared successfully on semi-artificial diets
2. Only about two dozen species have been successfully reared on completely artificial diets for several generations
3. Large-scale mass-rearing on artificial media has been developed for less than 20 species of insects
4. Quality control is essential, since there can be dietary effects on all critical performance traits of the mass-reared insect and also on the parasitoid or predator produced on a host that was mass-reared on an artificial medium
5. Good bioassays are important for determining the ultimate effect of the diet on the reared insect.

The last step in trying to reduce rearing costs is the search for ways to rear the natural enemy on an artificial diet. This is possible for several endo- and ectoparasitoids and a few predators. The technology is, however, far less developed than that for rearing Lepidoptera, Coleoptera and Diptera (Singh, 1984). House (1977) has reviewed nutrition requirements for natural enemies.

The following problems limit mass production:

1. The production of qualitatively good natural enemies at economic costs (Beirne, 1974). However, when the number of efficient chemical pesticides decreases and the costs per unit of volume increases—a trend which started about 10 years ago—this aspect will be of less importance
2. Lack of effective techniques for mass-producing natural enemies on artificial diets (Beirne, 1974)
3. Superparasitism in parasitoids may lead to very expensive rearing procedures in order to prevent high parasitoid mortalities, development of small and weak adults, as well as strongly male-biased sex ratios (see Chapter 3)
4. Reduced vigour of natural enemies as the result of rearing on unnatural hosts or on an unnatural diet even when the host itself remains apparently unaffected
5. Infection by pathogens is often encountered in insect rearing. Pathogens and microbial contaminants can lead to high mortality, prolonged development, diminutive adults, wide fluctuations in the quality of insects and direct pathological effects (Goodwin, 1984; Shapiro, 1984; Sikorowski, 1984). Field collected insects used to start a laboratory colony are a major source for microbial contaminants and dietary ingredients are the second main source. The cause of contamination is usually found rapidly but elimination of pathogens is difficult (Bartlett, 1984b).

B. Genetic Changes in Laboratory Populations

Two groups of problems may influence the genetic composition of laboratory-reared populations:

1. Lack of techniques that prevent selection pressures leading to genetic deterioration (Mackauer, 1972) which may result in a loss of effectiveness (Boller and Chambers, 1977)
2. Conflicting requirements for natural enemies in mass-rearing

programmes *versus* field performance, which may lead to selection of ineffective parasitoid strains. In the mass-rearing programmes one opts for intensive parasitism at high host densities and for minimal dispersal, but in the field opposite requirements are essential.

These obstacles, together with those mentioned in the previous section, lead to the conclusion that a good quality control test is essential in each mass production programme.

C. Quality Control in General

Most producers of entomophagous insects start quality control after having experienced serious problems and usually after they have heard from their clients that the insects did not control the pest sufficiently. It would, of course, be better to make an inventory of the changes we can expect when mass-rearing is started and check certain characteristics so as to prevent problems arising. Bartlett (1984a) recently stated: "I believe an unappreciated element of this problem is that the genetic changes taking place when an insect colony is started are natural ones that occur whenever any biological organism goes from one environment to another. These processes have been very well studied as evolutionary events and involve such concepts as colonization, selection, genetic drift, effective population numbers, migration, genetic revolutions, and domestication theory". In another article Bartlett (1984b) discussed what happens to genetic variability in the process of domestication. In the laboratory insects are selected to survive in the new environment: a process called winnowing by Spurway (1955) or, less appropriately, forcing insects through a bottleneck (Boller, 1979). Table III lists changes which a field population may undergo when introduced into the laboratory.

Variability is usually abundantly present in natural populations (Prakash, 1973), and can remain large even in inbred populations (Yamazaki, 1972). When a part of the "open population", where gene migration can occur and environmental diversity is large, is brought into the laboratory, it becomes a "closed population" and all the genetic changes will be made from the limited genetic variation of the founders (Bartlett, 1984b). Thus the size of the founder population will directly affect how much variation will be taken from the gene pool. Further, fitness characteristics for the field will be different from those for the laboratory, so laboratory selection forces may produce a genetic revolution (Mayr, 1970) and a new balanced gene system will be selected (Lopez-Fanjul and Hill, 1973).

The regular rejuvenation of the gene pool with wild insects (gene infusion) is often suggested to overcome or correct for genetic revolutions. However, if

TABLE III. Differences between field and laboratory situations and possible changes in field populations when introduced into the laboratory

1. Laboratory populations are kept in constant environments (light, temperature, wind, humidity) and constant biotic factors (food, no predation or parasitism), and there is no selection to overcome unexpected stresses. The result is a change in fitness and a modification of the whole genetic system (Lerner, 1958).
2. There is no interspecific competition in laboratory populations, which may result in a change in genetic variability (Lerner, 1958).
3. Laboratory conditions select for the average, sometimes for the poorest, genotype. No choice is possible as all individuals are confined to the same environment. The result is a possible decrease in genetic variability (Lerner, 1958).
4. Density dependent behaviour (e.g. searching efficiency) may be affected in laboratory situations (Bartlett, 1984b).
5. Mate-selection processes may be changed because unmated or previously mated females will have restricted means of escape (Bartlett, 1984b).
6. Dispersal characteristics, specifically adult flight behaviour and larval dispersal, may be severely restricted by laboratory conditions (Bush *et al.*, 1976).

the rearing conditions remain the same in the laboratory, the introduced wild individuals will be subjected to the same process of genetic revolution. If genetic differentiation between laboratory and field population has occurred, leading to genetic isolation [Oliver, 1972; positive correlations have been found between the incompatibility of races and the difference between the environments where the races occur (Jaenson, 1978; Jansson, 1978) and for the length of time two populations have been isolated], introduction of wild individuals will be useless. If one wants to introduce genes from the field it should be done regularly and from the start of laboratory rearing and should not be delayed until problems occur. The risk of introducing wild insects is concurrent introduction of parasitoids, predators or pathogens (Bartlett, 1984b).

Another effect of laboratory colonization can be inbreeding. The inbreeding coefficient is directly related to the size of the founder population and, because of artificial selection in the laboratory which results in an even smaller population size, the rate of inbreeding will increase and the result is often a definite and rapid change in genetic composition (Bartlett, 1984b). Inbreeding can be prevented by different methods to maintain genetic variability (Joslyn, 1984).

As a "precolonization" method, selection and pooling of founder insects from throughout the range of the species will provide a wide representation of the gene pool resulting in a greater fitness of the laboratory material. After colonization, variation of laboratory environments in time and space to maintain variability, although a simple concept, is difficult to put into practice. For example, the investment required for rearing facilities with varying temperatures, humidities and light regimes, for the creation of possibilities to choose from diets (hosts), and for the provision of space for dispersal etc., is very high.

Very few data are available about the effective population size to prevent inbreeding (Messenger *et al.*, 1976). Joslyn (1984) stated that to maintain sufficient heterogeneity, a colony should not decline below the number of founder insects. The larger the colony the better and he suggests a minimum of 500 individuals.

A list of criteria to be considered before mass-rearing is started is given in Table IV.

TABLE IV. Criteria to be considered before starting a mass-rearing programme (after Bartlett, 1984b)

1. The effective number of parents at the start of mass breeding is much lower than the number of founder individuals, so start with a large population.
2. Compensate for density dependent phenomena (large cages).
3. Create a proper balance of competition, but no overcrowding.
4. Set environmental conditions for the best genotype, use fluctuating abiotic conditions.
5. Maintain separate laboratory strains and cross them systematically to increase F_1 variability.
6. Measure frequencies of biochemical and morphological markers in founder populations and monitor changes.
7. Develop morphological and biochemical genetic markers for population studies
8. Determine the standards that apply to the intended use of the insects, and then adapt rearing procedures to maximize those values in the domesticated strain.

Quality control is placed in a wider perspective by Chambers and Ashley (1984), who approach it from the industrial side and consider three elements as essential: control of product, process and production. However, these elements are not used in all mass-rearing programmes. Mass-rearing, usually done by small private companies, is developed by trial and error; knowledge about mass-rearing techniques is limited in such organizations and producers are generally more than happy if they can meet demands.

D. Genetic Improvement of Parasitoids

Parasitic insects, especially parasitic Hymenoptera, are most often used in biological control programmes (van Lenteren, 1983) and are thus mass-reared more frequently than predators. Research on genetic improvement of hymenopteran parasitoids has been summarized by Roush (1979). There has been interest in selective breeding for about 80 years because adaptive weaknesses were noted and races with different characteristics were observed. Successes have been obtained in the laboratory: for example, Allen (1954) succeeded in changing the host preference of the parasitoid *Horogenes molestae* (Uchida) towards an unnatural host, the potato tuberworm, *Phthorimaea operculella* (Zeller), which made it possible to mass rear *H. molestae* since the natural host, the oriental fruit moth *Cydia molesta* (Busk), is very difficult to rear. No-one has been able to demonstrate increased effectiveness in the field (see DeBach, 1964; DeBach and Hagen, 1964; Wilson, 1965; Messenger and van den Bosch, 1971; Mackauer, 1972, 1976; Hoy, 1976; Messenger *et al.*, 1976 for reviews of this topic). Roush (1979) also summarized the arguments against success:

1. The features which must be selected cannot be satisfactorily identified
2. Natural selection in the field will act against artificially introduced attributes, and the population will revert to its wild state
3. Laboratory selection programmes are inherently incapable of success because they reduce genetic variability and introduce correlated deleterious pleiotropic effects.

He comments that the first argument denies scientific ingenuity and reasons that the kind of genetic improvement at which one aims is related to the kind of release programme one uses and he suggests: "Genetic selection can possibly be used to lower rearing costs, improve searching behaviour, produce winglessness, increase the proportion of females produced, improve seasonal adaptation, and alter host preferences." He then discusses possibilities for genetic improvement of parasitoids for use in different biological control programmes and considers the greatest single constraint to progress is psychological: that there has been no solid success to encourage further efforts.

My own experience is limited. We tried to increase the average number of ovarioles of *Encarsia formosa* through selection but the number appeared to be determined mainly by environmental factors, i.e. by the size of the host on which the parasitoid is reared (van Vianen and van Lenteren, 1982).

E. Behavioural Changes

Behavioural changes in natural enemies may occur as a result of rearing under unnatural conditions, on unnatural hosts or on artificial media. Data in support are scarce but some information is available on conditioning (for a more extensive discussion see van Lenteren, 1986). Experiments designed to test the hypothesis of *pre-imaginal conditioning* have shown that it may result in a changed preference pattern. However, long-term rearing (30 generations) on different hosts did not result in a change in dependence on, or preference for, these hosts [viz. Salt (1935) for *Trichogramma evanescens* Westwood and Samson-Boshuizen and Van Lenteren (unpublished) for *Leptopilina heterotoma* (Thomson)]. Thus, learning during pre-imaginal development seems to be of minor importance (Vet, 1983; Vet and van Opzeeland, 1984).

Imaginal learning in Hymenoptera, however, was demonstrated long ago (Tinbergen, 1932) and learning by adult insects seems to be commonplace (Jaenike, 1983; Prokopy *et al.*, 1982). Hymenopterous parasitoids can learn to associate physical or chemical cues with suitable host species (Arthur, 1971; Vinson *et al.*, 1977; Gardner *et al.*, unpublished; Vet, 1983, 1985, in press) thereby enhancing their efficiency of host habitat and host location. A series of conditioning experiments during oviposition (associative learning; Vet, 1984; and see Chapter 2) clearly showed the importance of associative learning in different Hymenoptera taxa. Conditioning during oviposition may firstly lead to acquisition of a long-distance host habitat location capacity (volatile host cues could be used only after having experienced a host and its food); secondly, result in modification of the long-distance host habitat location behaviour (parasitoid experience with hosts in different habitats modifies the habitat-odour preference pattern), and thirdly, lead to acquisition of a short-distance host location capacity (probably through association of the host and its contact kairomone(s) during oviposition).

As olfaction is probably the most important sense in host habitat searching in many parasitic Hymenoptera, learning processes occurring after emergence may strongly influence efficiency in the field. It is of high priority to obtain insight on the effect of rearing on unnatural hosts. An important question is whether specialists (mono- or oligophagous species) are likely to respond to stimuli which are specific to their host species, and generalists (polyphagous species) more often show associative learning and in this way become "individual specialists" (Vet, in press). Such information could be used to improve natural enemy performance through pre-release stimulation of parasitoids, as has been shown for *Trichogramma* (Gross, 1981).

F. Storage of Parasitoids

Problems related to good planning of production and the difficulty of predicting demand make it necessary to have storage methods and facilities available. For many beneficials short-term storage methods have been developed. They usually involve placing immature parasitoids, most often pupae, in low temperatures. Storage only lasts several weeks but even then a reduction in fitness is the rule. Data on long-term storage are limited but scientists, particularly in Finland (e.g. Forsberg, 1980) and Czechoslovakia (e.g. Havelka, 1980), have studied the possibility of storing beneficials in a diapausing stage. This work has not yet led to practical application, because of rather high mortalities which occurred during artificially introduced diapause. However, I am convinced that solutions for long-term storage will be found through the study of diapause mechanisms. Suggestions for research in this field can be found in papers by Tauber and Tauber (1976, 1981).

G. Shipment and Release of Parasitoids

Recently I discussed technical problems related to shipment and release of parasitoids (van Lenteren, 1986). Here I concentrate on the timing of releases of parasitoids.

Usually parasitoids are released when pests have been observed, or, when oviposition can be expected. If the pest is univoltine, or when generations are discrete, proper timing is essential so as to have the beneficials available when the preferred host stages are present (van Lenteren, 1980, 1983; van Lenteren *et al.*, 1980). Sometimes pest density has already increased to an intolerable level so that chemical treatment is necessary to achieve an initial reduction of pest numbers. Several authors (e.g. Hussey and Scopes, 1977) have discussed the possibility of introducing the pest before (pest-in-first method) or concurrent with the natural enemy to ensure that the beneficials can survive and reproduce to overcome the need for extensive sampling or to establish a precise and predictable interaction between parasitoid and host. Although these methods proved to be feasible and gave equally good control results as the introduction-after-pest-is-seen method, their acceptance seems unlikely because growers are reluctant to introduce the pest into their crop. Eggenkamp-Rotteveel Mansveld *et al.* (1982) reviewed the alternatives and concluded that the introduction-after-pest-is-seen method is most popular in greenhouse biological control.

Determining the dosage, distribution and frequency of releases are very difficult problems encountered in inundative and seasonal inoculative release programmes. Most programmes are based on empirical data rather than on calculations based on the dispersal behaviour and reproductive capacity, although one would expect, especially in this area of biological control, an important contribution from simulation models. Release ratios are not critical in inundative release programmes so long as it is possible to release an excess of natural enemies; however, the costs of mass production may be a limiting factor. In seasonal inoculative programmes the determination of release ratios is more critical because if too few beneficials are released, control is obtained after the pest has crossed the economic threshold: if too many, one risks extermination of the pest and consequently also, of the natural enemy; pest resurgence is a likely consequence and a serious threat to biological control. Also, in seasonal inoculative programmes, release ratios are usually determined by trial and error but Sabelis (1983) recently attempted to base ratios on biological data.

V. QUALITY CONTROL IN GREENHOUSE SYSTEMS: A CASE STUDY

One Dutch producer of natural enemies (Koppert Company) rears several beneficial organisms on their natural hosts and, even during the season when few natural enemies are sold, a minimal population size of several hundred thousands is normal. Further, they are reared under conditions similar to those in the greenhouse. They produce *Encarsia formosa* (estimated production in 1985: one billion), *Dacnusa sibirica* Telenga (one million), both for application in greenhouses; and *Trichogramma evanescens* (one billion). The main differences from the situation in which they are used are: high host density, pesticides are not applied, host plants are different and rearing temperatures are higher.

The first factor has not led to detectable deterioration of the two natural enemies used in greenhouses. The numbers released and frequency of releases during the growing season has tended to decrease during the 17 years of experience, without a decrease in control. The second factor could lead to a loss of resistance which would result in problems of integrating natural enemies into the system. No detrimental effect due to the other two differences is known. The manipulation of adult predators and parasitoids during collection, package and shipment may also influence field performance. Since 1980 shipment and release of natural enemies has been carried out on artificial substrates. Besides, beneficial organisms are not always shipped immediately but stored, usually at low temperatures.

The founder population of *Encarsia formosa* was imported from Switzerland in 1970 (Maag Ltd., 1500 individuals) and England [Glasshouse Crops Research Institute (GCRI), 400 individuals]. The Swiss strain probably originated from the English GCRI strain. This dates back to the rearing of *Encarsia formosa* in England by Speyer (1927) at the Cheshunt Experimental Research Station (since transferred to GCRI). The detailed history before importation into Holland is not known exactly but the greenhouse performance of *E. formosa* is excellent. *Dacnusa sibirica* was collected in greenhouses in the Netherlands and is native. Rearing was started with several hundreds of parasitoids.

Cultures are always maintained using hundreds or thousands of individuals under natural greenhouse climatic conditions, although sometimes temperatures 2–5°C higher are used. For these reasons the producer applies no quality control during the rearing process. Each year at the end of the growing season new parasitoid and host material is collected from a number of greenhouses to reinforce or completely replace the mass-production colonies, so that greenhouse-collected strains which were exposed to practical cropping procedures are used for natural enemy production in the following year. In this way the producer expects to maintain a good overall quality (vigour, searching capacity, pesticide resistance), but quality is only indirectly checked through the performance in the commercial greenhouses in the next season.

Specific quality control procedures were only introduced when problems were experienced during storage, packing, shipment and other technical manipulations. Thus, quality control for *E. formosa* had to be developed when a change in the procedure for "harvesting" and shipping this parasitoid took place in 1979. Until then the parasitoids were shipped as pupae on host plant leaves. This had the following disadvantages: (1) the degree of parasitism on each leaf had to be checked before shipment; (2) the leaf should not contain many whiteflies; (3) during transport leaves started to wilt or deteriorate resulting in high pupal mortality; (4) similarly when leaves were stored for short periods; (5) distribution was laborious and (6) there was always the risk of introducing pests or diseases from the insectary. Now the parasitized pupae are removed from the leaf, glued on cards and shipped for release so that storage is easier, transport and introduction is more convenient and there is no risk of infection of the crop with unwanted pests or diseases. The possible side-effects of the removal procedure and of glueing pupae on cards on the survival and quality of the parasitoid were:

1. Mechanical damage which can easily be corrected by adding extra pupae to a card to obtain a sufficiently high number of healthy pupae
2. The influence of the glue; but natural glues are used which have no effect on parasitoid development and performance
3. Influence of glueing parasitoids in an abnormal position on a card and pupae partially covered with glue.

This last point resulted in most problems. We have developed a quality control method to trace differences between parasitoids on leaves, *vs* those removed during the pupal stage from the same leaf and glued to a card. We compared:

1. The orientation of the pupae
2. Emergence of adults according to orientation during time
3. Head width and abdomen length
4. Oviposition frequency during the first days after emergence.

After changing to the new shipment procedure we observed the orientation of pupae was 25% normal, 50% on one side and 25% upside down, while on the leaf all pupae were oriented in a normal position. The emergence of adults on cards occurred later than on the leaf and the percentage emergence was also lower. The head width of both populations was the same, but the abdomen length of parasitoids from the cards was shorter. The adults had difficulties with crawling out of the glue-covered or wrongly oriented pupae, especially when pupae were upside down. This may explain the delayed emergence as well as the shorter abdomens of females from the card population. Many carded females were visibly in a poorer condition after emergence than those from leaves since they did not search for hosts and their oviposition frequency was very low—many females did not oviposit at all. The high percentage mortality was caused by puncturing of pupae by the brushes used for removing the pupae from the leaf (van Lenteren and van Eck, unpublished). We have not checked the long-distance dispersal capacity.

Addition of honey to the cards appeared to be very important because when parasitoids are not able to feed during the first two days after emergence their life span is reduced from a month or more to about a week and their fertility drops from 400 to less than 100 eggs (van Lenteren *et al.*, in press).

In response to these findings the Koppert Company improved the procedure of removal and glueing pupae on the cards and applies a honey-water mixture to the cards. As a result there are now insignificant differences for the various characteristics discussed, except for pupal orientation (Ravensberg, personal communication).

VI. CONCLUSIONS

Biological pest control in greenhouses through the introduction of parasitoids is a viable and valuable pest control method; development is cheaper than for chemical control, production and application are safer, there are no environmental risks, the chance of resistance is much less and once developed a biological control method can be used indefinitely. The only drawback—which is also one of the most important points in its favour—is its extreme specificity. Developments in chemical pest control will result in a need to search for other pest control methods and biological control is a main candidate. Parasitoids can be used in several types of biological control programmes, with a long- and short-term effect. To be successful in these different circumstances, parasitoids should satisfy different conditions. The main problem in evaluating the control capacity of parasitoids prior to introduction is the lack of good evaluation criteria and the limited insight on the working mechanism of biological control. Strong input is needed, both in the area of pure research—e.g. development and testing of population dynamic and optimal foraging theory—and in applied research—development and testing of criteria for pre-introductory evaluation. Such developments might result in a more efficient search for natural enemies than at present, thus reducing the costs of developing biological control programmes.

Development of mass-rearing techniques is necessary for programmes with a short-term control effect. Only some 15 natural enemy species are presently reared on a large scale. The technology for rearing natural enemies on unnatural hosts and host plants, or on artificial diets, is little developed. Physiological, ethological, ecological and genetic problems hamper developments. Conflicting requirements for natural enemies in mass-rearing and field performance, combined with risks from rearing on artificial hosts or diets, make good quality control a necessity. They are, however, available but are applied on a very limited scale. Innovations in long-term storage, shipment and release methods may lead to a further reduction in costs. Little success has been achieved in attempting genetic improvement of parasitoids.

It is about a hundred years since the first biological control success was obtained [*Icerya purchasi* Maskell with *Rodolia cardinalis* (Mulsant) in the USA] and in greenhouses the first success dates back to the 1920s. The future of biological control has never been so bright since the large-scale introduction of synthetic organic insecticides after the Second World War.

Acknowledgements

The development of pre-introductory evaluation criteria for natural enemies is a common effort of the Department of Population Biology (University of Leiden) and Entomology (Agricultural University Wageningen). M. Bergeman, O. P. J. M. Minkenberg, G. A. Pak, M. W. Sabelis and L. E. M. Vet assisted in improving this article. M. Koopman, T. de Vries and T. Wildeman are thanked for typing the manuscript.

REFERENCES

Allen, H. W. (1954). Propagation of *Horogenes molestae*, an Asiatic parasite of the Oriental fruit moth on the potato tuberworm. *Journal of Economic Entomology* **47,** 278–281.

Andrewartha, H. G. (1961). "Introduction to the Study of Animal Populations." Methuen, London.

Anonymous (1981). Report of the 2nd Session of the FAO Panel of Experts on Pest Resistance to Pesticides and Crop Loss Assessment. FAO, Rome.

Arthur, A. P. (1971). Associative learning by *Nemeritis canescens* (Hymenoptera: Ichneumonidae). *Canadian Entomologist* **103,** 1137–1141.

Askew, R. R. (1971). "Parasitic Insects." Heinemann, London.

Bartlett, A. C. (1984a). Establishment and maintenance of insect colonies through genetic control. *In* "Advances and Challenges in Insect Rearing" (E. G. King and N. C. Leppla, eds.), p. 1. USDA/ARS, New Orleans.

Bartlett, A. C. (1984b). Genetic changes during insect-domestication. *In* "Advances and Challenges in Insect Rearing" (E. G. King and N. C. Leppla, eds.), pp. 2–8. USDA/ARS, New Orleans.

Beirne, B. P. (1974). Status of biological control procedures that involve parasites and predators. *In* "Proceedings of the Summer Institute on Biological Control of Plant Insects and Diseases" (F. G. Maxwell and F. A. Harris, eds.), pp. 69–76. University Press of Mississippi, Jackson.

Beirne, B. P. (1980). Biological control: benefits and opportunities. *In* " Perspectives in World Agriculture" pp. 307–321. Commonwealth Agricultural Bureaux, Farnham Royal, England.

Besemer, A. F. H. (1985). "Veertig Jaar Volhardend Zoeken: het "Ideale" Bestrijdingsmiddel een Utopie?" Afscheidscollege Landbouwhogeschool, Wageningen.

Boller, E. F. (1979). Behavioral aspects of quality in insectary production. *In* "Genetics in Relation to Insect Management" (M. A. Hoy and J. J. McKelvey, eds.), pp. 153–160. Rockefeller Foundation, New York.

Boller, E. F., and Chambers, D. L. (1977). Quality aspects of mass-reared insects. *In* "Biological Control by Augmentation of Natural Enemies" (R. L. Ridgway and S. B. Vinson, eds.), pp. 219–236. Plenum, New York.

Bush, G. L., Neck, R. W., and Kitto, G. B. (1976). Screwworm eradication: inadvertent selection for noncompetitive ecotypes during mass rearing. *Science* **193,** 491–493.

Buys, M. J., Pirovano, I., and van Lenteren, J. C. (1981). *Encarsia pergandiella*, a possible biological control agent for the greenhouse whitefly, *Trialeurodes vaporariorum*: a study on intra- and interspecific host selection. *Mededelingen Faculteit Landbouwwetenschappen Rijksuniversiteit Gent* **46,** 465–574.

Chambers, D. L., and Ashley, T. R. (1984). Putting the control in quality control in insect rearing. *In* "Advances and Challenges in Insect Rearing" (E. G. King and N. C. Leppla, eds.), pp. 256–260. USDA/ARS, New Orleans.

Christochowitz, E. E., van der Fluit, N., and van Lenteren, J. C. (1981). Rate of development and oviposition frequency of *Trialeurodes vaporariorum*, *Encarsia formosa* (two strains) and *E. tricolor* at low glasshouse temperatures. *Mededelingen Faculteit Landbouwwetenschappen Rijksuniversiteit Gent* **46,** 477–485.

Clausen, C. P. (1978). "Introduced Parasites and Predators of Arthropod Pests and Weeds: a World Review." USDA/ARS, Washington.

Coppel, H. C., and Mertins, J. W. (1977). "Biological Insect Suppression." Springer, Berlin.

DeBach, P. (1964). Successes, trends, and future possibilities. *In* "Biological Control of Insect Pests and Weeds" (P. DeBach, ed.), pp. 673–713. Chapman and Hall, London.

DeBach, P. (1971). The use of imported natural enemies in insect pest management ecology. *Proceedings of the Tall Timbers Conference on Ecological Animal Control and Habitat Management* **3,** 211–234.

DeBach, P. (1974). "Biological Control by Natural Enemies." Cambridge University Press, Cambridge.

DeBach, P., and Hagen, K. S. (1964). Manipulation of entomophagous species. *In* "Biological Control of Insect Pests and Weeds" (P. DeBach, ed.), pp. 429–458. Chapman and Hall, London.

Dover, M., and Croft, B. (1984). Getting tough: public policy and the management of pesticide resistance. *World Resources Institute Study* **1,** 80.

Eggenkamp-Rotteveel Mansveld, M. H., Ellenbroek, F. J. M., van Lenteren, J. C. (1982). The parasite-host relationship between *Encarsia formosa* (Hymenoptera: Aphelinidae) and *Trialeurodes vaporariorum* (Homoptera: Aleyrodidae). XII. Population dynamics of parasite and host in a large, commercial glasshouse and test of the parasite introduction method used in the Netherlands. *Zeitschrift für angewandte Entomologie* **93,** 113–130; 258–279.

Ehler, L. E. (1982). Foreign exploration in California. *Environmental Ecology* **11,** 525–530.

van Emden, H. F. (1974). "Pest Control and its Ecology." Edward Arnold, London.

Finney, G. L., and Fisher, T. W. (1964). Culture of entomophagous insects and their hosts. *In* "Biological Control of Insect Pests and Weeds" (P. DeBach, ed.), pp. 329–355. Chapman and Hall, London.

Flanders, S. (1957). Principles and practices of biological control utilizing entomophagous insects. Division of Biological Control, University of California, Riverside.

Forsberg, A. (1980). Possibilities of using the diapause of *Aphidoletes aphidimyza* (Rond.) (Diptera: Cecidomyiidae) in its mass production. *Bulletin OILB/SROP* **III/3,** 35–39.

Gardner, S. M., van Dissevelt, M., and van Lenteren, J. C. (unpublished). Behavioural adaptations in host finding by *Trichogramma evanescens* (Westwood): the influence of oviposition experience on response to host contact kairomones.

Goodwin, R. H. (1984). Recognition and diagnosis of diseases in insectaries and the effects of disease agents on insect biology. *In* "Advances and Challenges in Insect Rearing" (E. G. King and N. C. Leppla, eds.), pp. 96–129. USDA/ARS, New Orleans.

Gross, H. R. (1981). Employment of kairomones in the management of parasitoids. *In* "Semiochemicals: Their Role in Pest Control" (D. A. Nordlund, R. L. Jones and W. J. Lewis, eds.), pp. 137–150. John Wiley, New York.

Hassan, S. A. (1981). Mass-production and utilization of *Trichogramma.* 2. Four years successful biological control of the European corn borer. *Mededelingen Faculteit Landbouwwetenschappen Rijksuniversiteit Gent* **46,** 417–428.

Hassell, M. P., and May, R. M. (1973). Stability in insect host-parasite models. *Journal of Animal Ecology* **42,** 693–726.

Hassell, M. P., and Rogers, D. J. (1972). Insect parasite responses in the development of population models. *Journal of Animal Ecology* **41,** 661–676.

Havelka, J. (1980). Some aspects of photoperiodism of the aphidophagous gallmidge *Aphidoletes aphidimyza* Rond. *Bulletin OILB/SROP* **III/3,** 75–81.

Hendrikse, A. (1980). A method for mass rearing two braconid parasites (*Dacnusa sibirica* and *Opius pallipes*) of the tomato leafminer (*Liriomyza bryoniae*). *Mededelingen Faculteit Landbouwwetenschappen Rijksuniversiteit Gent*, **45,** 563–572.

Hendrikse, A., Zucchi, R., van Lenteren, J. C., and Woets, J. (1980). *Dacnusa sibirica* Telenga and *Opius pallipes* Wesmael, (Hym., Braconidae) in the control of the tomato leafminer *Liriomyza bryoniae* Kalt. *Bulletin OILB/SROP* **III/3,** 83–98.

House, H. L. (1977). Nutrition of natural enemies. *In* "Biological Control by Augmentation of

Natural Enemies" (Ridgway, R. L., and Vinson, S. B., eds.), pp. 151–182. Plenum, New York.

Hoy, M. A. (1976). Genetic improvement of insects: fact or fantasy. *Environmental Entomology* **5,** 833–839.

Huffaker, C. B. (1958). Experimental studies on predation: dispersion factors and predator-prey oscillations. *Hilgardia* **28,** 343–383.

Huffaker, C. B. (1976). An overview of biological control, with particular commentary on biological weed control. *In* Proceedings of the 4th International Symposium on Biological Control of Weeds (T. D. Freeman, ed.), Center for Environmental Progress, University of Florida at Gainesville, 3–12.

Huffaker, C. B., and Messenger, P. S. (1976). "The Theory and Practice of Biological Control." Academic Press, New York.

Huffaker, C. B., and Rabb, R. L. (1984). "Ecological Entomology." John Wiley, New York.

Huffaker, C. B., Messenger, P. S., and DeBach, P. (1971). The natural enemy component in natural control and the theory of biological control. *In* "Biological Control" (C. B. Huffaker, ed.), pp. 16–67. Plenum, New York.

Huffaker, C. B., Simmonds, F. J., and Laing, J. E. (1976). The theoretical and empirical basis of biological control. *In* "Theory and Practice of Biological Control" (C. B. Huffaker and P. S. Messenger, eds.), pp. 42–80. Academic Press, New York.

Hulspas-Jordaan, P. M., and van Lenteren, J. C. (1978). The relationship between host-plant leaf structure and parasitization efficiency of the parasitic wasp *Encarsia formosa* Gahan (Hymenoptera: Aphelinidae). *Mededelingen Faculteit Landbouwwetenschappen Rijksuniversiteit Gent*, **43,** 431–440.

Hulspas-Jordaan, P. M., Christochowitz, E. E., Woets, J., and van Lenteren, J. C. (in press). The parasite-host relationship between *Encarsia formosa* (Hym.: Aphelinidae) and *Trialeurodes vaporariorum* (Hom.: Aleyrodidae). XXVII Effectiveness of *E. formosa* at low greenhouse temperatures. *Zeitschrift für angewandte Entomologie.*

Hussey, N. W., and Scopes, N. E. A. (1977). The introduction of natural enemies for pest control in glasshouses: ecological considerations. *In* "Biological Control by Augmentation of Natural Enemies" (R. L. Ridgway and S. B. Vinson, eds.), pp. 183–217. Plenum, New York.

Hussey, N. W., and Scopes, N. E. A. (1985). "Biological Pest Control, the Glasshouse Experience." Blandford Press, Poole, Dorset.

Jaenike, J. (1983). Induction of host preference in *Drosophila melanogaster. Oecologia* **58,** 320–325.

Jaenson, T. G. T. (1978). Mating behavior of *Glossina pallides* Austen (Diptera, Glossinidae): genetic differences in copulation time between allopatric populations. Entomologia Experimentalis et Applicata **24,** 100–108.

Jansson, A. (1978). Viability of progeny in experimental crosses between geographically isolated populations of *Arctocorisa carinata* (C. Sahlberg) (Heteroptera, Corixidae). *Annales Zoologici Fennici* **15,** 77–83.

Joslyn, D. J. (1984). Maintenance of genetic variability in reared insects. *In* "Advances and Challenges in Insect Rearing" (E. G. King and N. C. Leppla, eds.), pp. 20–29. USDA/ARS, New Orleans.

Kajita, H., and van Lenteren, J. C. (1982). The parasite-host relationship between *Encarsia formosa* (Hymenoptera: Aphelinidae) and *Trialeurodes vaporariorum* (Homoptera: Aleyrodidae). XIII. Effects of low temperatures on egg maturation of *Encarsia formosa. Zeitschrift für angewandte Entomologie* **93,** 430–439.

King, E. G., and Morrison, R. K. (1984). Some systems for production of eight entomophagous arthropods. *In* "Advances and Challenges in Insect Rearing" (E. G. King, and N. C. Leppla, eds.), pp. 206–222. USDA/ARS, New Orleans.

Krebs, C. J. (1972). "Ecology, the Experimental Analysis of Distribution and Abundance." Harper and Row, New York.

van der Laan, E. M., Burgraaf-van Neirop, Y. D., and van Lenteren, J. C. (1982). Oviposition frequency, fecundity and life-span of *Encarsia formosa* (Hymenoptera: Aphelinidae) and *Trialeurodes vaporariorum* (Homoptera: Aleyrodidae) and migration capacity of *E. formosa* at low greenhouse temperatures. *Mededelingen Faculteit Landbouwwetenschappen Rijksuniversiteit Gent* **47,** 511–521.

Laing, J. E., and Hamai, J. (1976). Biological control of insect pests and weeds by imported parasites, predators and pathogens. *In* "Theory and Practice of Biological Control" (C. B. Huffaker and P. S. Messenger, eds.), pp. 42–80. Academic Press, New York.

van Lenteren, J. C. (1980). Evaluation of control capabilities of natural enemies: does art have to become science? *Netherlands Journal of Zoology* **30,** 369–381.

van Lenteren, J. C. (1983). The potential of entomophagous parasites for pest control. *Agriculture, Ecosystems and Environment* **10,** 143–158.

van Lenteren, J. C. (1986). Evaluation, mass production, quality control and release of entomophagous insects. *In* "Biological Plant and Health Protection" (J. M. Franz, ed.), pp. 31–56. Fisher Verlag, Stuttgart.

van Lenteren, J. C., and Hulspas-Jordaan, P. M. (1983). Biological control of the greenhouse whitefly, *Trialeurodes vaporariorum* (Westwood) at low greenhouse temperatures: a summary. *Proceedings of the International Conference on Integrated Plant Protection* **3,** Budapest, 1–17.

van Lenteren, J. C., and van der Schaal A. W. J. (1981). Temperature thresholds for oviposition of *Encarsia formosa, E. tricolor* and *E. pergandiella* in larvae of *Trialeurodes vaporariorum. Mededelingen Faculteit Landbouwwetenschappen Rijksuniversiteit Gent* **46,** 457–464.

van Lenteren, J. C., Nell, H. W., and Sevenster-van der Lelie, L. A. (1980). The parasite-host relationship between *Encarsia formosa* (Hymenoptera: Aphelinidae) and *Trialeurodes vaporariorum* (Homoptera: Aleyrodidae). IV. Oviposition behaviour of the parasite, with aspects of host selection, host discrimination and host feeding. *Zeitschrift für angewandte Entomologie* **89,** 442–454.

van Lenteren, J. C., van Vianen, A., Gast, H. P., and Kortenhoff, A. (in press). The parasite-host relationship between *Encarsia formosa* Gahan (Hymenoptera: Aphelinidae) and *Trialeurodes vaporariorum* (Westwood) (Homoptera: Aleyrodidae). XVI. Feeding-frequency and food-quality effects on oogenis, oviposition, life-span and fecundity of *Encarsia formosa* and other hymenopterous parasites. *Zeitschrift für angewandte Entomologie.*

Lerner, I. (1958). "The Genetic Basis of Selection." John Wiley, New York.

Lewontin, R. C. (1965). Selection for colonizing ability. *In* "The Genetics of Colonizing Species" (H. G. Baker and G. L. Stebbins, eds.). Academic Press, London.

Lopez-Fanjul, C., and Hill, W. G. (1973). Genetic differences between populations of *Drosophila melanogaster* for a quantitative trait. II. Wild and laboratory populations. *Genetical Research* **22,** 69–78.

Mackauer, M. (1972). Genetic aspects of insect control. *Entomophaga* **17,** 27–48.

Mackauer, M. (1976). Genetic problems in the production of biological control agents. *Annual Review of Entomology* **21,** 369–385.

Madueke, E. D. N. N. (1979). "Biological Control of *Trialeurodes vaporariorum.*" PhD Thesis, Cambridge.

Mayr, E. (1970). "Populations, Species, and Evolution." Harvard University Press, Cambridge, Massachusetts.

Messenger, P. S., and van den Bosch, R. (1971). The adaptability of introduced biological control agents. *In* "Biological Control" (C. B. Huffaker, ed.), pp. 68–92. Plenum, New York.

Messenger, P. S., Wilson, F., and Whitten, M. J. (1976). Variation, fitness, and adaptability of natural enemies. *In* "Theory and Practice of Biological Control," (C. B. Huffaker and P. S. Messenger, eds.), pp. 233–254. Academic Press, New York.

Metcalf, R. L. (1980). Changing role of insecticides in crop protection. *Annual Review of Entomology* **25,** 219–256.

Minks, A. K. and Gruys, P. (eds.) (1980). "Integrated Control of Insect Pests in the Netherlands." Pudoc, Wageningen.

Morrison, R. K., and King, E. G. (1977). Mass production of natural enemies. *In* "Biological Control by Augmentation of Natural Enemies" (R. L. Ridgway and S. B. Vinson, eds.), pp. 183–217. Plenum, New York.

Murdoch, W. W., Reeve, J. D., Huffaker, C. B., and Kennett, C. E. (1984). Biological control of olive scale and its relevance to ecological theory. *American Naturalist* **123,** 371–392.

Murdoch, W. W., Chesson, J., and Chesson, P. L. (1985). Biological control in theory and practice. *American Naturalist* **125,** 344–366.

Oliver, C. G. (1972). Genetic and phenotypic differentiation and geographic distance in four species of Lepidoptera. *Evolution* **26,** 221–241.

de Ponti, O. M. B., and van Lenteren, J. C. (1981). Resistance and glabrousness: different approaches to biological control of two cucumber pests. *Tetranychus urticae* and *Trialeurodes vaporariorum. Bulletin OILB/SROP.* **VI/1,** 109–113.

Prakash, S. (1973). Patterns of gene variation in central and marginal populations of *Drosophila robusta. Genetics* **75,** 347–369.

Prokopy, R. J., Averill, A. L., Cooley, S. S., and Roitberg, C. A. (1982). Associative learning in egglaying site selection by apple maggot flies. *Science* **218,** 76–77.

Pschorn-Walcher, H. (1977). Biological control of forest insects. *Annual Review of Entomology* **22,** 1–22.

Ridgway, R. L., and Vinson, S. B. (1977). "Biological Control by Augmentation of Natural Enemies: Insect and Mite Control with Parasites and Predators." Plenum Press, New York.

Roush, R. T. (1979). Genetic improvement of parasites. *In* "Genetics in Relation to Insect Management" (M. A. Hoy and J. J. McKelvey, eds.), pp. 97–105. Rockefeller Foundation, USA.

Sabelis, M. W. (1983). Experimental validation of a simulation model of the interaction between *Phytoseiulus persimilis* and *Tetranychus urticae* on cucumber. *Bulletin OILB/SROP* **VI/3,** 207—229.

Salt, G. (1935). Experimental studies in insect parasitism. III. Host selection. *Proceeding Royal Society London* **117,** 413–435.

Shapiro, M. (1984). Micro-organisms as contaminants and pathogens in insect rearing. *In* "Advances and Challenges in Insect Rearing" (E. G. King and N. C. Leppla, eds.), pp. 130–142. USDA/ARS, New Orleans.

Sikorowski, P. P. (1984). Microbial contamination in insectaries. *In* "Advances and Challenges in Insect Rearing" (E. G. King and N. C. Leppla, eds.), pp. 143–153. USDA/ARS, New Orleans.

Simmonds, F. J., Franz, J. M., and Sailer, R. I. (1976). History of biological control. *In* "Theory and Practice of Biological Control" (C. B. Huffaker and P. S. Messenger, eds.), pp. 17–41. Academic Press, New York.

Singh, P. (1984). Insect diets. Historical developments, recent advances, and future prospects. *In*

"Advances and Challenges in Insect Rearing" (E. G. King and N. C. Leppla, eds.), pp. 32–44. USDA/ARS, New Orleans.

Smith, C. N. (1966). "Insect Colonization and Mass Production". Academic Press, New York.

Smith, H. S. (1919). On some phases of insect control by the biological method. *Journal of Economic Entomology* **12,** 288–292.

Speyer, E. R. (1927). An important parasite of the greenhouse white-fly *Trialeurodes vaporariorum* (Westwood). *Bulletin of Entomological Research* **17,** 301–308.

Spurway, H. (1955). The causes of domestication: an attempt to integrate some ideas of Konrad Lorenz with evolution theory. *Journal of Genetics* **53,** 325–362.

Sweetman, H. L. (1958). "The Principles of Biological Control." Brown, Dubuque.

Takada, H., and Kamijo, K. (1979). Parasite complex of the gardenpea leafminer, *Phytomyza horticola* Gonzea, in Japan. *Kontyû* **47,** 18–37.

Tauber, C. A., and Tauber, M. J. (1981). Insect seasonal cycles: genetics and evolution. *Annual Review of Ecology and Systematics* **12,** 281–308.

Tauber, M. J., and Tauber, C. A. (1976). Insect seasonality: diapause maintenance, termination, and post-diapause development. *Annual Review of Entomology* **21,** 81–107.

Tinbergen, N. (1932). Über die Orientierung des Bienenwolfes (*Philanthus triangulum* Fabr.). *Zeitschrift für vergleichende Physiologie* **16,** 305–335.

Varley, G. C. (1951). Ecological aspects of population regulation. *Transactions of the 9th International Congress of Entomology in Amsterdam* **2,** 210–214.

Varley, G. C., Gradwell, G. R., and Hassell, M. P. (1973). "Insect Population Ecology, an Analytical Approach." Blackwell, Oxford.

Vet, L. E. M. (1980). Laboratory studies on three *Encarsia* spp. and one *Eretmocerus* sp. (Hymenoptera: Aphelinidae), parasites of the greenhouse whitefly *Trialeurodes vaporariorum* (Westw.) to assess their efficiency as biological control agents under low temperature conditions. *Mededelingen Faculteit Landbouwwetenschappen Rijksuniversiteit Gent* **45,** 555–561.

Vet, L. E. M. (1983). Host-habitat location through olfactory cues by *Leptopilina clavipes* (Hartig) (Hym.: Eucoilidae), a parasitoid of fungivorous *Drosophila*: the influence of conditioning. *Netherlands Journal of Zoology* **33,** 225–248.

Vet, L. E. M. (1984). "Comparative Ecology of Hymenopterous Parasitoids." PhD Thesis, University of Leiden.

Vet, L. E. M. (1985). Response to kairomones by some alysiine and eucoilid parasitoid species (Hymenoptera). *Netherlands Journal of Zoology* **35,** 486–496.

Vet, L. E. M. (in press a). Olfactory microhabitat location in some eucoilid and alysiine species (Hymenoptera), larval parasitoids of Diptera. *Netherlands Journal of Zoology.*

Vet, L. E. M., and van Lenteren, J. C. (1981). The parasite-host relationship between *Encarsia formosa* (Hymenoptera: Aphelinidae) and *Trialeurodes vaporariorum* (Homoptera: Aleyrodidae). X. A comparison of three *Encarsia* spp. and one *Eretmocerus* sp. to estimate their potentialities in controlling whitefly on tomatoes in greenhouses with a low temperature regime. *Zeitschrift für angewandte Entomologie* **91,** 327–348.

Vet, L. E. M., and van Opzeeland, K. (1984). The influence of conditioning of olfactory microhabitat and host location in *Asobara tabida* (Nees) and *A. rufescens* (Foerster) (Braconidae: Alysiinae), larval parasitoids of Drosophilidae. *Oecologia* **63,** 171–177.

Vet, L. E. M., van Lenteren, J. C., and Woets, J. (1980). The parasite-host relationship between *Encarsia formosa* (Hymenoptera: Aphelinidae) and *Trialeurodes vaporariorum* (Homoptera: Aleyrodidae). IX. A review of the biological control of the greenhouse whitefly with suggestions for future research. *Zeitschrift für angewandte Entomologie* **90,** 26–51.

van Vianen, A., and van Lenteren, J. C. (1982). Increasing the number of ovarioles of *Encarsia formosa*—a possibility to improve the parasite for biological control of the greenhouse whitefly, *Trialeurodes vaporariorum*? *Mededelingen Faculteit Landbouwwetenschappen Rijksuniversiteit Gent* **47,** 523–531.

van Vianen, A., and van Lenteren, J. C. (in press). The parasite-host relationship between *Encarsia formosa* (Hymenoptera: Aphelinidae) and *Trialeurodes vaporariorum* (Homoptera: Aleyrodidae). XIV. Genetic and environmental factors influencing body size and number of ovarioles of *Encarsia formosa*. *Zeitschrift für angewandte Entomologie*.

Vinson, S. B., Barfield, C. S., and Henson, R. D. (1977). Oviposition behaviour of *Bracon mellitor*, a parasitoid of the boll weevil (*Anthonomus grandis*). II. Associative learning. *Physiological Entomology* **2,** 157–164.

Waage, J. K., and Hassell, M. P. (1982). Parasitoids as biological control agents—a fundamental approach. *Parasitology* **84,** 241–268.

Way, M. J. (1973). Objectives, methods and scope of integrated control. *In* "Insects: Studies in Population Management" (W. P. Geier, L. R. Clark, D. J. Anderson and H. A. Nix, eds.), *Ecological Society of Australia Memoirs* **1,** 1–17.

Wilson, F. (1965). Biological control and the genetics of colonizing species. *In* "The Genetics of Colonizing Species" (H. G. Baker and G. L. Stebbins, eds.), pp. 307–329. Academic Press, New York.

Yamazaki, T. (1972). Detection of single gene effect by inbreeding. *Nature: New Biology* **240** (97), 53–54.

Zwölfer, H., Ghani, M. A., and Rao, V. P. (1976). Foreign exploration and importation of natural enemies. *In* "Theory and Practice of Biological Control" (C. B. Huffaker and P. S. Messenger, eds.), pp. 189–207. Academic Press, New York.

Index

A

D

I

P

T

U

V

W

X

Z